Gleason's

P L A N T S

of Michigan

Richard K. Rabeler
University of Michigan Herbarium

First Edition

EDITOR: VIVIENNE N. ARMENTROUT
ILLUSTRATIONS BY ELISE C. BUSH

The University of Michigan Press
Ann Arbor

Copyright © 2007 by the University of Michigan
Published by the University of Michigan Press 2007
First published by Oakleaf Press 1998
All rights reserved
Published in the United States of America by
The University of Michigan Press
Manufactured in the United States of America
♾ Printed on acid-free paper

2010 2009   4 3 2

U.S. CIP data applied for.

ISBN-13: 978-0-472-03246-4
ISBN-10: 0-472-03246-1

EDITOR'S NOTE

This book is a substantial revision and expansion of *The Plants of Michigan* by H.A. Gleason, first published by George Wahr Publishing Co. in 1918. Although the basic structure of the original has been retained, most of the keys have been completely rewritten and new material has been added. Family descriptions have been rewritten and expanded. Information on likely habitats and on distribution has been added to the keys to facilitate identification. Illustrations have been provided for some key families and to illustrate terms. Short boxed notes throughout the text discuss topics ranging from taxonomic issues, to changes in plant distributions, to related cultivated plants. Instructions for using the keys for plant identification are included in the Introduction. A section on botanical terminology and a glossary of terms have been provided. A comprehensive index lists all Latin and common plant names. Key bibliographic references to certain plant groups and to Michigan plants have been included. See the Preface for a discussion of H.A. Gleason and his contributions to Michigan botany.

Keys to flowering plants, including trees and shrubs, and to gymnosperms, are included. The keys have been revised to reflect current interpretations of the classification of Michigan plants. This edition of *Gleason's Plants of Michigan* follows the arrangement and definition of families used in *Michigan Flora* (1972, 1985, 1996). Nomenclature and information on plant distribution are also based on those provided in *Michigan Flora*. Plant species collected only rarely in Michigan have been omitted.

Every effort has been made to make this revised and expanded edition of *Plants of Michigan* accurate and current. Plant heights and other characters were checked against the plant descriptions in Gleason and Cronquist (1991) and the keys were compared with characters cited by Voss (1972, 1985). Collections in the University of Michigan Herbarium were also consulted. Dr. Anton Reznicek provided additional information in the course of his review of the unpublished manuscript.

NOTES ON THE SECOND PRINTING

With this second printing of *Gleason's Plants of Michigan*, typographical errors have been corrected and page numbers provided in the Group Keys where a line leads to a plant species or family. Some changes have been made in the Group 4 (Dicots) Key for better identification of some families. A Quick Index to Plant Families has been added to the end of the book. Special thanks are due to Dr. James Doyle for suggesting this index and for preparing the first draft.

Additional useful references for aquatic plants:

Chadde, Steve W. 1998. A Great Lakes Wetland Flora: A Complete, Illustrated Guide to the Aquatic and Wetland Plants of the Upper Midwest. Pocketflora Press. 550 p. (ISBN 0-965-13852-6)

Crow, Garrett E., & C. Barre Hellquist. 2000. Aquatic and Wetland Plants of Northeastern North America. A Revised and Enlarged Edition of Norman C. Fassett's *A Manual of Aquatic Plants*. Vol. I: Pteridophytes, Gymnosperms, and Angiosperms: Dicotyledons. University of Wisconsin Press. 448 p. (ISBN 0-299-16330-X)

Crow, Garrett E., & C. Barre Hellquist. 2000. Aquatic and Wetland Plants of Northeastern North America. A Revised and Enlarged Edition of Norman C. Fassett's *A Manual of Aquatic Plants*. Vol II: Angiosperms: Monocotyledons. University of Wisconsin Press. 464 p. (ISBN 0-299-16280-X)

TABLE OF CONTENTS

ACKNOWLEDGMENTS

The author and editor wish to thank Dr. Edward G. Voss for making his unpublished manuscript of the third volume of *Michigan Flora* and his specimen data files available for preparation of this revision of *The Plants of Michigan*. Without his many years of meticulous and comprehensive research on the flora of Michigan, this work would not have been possible. Special thanks are also due to Dr. Voss for his review of the Preface and of the entire manuscript prior to publication.

We also wish to thank Dr. Anton A. Reznicek for reviewing the manuscript as sections were completed; his comments and suggestions have been invaluable in improving the keys. Dr. Reznicek also provided important information about habitats for individual Michigan plant species.

Thanks to Beverly S. Walters for reviewing the botanical terminology and glossary, and to Dr. William R. Anderson, the Director of the University of Michigan Herbarium, for permission to use the collections, library, and workrooms of the Herbarium in the course of preparation of this book.

The editor is grateful to Elizabeth K. Davenport of the George Wahr Publishing Company for her support and encouragement of this project. Mrs. Davenport initiated the plan to revise *The Plants of Michigan*, which was issued in its three original editions by George Wahr Publishing Company.

Finally, the author is indebted to his wife Karen for her support throughout the evolution of this work.

PREFACE
DR. HENRY ALLAN GLEASON: HIS MICHIGAN PERIOD

North American botanists may most readily associate Gleason's
name with The New York Botanical Garden, from using either *The
New Britton and Brown Illustrated Flora of the Northeastern United
States and Adjacent Canada* (1952) or the *Manual of Vascular
Plants of Northeastern United States and Adjacent Canada* (with
Arthur Cronquist, 1963, 1991). Before he wrote any of these botani-
cal classics, Dr. Henry Allan Gleason spent some years in Michigan,
during which he wrote a key to its flora, *The Plants of Michigan.*

Dr. Gleason came to Michigan in 1910, joining the University of
Michigan Botany Department as an Assistant Professor, being pro-
moted to the rank of Associate Professor six years later. His research
involved both plant taxonomy and plant ecology; the latter was a relatively new field of endeavor in which Gleason's work was to be
very influential for many years (Smith 1951). The first edition of
The Plants of Michigan (1918), his most extensive floristic work to
that date, appeared one year before he would leave the University.

Besides his duties in the Botany Department, Dr. Gleason became
involved in both teaching and administration at the Biological Station near Pellston. He taught during seven different summers at the
Station, offering one or more of the following courses: systematic
botany, field and forest botany, plant ecology, or plant anatomy.
After serving as Acting Director during 1913 and 1914, he was ap-
pointed Director of the Biological Station for the 1915 session
(Gates 1983). Several of his papers on Michigan plants arose from
his work at and around the Biological Station. Curiously, most of
these that I have encountered, including "The introduced vegetation
in the vicinity of Douglas Lake, Michigan" (1914, with Frank T.
McFarland), deal at least in part with the introduced rather than the
native flora of that area; an intriguing parallel to my own interests!

H. A. Gleason was also involved with the University of Michigan
Botanical Gardens. He served as Director of the "Botanical Gardens
and Arboretum" from 1915 until his departure from the University in
1919. During his tenure, he would have dealt with the move of the

Gardens from what is now the Nichols Arboretum to the Iroquois Street site, which preceded the present site on Dixboro Road east of Ann Arbor (Bartlett 1943).

After Dr. Gleason joined the staff of the New York Botanical Garden in 1919—Smith (1951) noted that Gleason was "prevailed upon" by Dr. N. L. Britton, Director of the NYBG and a key advisor to Gleason's earlier doctoral study there (1905–1906; Gleason 1976), to leave Michigan—his "Michigan connection" was not broken. He taught plant anatomy at the Biological Station for a final time in 1923 and spent several other summers conducting research at the station (Gates 1983, Smith 1951). Two more editions of *Plants of Michigan* appeared, the second in 1926, the third in 1939. Although titled "editions", the changes from his initial 1918 text were minor. In a 1925 letter from Gleason to George Wahr, the publisher, Gleason enclosed three pages "indicating the necessary changes for a second edition . . . only one line needs to be added."

The idea of revising the 3rd edition of *Plants of Michigan* dates back to 1950. After selling all of the ca.1000 copies of the 3rd edition printed in 1939, the publisher wrote to Gleason in May of 1950 suggesting that the work should be revised before being reprinted. Gleason liked the idea but withdrew from the project in December of 1950 because of poor health. Photolithoprinted copies of the 3rd edition (1939) were produced, beginning in 1951. Throughout the 1950's a total of 1900 copies were printed. The final reprinting appears to have been 500 copies produced in 1963.

Henry Allan Gleason published many works besides those already mentioned during his thirty-one-year career at the New York Botanical Garden; they included papers on ecology, systematics of South American plants (especially members of the family Melastomataceae), and a small volume that is strikingly similar to *The Plants of Michigan*. First appearing in 1935 (Gleason 1935), in a second "edition" (apparently identical except for correcting known errata in the 1935 work) twelve years later, (Gleason 1947), and as a "revised edition" (and enlarged) in 1962 (Gleason 1962), *Plants of the Vicinity of New York* arose from his interest in the local flora (Smith 1951) and is to serve for plant identification in "all the region within two hundred miles of New York City" (Gleason 1962). The layout of group keys leading to family treatments first used in *Plants of Michigan* is maintained; in fact, it is likely the model which Gleason followed for the introductory keys in his 1952 *The New Britton and Brown Illustrated Flora* and those in the later Gleason and Cronquist manuals (Voss 1996).

The major differences between Gleason's *Plants of the Vicinity of New York* and his *The Plants of Michigan* are the inclusion of line illustrations, a section discussing plant structure, and a key to Ferns and Fern Allies in the New York volumes. Would Gleason have added these sections to a subsequent revision of *The Plants of Michigan*? One can only suspect that, given the success of his edition of *The New Britton and Brown Illustrated Flora,* illustrations similar to those we have included might have appeared.

The preface which appears in each of H. A. Gleason's three editions of *The Plants of Michigan* was succinct. After stating that the book was "not intended for the expert botanist" nor for "the merely curious" and that "it is not a textbook", Gleason clarified its purpose in a single sentence:

> "Its mission is fully accomplished if, through its use, students, vacationists, and plant-lovers in general are able to recognize by name the plants about them."

I hope that, in this revised form, this book still fulfills the mission which Henry Allan Gleason foresaw some 80 years ago.

Richard K. Rabeler
1998

ADDITIONS AND CORRECTIONS

Page	Key	Additions and Corrections
20, 121, 127, 165, 226, 227, 237, 254, 262	various	These are all references to the Michigan "Christmas tree law" that prohibited removal of several wild plants from private property without permission. This law has been repealed.
55	Woody Plants	Line 89b. should lead to new couplet 89.5
55	Woody Plants	Add new couplet between 89 and 90: 89.5a. Leaves covered with silver or silver and brown scales beneath — *Elaeagnus* spp., in ELAEAGNACEAE, p. 240 89.5b. Leaves lack silver or silver and brown scales beneath — 90
56	Woody Plants	Line 109a. should lead to new couplet 109.5
56-57	Woody Plants	Add new couplet between 109 and 110: 109.5 a. Leaves covered with silver or silver and brown scales beneath — *Elaeagnus* spp., in ELAEAGNACEAE, p. 240 109.5b. Leaves lack silver or silver and brown scales beneath — 110
58	Woody Plants	Line 125b. should lead to line 127. Line 127b. should lead to line 128.
70	Dicots	Line 2b. should read: Leaves simple and entire, toothed, or lobed (even cleft deeply), but not dissected —15
70	Dicots	Line 8b. should read: Leaflets (or leaves) coarsely toothed or lobed — 12
71	Dicots	Line 14b. should read: Petals pink or blue (rarely white in albino flowers); stamens 5; ovary 1—14.5
71	Dicots	Add new couplet between 14 and 15: 14.5a. Petals pink; leaves pinnately dissected— *Erodium cicutarium*, in GERANIACEAE, p. 217 14.5b. Petals blue (rarely white in albino flowers); leaves deeply cleft, nearly dissected— *Viola* spp., in VIOLACEAE, p. 236
77	Dicots	Line 81b should lead to new couplet 87.5.
78	Dicots	Add new couplet between 87 and 88: 87.5a. Sepals greenish or white, petals absent— *Mollugo verticillata,* in MOLLUGINACEAE, p. 156 87.5b. Sepals usually green, petals present — 88
85	Dicots	Line 172a. should read "calyx of 5 sepals, the outer 2 much narrower than the *inner* 3"
191	Rosaceae	Line 3b. should read "each flower 10 mm or more wide"

Page	Key	Additions and Corrections
192	Rosaceae	Changes to couplet 8: 8a. Leaves pinnately compound (or if trifoliolate, then flowers pink); flowers pink or red, rarely white or yellow, 2–10 cm across; fruit of achenes enclosed in a fleshy receptacle (early summer) (***Rosa*** spp., **Rose**) — 9 8b. Leaves trifoliolate or palmately compound; flowers white, 1–3 cm across; fruit a cluster of fleshy drupelets (late spring) (***Rubus*** spp. in part, **Bramble**) — 19
240	Elaeagnaceae	Family description should read "Shrubs or small trees with opposite or alternate, simple, entire leaves covered with silvery and/or rusty scales."
272	Boraginaceae	Changes to couplet 3: 3a. Corolla tubular, 10mm long or more, the lobes erect, the tube distinctly longer than the calyx — 4 3b. Corolla funnelform or salverform, mostly less than 10 mm long, the lobes spreading, the tube equaling, shorter than, or occasionally longer than the calyx — 7
283	Labiatae	Changes to couplet 39: 39a. All flowers in terminal panicles; corolla two-lipped; occasional escape from cultivation (40-80 cm high; summer) — **Oregano**, ***Origanum vulgare*** 39b. All flowers in axillary whorls, corolla almost regular or two-lipped — 39.5
283	Labiatae	Add new couplet between 39 and 40: 39.5a. Corolla almost regular; moist areas (20-80 cm high) —**Wild Mint**, ***Mentha arvensis*** 39.5b. Corolla two-lipped; fields, roadsides, railroad rights-of-way, etc. (30–100 cm high, summer and autumn) — **Catnip**, ***Nepeta cataria***
290	Scrophulariaceae	Line 16a. should lead to new couplet 16.5.
290	Scrophulariaceae	Add new couplet between 16 and 17: 16.5a. Corolla with a spur which protrudes between the lower two lobes of the calyx; railroad ballast, roadsides, etc. (10–50 cm high) — **Dwarf Snapdragon**, ***Chaenorrhinum minus*** 16.5 b. Corolla lacks a spur — 17
290	Scrophulariaceae	Line 20b. should read: Filaments four, all fertile; corolla two-lipped *or not* — 26
347	Glossary	**Dissected**: Leaf blade, usually of a pinnate leaf, which is extremely finely divided (*Fig. 8*)
349	Glossary	Insert new term: **Lacinate**: cut into narrow segments
352	Glossary	Insert new term: **Scarious**: thin, of dry texture, not green
397	Index	Add Virginia Creeper, p. 231 after Violet

INTRODUCTION

Gleason's Plants of Michigan has been designed to be a quick reference for the plant collector or naturalist for identification of the flowering plants and gymnosperms most likely to be found in Michigan. The diagnostic keys are not all-inclusive. Some species of very limited distribution and those known only in cultivation have been omitted, and the numbers of individual species listed for some difficult groups have been limited. For a comprehensive treatment of all flowering plant species reported in Michigan, consult the definitive *Michigan Flora* (Voss 1972, 1985, 1996). The ferns and the "fern allies" (horsetails, club mosses, etc.) are not discussed here; see Billington (1952) or Lellinger (1985) for information on these plants. References noted in the text are listed in the Bibliography and should be consulted for further information on various groups. For full plant descriptions, consult reference works cited in the Bibliography, such as Gleason and Cronquist (1991) or Fernald (1987), or one of the references cited for a specific group.

How to use this book The keys in this book are designed to be used in the field, thus the *characters* on which the keys are based will be visible with the naked eye or with a hand lens. The only other tool required is a centimeter rule. A centimeter scale has been provided in the endpapers for your convenience.

The first key, the "Key to Groups" will direct you to one of four keys to the families of flowering plants. These keys make direct reference to the second set of keys, those for the individual families. Begin with this first set of keys if you are uncertain about the likely family to which your plant belongs. If you know the family, starting directly at the family key is most convenient. As you become familiar with some of the more common families, identification will become much more rapid. See the Quick Index, end of the book.

Each family key begins with a description of the characteristics of that family, especially the floral morphology. In some cases, Michigan representatives of a plant family may have only a limited range of the morphological features found in the family worldwide. Family descriptions used here are limited to characteristics of Michigan representatives. Where the family is represented in Michigan by a single species, the description is of that species. Families of plants not found in Michigan are not included.

The understanding of plant family relationships and the grouping of plants in families has advanced significantly since the original *The Plants of Michigan* by H.A. Gleason was published (1918). The names, arrangement, and circumscription of the families here follow Engler and Prantl as modified by Voss (1972, 1985, 1996). Reviews of concepts regarding plant family systems may be found in Brummitt (1992) and Zomlefer (1994).

Each key consists of a number of *couplets*, each consisting of two *leads*, or statements; the pairing of statements makes the key *dichotomous*. Each lead either ends with a number of a subsequent couplet or a plant name. To use a key, read the first pair of leads and decide which one describes the plant in hand. If that statement ends in a number, go to that couplet and repeat the process. If the statement ends in a plant name, you have completed your mission. Botanical terms used in the keys are discussed in Botanical Terminology, which follows this introduction, and terms are also defined in the Glossary. All measurements follow the metric system. The height of the plant and/or flowering season (adapted from Gleason and Cronquist 1991, in some cases enhanced with additional information from Swink and Wilhelm 1994, Voss 1972 and 1985, and specimens at the University of Michigan herbarium) is given in parentheses for many plants.

Plant names All plants have a Latin name, or what is commonly known as a *scientific name*. The application of these names is governed by the International Code of Botanical Nomenclature. The name, or *binomial*, consists of two Latin words. The first word is the name of the *genus* to which the plant belongs, e.g., *Acer*, the genus for maple trees. The second word, or *epithet*, together with the first word provides a unique name for a *species* of plant; combining *Acer* with the epithet *saccharum* gives the name *Acer saccharum* (the name of a common Michigan tree, the sugar maple). Another element associated with a scientific name is the *author*, the name of the person who first described the plant; in our example, it would be *Acer saccharum* Marshall. Authors for plant names have been omitted in this book. Nomenclature used in this volume follows that in Voss (1972, 1985, 1996) unless otherwise specified in the text. In some cases, alternate scientific names have been included in brackets, with the initials "CQ" serving as reference to Gleason and Cronquist (1991). A few other specialized references to alternate names are also included.

Several of the genera represented in the Michigan flora include many, even dozens, of species. In an attempt to group related species within such large genera, names at the rank of *subgenus*, *section*, or *tribe* are sometimes employed. A few such names are included in the keys, e.g., *Polygonum* sect. *Persicaria*, where these "groups" can be easily recognized in the field.

Many plants have one or more *common names*. Some of these names may already be familiar to you; sugar maple is probably much more familiar to many people than the scientific name *Acer saccharum*. Usage of common names, while convenient, is not without major pitfalls. In contrast to the specific rules which govern the application of scientific names, there are no such rules for assigning common names to plants. Some are descriptive while others are nothing more than a translation of the scientific name. Some names imply relationships where there is none: for example, *Portulaca grandiflora*, commonly called a "moss rose", is not a member of the Rosaceae; the weedy purple loosestrife, *Lythrum salicaria*, is a member of the Loosestrife Family (Lythraceae), but the goose-necked loosestrife, *Lysimachia clethroides*, is a member of the Primrose Family (Primulaceae). Some plants may have ten or more common names, others lack even one. They often vary by region and local usage. As an example, plants which are called "indian paintbrush" in Michigan (species of *Castilleja*) are not related to the plant with the same common name in upstate New York (*Hieracium aurantiacum* = orange hawkweed in Michigan). For the most part, common names as used by Voss (1972, 1985, 1996) are used here, or the common name assigned by Gleason in the original edition is retained.

Names of all plants included in this book can be found in the Index which includes both scientific (alternative or otherwise) and common names of families and individual species.

Plant distributions Much information about the nature and distribution of the *native* or *presettlement vegetation* (the plants existing in Michigan before European settlement) is available from the extensive information compiled by the U.S. General Land Office Survey between 1816 and 1856. As surveyors platted the territory into 36 mile-square townships, they also mapped natural features, including rivers, lakes, wetlands, and large "witness trees". Comments on soil and vegetation were also noted (Comer et al. 1995). Fortunately for the vegetation record, this survey occurred before the onset of logging. The information compiled by the Land Office Survey has now been transcribed, interpreted into a series of digital maps, and imposed on the "regional landscape ecosystems" described by Albert (1995). The resulting regional descriptions encompass soil, climate, topography, and presettlement or native vegetation types (Comer et al. 1995).

The distribution of many Michigan plants has changed over the last 150 years, with habitat loss from conversion of areas first to agriculture and later to other more intensive uses, massive logging operations in the late 1800's and early 1900's, and competition from introduced plants. *Introduced plants* are species which now grow in Michigan but are not members of our native flora; they are, for the most part, seldom found in natural communities. Included among our introduced flora are most agricultural weeds, plants that have escaped from cultivation, and those that have arrived accidentally as soil or seed contaminants. The list of introduced plants is always changing as new species arrive and others disappear. *Endangered* or *threatened* plants, native species now found only in one or a few areas, are legally protected under Federal and/or Michigan endangered species legislation (Michigan Department of Natural Resources 1992, Chittenden 1996).

Michigan includes several phytogeographical zones, each with distinctive climate, soil, and plant communities. A striking difference in plant communities is seen north and south of the *tension zone*, which may be imagined as 50 km on either side of a line drawn horizontally across the southern Lower Peninsula between the cities of Saginaw and Muskegon (Medley & Harman, 1987). The line between the NM and SLP regions in Figure 1 approximates this line. North of the tension zone, soil is often a "podzol", with a thin layer of humus over coarse lower horizons with poor water-retention capabilities.

South of the tension zone, soils generally have a higher content of organic matter, with a deeper dark top horizon and good water retention.

Information on plant distributions is derived from Voss (1972, 1985, 1996), where complete maps showing the distribution by county of all plant species known in Michigan may be found. The distribution of some plant species, both native and introduced, is limited to certain regions of Michigan (Fig. 1). In these cases, the following abbreviations are used in the keys to indicate these regions:

SLP: Southern Lower Peninsula
LP: Lower Peninsula
UP: Upper Peninsula
NM: Northern portion of the Lower Peninsula, and the Upper Peninsula
WM: Western Michigan, including counties on the Lake Michigan coast
SE: Southeastern portion of the Southern Lower Peninsula
SW: Southwestern portion of the Southern Lower Peninsula
Straits: area adjacent to the Straits of Mackinac

Species of very limited distribution are generally omitted from this book; however, some with a very limited distribution but which may be locally conspicuous have been included. In these cases, the specific location is indicated, e.g., "Pellston" (the location of the University of Michigan Biological Station). Most species known in Michigan only from Isle Royale have been omitted; see Slavick and Janke (1993) for a checklist of the Isle Royale flora. Species that have been collected only one or a few times in the state but not in the last 50 years are likewise omitted.

Plant communities and habitats Many different approaches to terminology for distinctive plant communities have been used (e.g., Comer et al. 1995); defining them is outside the scope of this book. However, the concept is an important one. In recent years the intrinsic value of particular plant communities, such as wetlands, to the natural system as a whole has been more widely recognized. Others, such as prairies, are valued for their aesthetic qualities and the diversity of the plant species they support. In addition, the existence and survival of endangered and threatened plants at a site is probably

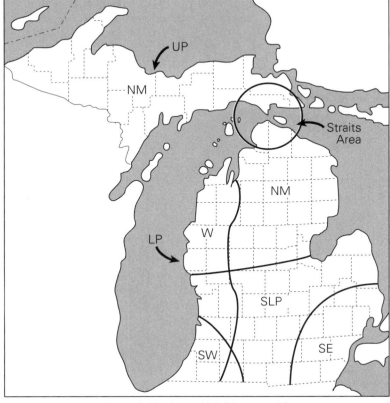

*Figure 1: Regions of Michigan used for
plant distribution information in keys*

linked to the survival of the entire plant community. Protection of
these special natural areas has emphasized the recognition of native
plants and the elimination of introduced or invasive plants. Vegetation analysis, in which plants are ranked for their value in a natural
landscape (Herman et al.1996, Swink & Wilhelm 1994), is becoming
a useful tool in making land use policy decisions.

The presettlement forests which once covered the northern Lower
Peninsula and eastern Upper Peninsula were chiefly hemlock-white
pine-northern hardwood (Braun 1950). Logging has shifted the bal-
ance in many areas of this region toward deciduous forests dominated by maples (Elliott 1953). Other dry northern sites have pine

forests, or oak-pine forests (Braun 1950). The major forest type in the southern Lower Peninsula is the beech-maple forest (Braun 1950), though highly drained (drier) areas are likely to be occupied by oak-hickory woods (Dodge & Harman 1984).

Two areas in Michigan with particularly distinct plant communities are the sand dunes along the shores of the Great Lakes (especially Lake Michigan and Lake Superior) and the western Upper Peninsula, particularly the shores of Lake Superior, the Keweenaw Peninsula, and Isle Royale. A number of plant species are found in this second area and also in the mountains of the Pacific Northwest (Marquis & Voss, 1981). These "western disjuncts" are in some cases species adapted to mountainous or arctic habitats.

Some general terms for plant community types are used in the keys to aid in identification. *Rich woods* include the beech-maple or beech-maple-hemlock communities. These are characterized by mesic conditions (abundant moisture, but the soil well-drained), and a rich soil such as loam or clay loam, with a nearly neutral pH. *Damp woods* are low spots with more moisture in these same communities. *Wet thickets* are dense, shrubby areas often found along streams or around swamps. *Dry woods* are oak-hickory, oak-pine, or mixed oak woods; the soil is often a sandy loam with faster drainage and an acidic pH. *Dry sandy woods* have soils containing more sand, and less organic matter; the moisture supply is very low and the pH is often more acidic. They are characterized by oak-pine or pine stands. *Upland habitats* may include meadows, fields, or sparse dry woodlands. *Swamps* are wet, periodically flooded, forested sites while *marshes* are wet areas which lack trees. *Bogs* often look like ponds, but are encircled by a floating mat of sphagnum moss; the moss gradually fills in the pond and helps create a habitat favorable for acid-loving plants such as members of the Ericaceae. *Fens* are calcareous (alkaline) wetlands, often meadow-like and dominated by sedges.

Some very general terms for characteristic habitats are also used in the keys. *Wet areas* indicates any place where the soil remains saturated or where there is standing water most of the time, such as low spots, marshes or swamps, and stream banks. *Moist areas* have soil which is noticeably moist to soggy most of the time. *Dry areas* occur where soil moisture is limited, such as sandy sites. *Disturbed areas* are where the soil surface has recently been broken, as in cultivated ground or construction sites. *Rocky areas* contain rock outcrops. *Calcareous areas* are derived from limestone, so that the pH is relatively alkaline.

Collecting a plant specimen   Try to gather fertile (flowering or fruiting) material if at all possible, since sterile specimens can be very difficult to identify. The specimen should be as complete as possible. With a small herbaceous plant, there is usually no problem in collecting the entire plant. With a shrub or tree, it is best to get a small branch with attached leaves and flowers and/or fruits if at all possible. Be sure to note whether it is a tree or shrub and the approximate height, information that cannot be gained from looking at just the specimen that you have collected. If you plan to press and dry your specimens for later study, be aware that while the green of the leaves remains, many floral parts don't retain their color. Making a note of the color of the petals, stamens, and/or fruit when you collect them will save guesswork later. A proper specimen is also labelled with the date and location of collection and any other useful information such as the habitat in which the plant was found.

Ethics of collecting   There are several legal and ethical considerations to keep in mind. Plant collecting is essential to the study of plants, but must not endanger the survival of local plant populations. Don't collect more individuals than necessary for your use. Some species which are legally protected in Michigan (see below) occur in large, but very local populations. Respect the rights of property owners by seeking permission to collect on any private property. Be aware that collecting specimens in state and national parks is not permitted. Permission for collecting on any public land should be obtained from the agency which manages it. Many privately held land preserves, such as those owned by the Nature Conservancy, also have restrictions on collecting.

Some plants in Michigan are legally protected from collection or sale without obtaining proper permission. Michigan Public Act 182 of 1962, the so-called Christmas Tree Act, makes collection of the following groups of plants illegal on private land without obtaining written permission or a bill of sale from the owner of the property:

"trailing arbutus, bird's foot violet, climbing bittersweet, club mosses, flowering dogwood, all Michigan holly, North American lotus, pipsissewa, and all native orchids, trilliums and gentians".

Under the Michigan Endangered Species Act, (Public Act 203 of 1974), a list of plants thought to be endangered, threatened, or of special concern on the recommendation of the Endangered Species Technical Advisory Committee is given legal protection. The list is periodically updated (see Michigan Department of Natural Resources 1992). The status of endangered species is not included here.

Cautions Some plants are known for their irritant properties, being either irritating to the touch or poisonous to eat. Become familiar with the most common poisonous or allergenic plants for your own safety. At a minimum, recognition of poison-ivy, poison sumac, and stinging nettle is highly recommended, as many people are sensitive to the chemicals which these plants contain.

Some plants have edible fruit or other parts. Information on edibility is not included in this text, nor are poisonous plants identified. Before sampling *any* plant, be certain of its identity, and consult a reference on edible plants. Don't rely on common names which may be suggestive, nor on the appearance of the plant.

Botany in Michigan There are a number of options for anyone who desires to learn more about the Michigan flora. Courses on the local flora may be available at a nearby high school, community college, county extension office, botanical garden, or nature center. Enrollment in a course at either the University of Michigan Biological Station near Pellston or Michigan State University's Kellogg Biological Station at Hickory Corners is a way to get an in-depth experience. The W.J. Beal Botanical Garden at Michigan State University in East Lansing includes a collection of labelled Michigan native plants.

A good way to become familiar with the flora is to join one of the organizations which sponsor the study of the flora of Michigan. The oldest of these is the Michigan Botanical Club (c/o Herbarium, 2001 North University Building, University of Michigan, Ann Arbor, MI 48109-1057), founded in 1940. The Club has five chapters (*Huron*

Valley, Ann Arbor, *Southeast*, the Detroit Metropolitan Area, *Red Cedar*, East Lansing, *White Pine*, Grand Rapids, and *Southwest*, Kalamazoo), each sponsoring a variety of lectures and field trips. Since 1962, the Club has sponsored the publication of the quarterly scientific journal, *The Michigan Botanist*, devoted to articles about the botany of the Upper Great Lakes region; a number of articles from that journal are cited in the Bibliography.

One of the most important resources for any flora is the herbarium. A herbarium is a collection of dried plants, each specimen pressed, dried, and mounted on paper with a label describing when and where the plant was found. These specimens are a permanent record of the flora and are regularly consulted by researchers interested in working on floras and in-depth studies of individual plant genera and species. The most important collections of the Michigan flora are at herbaria at the University of Michigan and Michigan State University.

BIBLIOGRAPHY

Books and articles cited in the text

Albert, Dennis A. 1995. Regional Landscape Ecosystems of Michigan, Minnesota, and Wisconsin: A Working Map and Classification. USDA Forest Service, North Central Forest Experiment Station General Technical Report NC-178. 250 p.

Bailey, Liberty H. 1949. Manual of Cultivated Plants, rev. ed. Macmillan Publ. Co., New York. 1116 p.

Bailey, Liberty H., & Ethel Zoe Bailey. 1976. Hortus Third. Revised and expanded by the staff of the Liberty Hyde Bailey Hortorium. Macmillan Publ. Co., New York. xiv + 1290 p.

Ballard, Harvey. 1994. Violets of Michigan. Michigan Bot. 33: 131–199.

Ballard, Harvey. 1995. Addenda and errata: Violets of Michigan. Michigan Bot. 34: 84.

Barnes, Burton V., & Warren H. Wagner, Jr. 1981. Michigan Trees. A Guide to the Trees of Michigan and the Great Lakes Region. University Mich. Press, Ann Arbor. viii + 383 p. (ISBN 0-472-0818-0)

Bartlett, Harley H. 1943. The Botanical Gardens. p. 506–512 in Wilfred B. Shaw, ed. The University of Michigan an Encyclopedic Survey, Part III (I). Univ. Michigan Press, Ann Arbor.

Bayer, Randall J., & G. Ledyard Stebbins. 1993. A synopsis with keys for the genus *Antennaria* (Asteraceae: Inuleae: Gnaphaliinae) of North America. Canad. J. Bot. 71:1589–1604.

Beales, Peter. 1992. Roses: An Illustrated Encyclopaedia and Grower's handbook of Species Roses, Old Roses and Modern Roses, Shrub Roses and Climbers. Holt and Co., New York. iv + 472 p. (ISBN 0-8050-2053-5)

Billington, Cecil. 1952. Ferns of Michigan. Bull. Cranbrook Inst. Sci. 32. vii + 240 pp.

Bliss, Margaret. 1986. The morphology, fertility, and chromosomes of *Mimulus glabratus* var. *michiganensis* and *M. glabratus* var. *fremontii* (Schrophulariaciaeae)[sic]. Amer. Midl. Naturalist 116: 125–131.

Braun, E. Lucy. 1950. Deciduous Forests of Eastern North America. Blakiston Co., Philadelphia. 566 p.

Brewer, Lawrence G. 1982. Present status and future prospect for the American chestnut in Michigan. Michigan Bot. 21: 117–128.

Brewer, Lawrence G. 1995. Ecology of survival and recovery from blight in American chestnut trees (*Castanea dentata* (Marsh.) Brkh.) in Michigan. Bull. Torrey Bot. Club 122: 40–57.

Brummitt, Richard K. 1992. Vascular Plant Families and Genera. Royal Botanic Gardens, Kew, England. 804 p. (ISBN 0-947643 435)

Case, Frederick W., Jr. 1987. Orchids of the Western Great Lakes Region, rev. ed. Bull. Cranbrook Inst. Sci. 48. xxi + 251 p. (ISBN 87737-036-2)

Case, Frederick W., Jr., & Paul M. Catling. 1982. The genus *Spiranthes* in Michigan. Michigan Bot. 22:79–92

Chittenden, Elaine M. 1996. Endangered and Threatened Plants in Michigan. W. J. Beal Botanical Garden, Michigan State Univ., East Lansing. 53 p.

Comer, P.J., D.A. Albert, H.A. Wells, B.L. Hart, J.B. Raab, D.L. Price, D.M. Dashian, R.A. Corner, & D.W. Schuen. 1995. Michigan's Native Landscape, as interpreted from the General Land Office Surveys 1816–1856. Report to the U.S. E.P.A. Water Division and the Wildlife Division, Michigan Department of Natural Resources. Michigan Natural Features Inventory, Lansing. 76 p.

Cronquist, Arthur. 1976. Dr. H. A. Gleason: an appreciation. Garden J. 26: 56–59.

Cruden, Robert W., Ann M. McClain, & Gokaran P. Shrivastava. 1996. Pollination biology and breeding system of *Alliaria petiolata* (Brassicaceae). Bull. Torrey Bot. Club. 123: 273–280.

Dietrich, Werner, Warren L. Wagner, & Peter H. Raven. 1997. Systematics of *Oenothera* sect. *Oenothera* subsection *Oenothera*. Syst. Bot. Monogr. 50: 234 p.

Dodge, Sheridan L. & Jay R. Harman. 1984. Woodlot composition and successional trends in south-central lower Michigan. Michigan Bot. 24: 43–54

Dore, William G., & John McNeill. 1980. Grasses of Ontario. Agriculture
Canada, Res. Branch Monogr. 26: 566 p. (ISBN 0-660-10452-0)

Elliott, Jack C. 1953. Composition of upland second growth hardwood stands
in the tension zone of Michigan as affected by soils and man. Ecol.
Monogr. 23: 271–288

Fassett, Norman C. 1957. A Manual of Aquatic Plants with revision appendix
by Eugene C. Ogden, rev. ed. Univ. Wisconsin Press, Madison. ix +
405 p. (ISBN 0-299-01450-9)

Garlitz, Russ. 1992. The spread of *Puccinellia distans* (reflexed saltmarsh
grass) in Michigan. Michigan Bot. 31: 69–74.

Gates, David M., ed. 1983. The University of Michigan Biological Station
1909–1983. Univ. of Michigan Biol. Sta., Ann Arbor. 116 p.

Gereau, Roy E., & Richard K. Rabeler. 1984. Eurasian introductions to the
Michigan Flora. II. Michigan Bot. 23: 51–56.

Gleason, Henry A. 1913. Some interesting plants from the vicinity of Douglas
Lake. Rep. Michigan Acad. Sci. 15: 147–149.

Gleason, Henry A. 1918. The Plants of Michigan. George Wahr, Ann Arbor.
xlvii + 158 p.

Gleason, Henry A. 1935. Plants of the Vicinity of New York. The New York
Botanical Garden, New York. lxxxvi + 198 p.

Gleason, Henry A. 1947. Plants of the Vicinity of New York. The New York
Botanical Garden, New York. lxxxvi + 198 p.

Gleason, Henry A. 1952. The New Britton and Brown Illustrated Flora of the
Northeastern United States and Adjacent Canada. The New York
Botanical Garden, New York. 1: lxxv + 482 p.; 2: iv + 655 p.; 3: iii +
595 p.

Gleason, Henry A. 1962. Plants of the Vicinity of New York, 3rd ed. The New
York Botanical Garden, New York. 307 p.

Gleason, Henry A., & Arthur Cronquist. 1963. Manual of Vascular Plants of
Northeastern United States and Adjacent Canada. Van Nostrand Rein-
hold Co., New York. li + 810 p.

Gleason, Henry A., & Arthur Cronquist. 1991. Manual of Vascular Plants of Northeastern United States and Adjacent Canada, 2nd ed. The New York Botanical Garden, Bronx, NY. lxxv + 910 p. (ISBN 0-89327-365-1)

Gleason, Henry A., & Frank T. McFarland. 1914. The introduced vegetation in the vicinity of Douglas Lake, Michigan. Bull. Torrey Bot. Club 41: 511–521.

Harlow, William M. 1959. Fruit Key and Twig Key to Trees and Shrubs. Dover Publications, Inc., New York. vii + 50 + vi + 56 + vi p.

Herman, Kim D., Linda A. Masters, Michael R. Penskar, Anton A. Reznicek, Gerould S. Wilhelm, and William W. Brodowicz. 1996. Floristic quality assessment with wetland categories and computer applications programs for the State of Michigan. Michigan Department of Natural Resources, Wildlife Division, Natural Heritage Program, Lansing. iii + 21 p. + Appendices.

Kartesz, John T. 1994. A Synonymized Checklist of the Vascular Flora of the United States, Canada, and Greenland, 2nd ed. Timber Press, Portland, OR. 1: lxi + 622 p.; 2: vi + 816 p. (ISBN 0-88192-204-8)

Large, E. C. 1962. Advance of the Fungi. Dover Publications, Inc., New York. 488 p.

Lellinger, David B. 1985. A Field Manual of the Ferns & Fern-allies of the United States & Canada. Smithsonian Institution Press, Washington, DC. ix + 389 p.

Luken, James O., & John W. Thieret. 1995. Amur honeysuckle (*Lonicera maackii*; Caprifoliaceae): its ascent, decline, and fall. Sida 16: 479–503.

Marquis, Robert J., & Edward G. Voss. 1981. Distributions of some western North American plants disjunct in the Great Lakes region. Michigan Bot. 20: 53–82.

Medley, Kimberly E. & Jay R. Harman. 1987. Relationships between the vegetation tension zone and soils distribution across central lower Michigan. Michigan Bot. 26: 78–87

Michigan Department of Natural Resources. 1992. Michigan's Special Plants. Wildlife Division, Lansing. iii + 16 p.

Nuzzo, Victoria A. 1993. Current and historic distribution of garlic mustard (*Alliaria petiolata*) in Illinois. Michigan Bot. 32: 23–33.

Rabeler, Richard K. 1988. Eurasian introductions to the Michigan flora. IV. Two additional species of Caryophyllaceae in Michigan. Michigan Bot. 27: 85–88.

Rabeler, Richard K., & Roy E. Gereau. 1984. European introductions to the Michigan flora. I. Michigan Bot. 23: 39–47.

Reznicek, Anton A. 1980. Halophytes along a Michigan roadside with comments on the occurrence of halophytes in Michigan. Michigan Bot. 19: 23–30.

Ronse Decraene, Louis-Philippe, & J. R. Akeroyd. 1988. Generic limits in *Polygonum* and related genera (Polygonaceae) on the basis of floral characters. Bot. J. Linn. Soc. 98: 321–371.

Slavick, Allison D., & Robert A. Janke. 1993. The Vascular Flora of Isle Royale National Park. Isle Royale Natural History Association, Houghton. vi + 50 p. (ISBN 0-935289-05-4)

Smith, A. C. 1951. Dr. H. A. Gleason retires from The New York Botanical Garden. Garden J. 1: 53, 56, 64.

Stephenson, Stephen N. 1984. The genus *Dichanthelium* (Poaceae) in Michigan. Michigan Bot. 23: 107–119.

Swink, Floyd, & Gerould Wilhelm. 1994. Plants of the Chicago Region, 4th ed. Indiana Academy of Science, Indianapolis. xiv + 921 p.

Voss, Edward G. 1967. A vegetative key to the genera of submersed and floating aquatic vascular plants of Michigan. Michigan Bot. 6: 35–50.

Voss, Edward G. 1972. Michigan Flora. Part I. Gymnosperms and Monocots. Bull. Cranbrook Inst. Sci. 55 and Univ. Michigan Herbarium. xv + 488 p. (ISBN 87737-032)

Voss, Edward G. 1985. Michigan Flora. Part II. Dicots (Saururaceae–Cornaceae). Bull. Cranbrook Inst. Sci. 59 and Univ. Michigan Herbarium. xix + 724 p. (ISBN 87737-037-0)

Voss, Edward G. 1996. Michigan Flora. Part III. Dicots (Pyrolaceae–Composi-
    tae). Bull. Cranbrook Inst. Sci. 61 and Univ. Michigan Herbarium. xix
    + 622 p. (ISBN 87737-040-0)

Voss, Edward G., & Mark W. Böhlke. 1978. The status of certain hawkweeds
    (*Hieracium* subgenus *Pilosella*) in Michigan. Michigan Bot. 17:
    35–47.

Widrlechner, Mark P. 1983. Historical and phenological observations on the
    spread of *Chaenorrhinum minus* across North America. Canad. J. Bot.
    61: 179–187.

Zelmer, Carla D., & R. S. Currah. 1995. Evidence for a fungal liaison between
    Corallorhiza trifida (Orchidaceae) and *Pinus contorta* (Pinaceae).
    Canad. J. Bot. 73: 862–866.

Zomlefer, Wendy B. 1994. Guide to Flowering Plant Families. Univ. North
    Carolina Press, Chapel Hill. xiv + 430 p. (ISBN 0-8078-4470-5)

Other Books Useful in Identification of Michigan Plants

Billington, Cecil. 1949. Shrubs of Michigan, 2nd ed. Bull. Cranbrook Inst.
    Sci. 20: vi + 339 p.

Fassett, Norman C. 1976. Spring Flora of Wisconsin, 4th ed. rev. by Olive S.
    Thomson. Univ. Wisconsin Press, Madison. ix + 413 p. (ISBN 0-299-
    06750-5/0-299-06754-8)

Fernald, Merritt L. 1987 [facsimile reprint of 1970 printing of 1950 work].
    Gray's Manual of Botany, 8th ed. Dioscorides Press, Portland, OR.
    lxiv + 1632 p. (ISBN 0-931146-09-7)

Hoagman, Walter J. 1994. Great Lakes Coastal Plants. Michigan State Univ. &
    Michigan Dept. Nat. Resources. 135 p. (ISBN 1-56525-008-7)

Holmgren, N.H. 1998. Illustrated Companion to Gleason and Cronquist's
    Manual: Illustrations of the Vascular Plants of Northeastern United
    States and Adjacent Canada. New York Botanical Garden, Bronx NY
    xvi + 937 p. (ISBN 0-89327-399-6)

Lund, Harry C. 1998. Michigan Wildflowers in Color. Thunder Bay Press,
    Holt. 144 p. (ISBN 1-882376-56-0)

Smith, Helen V. 1966. Michigan Wildflowers, rev. ed. Bull. Cranbrook Inst.
    Sci. 42: xii + 468 p. (ISBN 87737-019-2)

TERMINOLOGY USED IN PLANT DESCRIPTIONS

This is a review of terms used in these keys for identifying plants. Terms are also defined in the Glossary; some specialized terms are defined within the family descriptions at the beginning of each family.

Flowers

The foundation of plant taxonomy is the classification of plants according to their floral structure. Plants are most easily identified in this or any other key if they are in flower or fruit. A description of the typical floral structure is included in each family description.

Flowers may include the following parts, from the inside out: *pistil* (or pistils), *stamens, petals*, and *sepals* (Fig. 2). These are attached to the *receptacle*. The stalk which supports the flower is the *pedicel*. When both stamens (containing male gametes) and pistil (containing female gametes) are in the same flower, the flower is *perfect*; when only one of these organs is present, the flower is *imperfect*. If flowers are imperfect but flowers of both sexes (both *staminate* and *pistillate* flowers) are present on one plant, the plant is *monoecious*. When a single sex is present on one plant (*i.e.*, two plants of different sexes are necessary for pollination), the plant is *dioecious*.

A *pistil* consists of an *ovary* containing one or more *ovules* (which develop into seeds), the (often) slender *style*, and the *stigma* or receptive surface at the tip of the style. The pistil is formed from one or more ovule-bearing units known as *carpels.* The number of carpels included in a pistil often can be determined by counting the number of styles and/or stigmas arising from the ovary. The interior of the ovary is often divided into chambers or "cells" known as *locules*. These are noted in family descriptions as, for example, "2-celled". The number of locules is related to the number of carpels, although the loss of internal partitions between adjacent carpels may reduce the number of locules. A flower may have more than one pistil; in such flowers, each pistil consists of a single carpel and each style can be traced to a different ovary.

The position of the ovary with respect to the receptacle is also an important character. The ovary may be above the receptacle, so that the petals and sepals are attached beneath the ovary (Fig. 12); this is a *superior ovary*. When the ovary is enclosed by the floral tissue, so

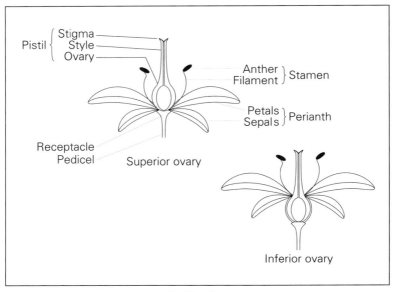

Figure 2: Floral structure

that the petals and sepals are attached at the top of the ovary, it is an *inferior ovary* (Fig. 39).

These keys have been designed to require a minimum of dissection. However, to determine whether the ovary of a flower is superior or inferior, it may be helpful to use a knife to make a longitudinal cut through the center of the flower. Use a hand lens if necessary.

A *fertile stamen* consists of a slender *filament* and the *anther*, which contains pollen. Some stamens may be *sterile*, meaning that a stamen-like structure exists without a pollen-producing anther. This sterile stamen (or *staminode*) may also be petal-like (*petaloid*) in appearance, *i.e.*, expanded and colored. The fusion of stamen filaments with each other (so that they are *connate*) or the other floral parts can be significant in identification.

The *sepals* and *petals* together are the *perianth* of the flower. The number of sepals (together called the *calyx*) and petals (the *corolla*) is often an important character; a plant description might indicate "parts in 4s" or that the "calyx and corolla are 4-merous". While the sepals of most plants are green, they may sometimes be petaloid. In some flowers, it is difficult to distinguish sepals from petals, thus the plant description may simply refer to *tepals*. The sepals and petal may be separate (*free*) or more or less fused to one another (*united*). Fusion of parts is noted in family descriptions; if no mention is made, it should be assumed that the individual sepals and petals are separate.

The flower may be subtended by specialized leaves known as *bracts*. In some families, such as the Euphorbiaceae, the bracts may be colored or otherwise prominent and resemble petals (Fig. 26). A tightly organized arrangement of bracts called the *involucre* forms the base of the head of the Compositae and is also present in flowers of some other families. Flowers may be *regular* with all parts of a whorl being the same size and shape. Regular flowers (Figs. 13, 19, 34) are usually radially symmetrical; a cut through more than one direction will produce two similar halves. *Irregular* flowers are most often bilaterally symmetrical, where floral parts are grouped or fused so that a cut through only one plane will produce two equal halves (Figs. 23, 36, 39), or they may be entirely asymmetrical.

Inflorescences

Flowers occur in arrangements which are often characteristic of the plant species. The entire flowering portion of the plant, including flowers, "stems" (pedicels), and bracts, is referred to as the *inflorescence* (Fig. 3). When the flower or inflorescence is at the top or end of the stem, it is *terminal*. When the inflorescence or flower is located in an *axil* (usually subtended by leaves), it is *axillary* (or *lateral*). *Solitary* flowers are often terminal, sometimes on a *scape* arising from a rosette of leaves. Flowers on pedicels along a single unbranched axis, erect or drooping, form a *raceme* (Fig. 23). When pedicels are not present, the flowers are *sessile* and the inflorescence is a *spike*. The basic branched inflorescence is the *panicle*, in which pedicels are located along several branches arising from a central axis. In an *umbel*, all pedicels arise from a central point, with the flowers arranged in a flat plane or a convex rounded cluster (Fig. 14). A *corymb* has a similar appearance, but the pedicels arise at different points along the stem. Both the umbel and corymb may be compound, with different tiers of pedicels within the inflorescence

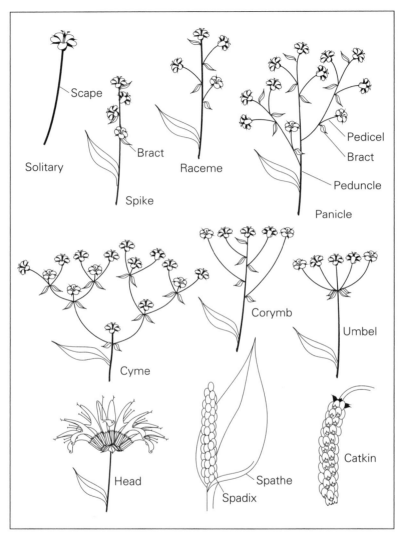

Figure 3: Inflorescence types

(Fig. 28, compound umbel). Combinations of two inflorescence types may occur, such as a panicle of racemes (Fig. 15). The *cyme* may take many forms, but is always characterized by a single central flower which matures before the rest of the inflorescence (Figs. 22,

34). The *head* is composed of flowers tightly arranged on a flat or discoid axis (Fig. 41). The typical inflorescence of the Araceae is the *spadix*, a fleshy spike, subtended by a large bract, the *spathe*. *Catkins* are typical of many woody species; they are spikes or racemes of unisexual flowers.

Fruits

The type of fruit produced is often a key character in plant identification; an ideal specimen includes both flowers and fruits. Fruits are the ripened ovaries of flowers, contain the seeds, and may include accessory structures. Some types of fruit are *dry* at maturity, while others have a *fleshy* fruit wall.

Indehiscent dry fruits do not open to release seeds at maturity. The fruit wall and the seed coat are fused together in the *grain*, the characteristic fruit of the Gramineae. The *achene* is also a small single-seeded fruit, with a dry thin fruit wall closely appressed to the seed coat. Wings or other appendages may be attached to the achene (Fig. 15); a *samara* is a winged achene typical of some woody plants. A *utricle* is a small single-seeded fruit resembling an achene, but with the fruit wall loose from the seed coat. *Nuts* and *nutlets* are dry indehiscent fruits with hard fruit walls. Sometimes there is an involucre attached to the nut, as in the *acorn* (the characteristic fruit of *Quercus* spp., Oaks).

Dehiscent dry fruits open to release seeds. The *capsule* develops from an ovary with more than one carpel, so that internal partitions between locules ("cells") are often visible. Seeds are released when the capsule opens at the top or sides. A *follicle* forms from one carpel and splits along one seam (Fig. 32). A *legume* is also formed from one carpel but splits along two seams; it is the typical fruit of the Leguminosae. The *silicle* and *silique* each split along both sides, leaving a membranous partition in the middle; they are typical of the Cruciferae (Figs.19, 20). Pieces of the dry fruit wall which separate from one another during dehiscence are called *valves*. A *schizocarp* arises from an ovary where the mature carpels split away from one another, so that each unit (a *mericarp*) resembles an entire indehiscent fruit.

Fleshy fruits are all indehiscent. The *berry* is several-celled (though partitions may not be visible at maturity) and many-seeded. A *drupe* has a single, usually large seed. A *pome* is typical of some Rosaceae (such as *Malus*, apple), in which the swollen, fleshy receptacle encloses the papery fruit wall. Some plants have multiple fruits which develop from the entire inflorescence.

Leaves

The leaf consists of a *petiole* (usually stalk-like) and an expanded *blade* (Fig. 4). The leaf is attached to the stem at the *node*, while the length of the stem between leaf attachments is the *internode*. Either the petiole or the blade may be modified or missing; the leaves of some plants lack blades and consist only of petioles, flattened or not. *Sessile* leaves only have a blade. In other leaves, the petiole may be flattened to form a

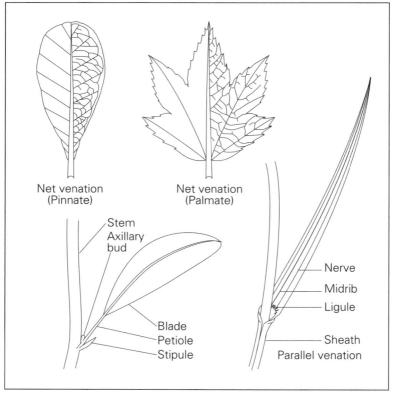

Figure 4: Leaf structure

sheath, which encircles the stem; a *ligule*, a ridge of tissue at the top of the sheath, may be present. Small (usually) leaflike structures, the *stipules*, are often present at each side of the point of leaf attachment. The space described by the angle between the upper side of the leaf attachment and the stem is referred to as the *axil*. The *axillary bud* may eventually expand to produce a branch or a flower.

Venation is an important character for identification (Fig. 4). The central vein (the *midrib*) is usually the most prominent. When other especially prominent veins are visible, they are called *nerves*. In leaves with *net venation*, two patterns of venation are possible. Secondary veins arise in two rows along the midrib in a leaf with *pinnate* venation, while secondary veins arise from a central point near the petiole in *palmate* venation. In leaves with *parallel venation*, all major veins are parallel to the midrib.

Leaf arrangement (Fig. 5) may be *alternate*, in which one leaf is attached at a node, *opposite*, with a pair of leaves at a node, or *whorled*, with three or more leaves at a node.

Leaves may have an undivided blade (a *simple* leaf), or may be divided into *leaflets* (a *compound* leaf). *Pinnately compound* leaves are based on pinnate venation, with parts of the midrib exposed between individual leaflets (Fig. 5). *Palmately compound* leaves are based on palmate venation, with leaflets meeting at a central point. A *ternately compound* leaf has three leaflets, based on palmate venation. A *trifoliolate* leaf has three leaflets, based on pinnate venation. Leaves may also show two or even three levels of compoundness; a twice-pinnately-compound leaf is referred to as *bipinnate*. A simple rule to distinguish leaflets from leaves is to look for an axillary bud; leaves have them, leaflets do not.

Leaf shapes are shown in Fig. 6. Terms for the shape of the leaf apex and leaf base are illustrated in Fig. 7. *Peltate* leaves are often round, with a centrally attached petiole. *Perfoliate* leaves are sessile, with the leaf base entirely surrounding the stem. *Decurrent* leaf bases continue down the stem as ridges of tissue. Leaf margins (Fig. 8) may be *entire* (lacking teeth or indentations), *toothed*, or *lobed*. Leaves may be *pinnately lobed* (based on pinnate venation) or *palmately lobed*. Deeply lobed leaves, with sinuses extending three-quarters of the distance from the margin to the midrib or more, are

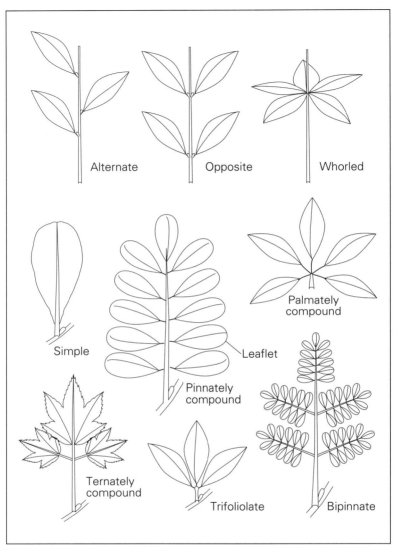

Figure 5: Leaf arrangement; compound leaves

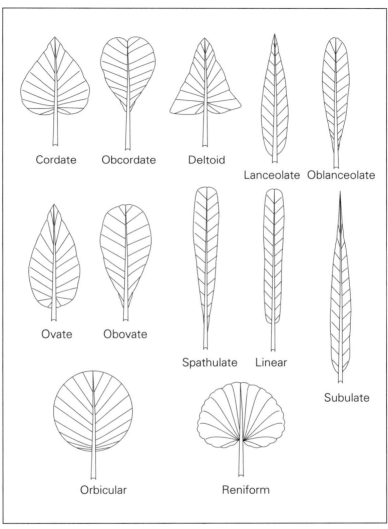

Figure 6: Leaf shapes

divided. Leaves divided into many narrow segments, producing a feathery appearance, are *dissected* (Fig. 8). Note that lobed, divided, and dissected leaves still retain some blade tissue along the midrib, in contrast to compound leaves. Terms for different types of toothed margins are illustrated in Fig. 8.

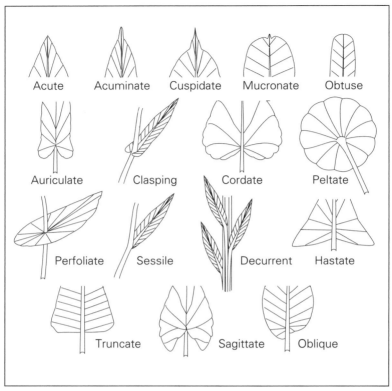

Figure 7: Leaf apices; leaf bases

Surfaces

Stems, leaves, and floral parts may have distinctive surface structures or appearance which are useful in identification. A magnifying glass or hand lens is sometimes useful to distinguish them. A surface without hairs is *glabrous*. A hairy surface is *pubescent* and is covered by *pubescence*, which can be quite variable in length, shape, and texture. For example, a *pilose* surface is covered with fine, thin hairs, while a *tomentose* surface is densely hairy. Hairs may have special shapes, such as hooked or *stellate* (star-shaped). Hairs which point down towards the base of the plant are *retrorse*. A *glaucous* surface is noticeably waxy, with a bluish or whitish bloom; a *glan-*

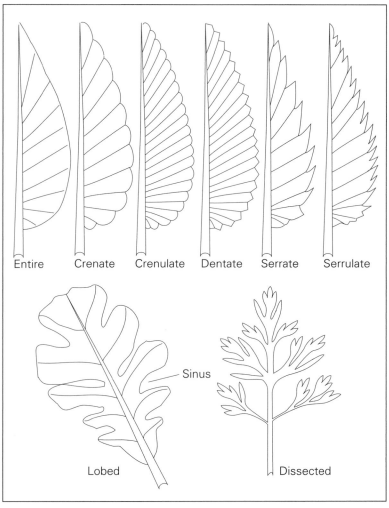

Figure 8: Leaf margins

dular surface has visible, often darker, gland cells. Surfaces may be *punctate*, with tiny depressions, or *tuberculate*, with small protrusions. *Prickles* are sharp pointed extrusions and may be found on any surface, while *thorns* are hard, sharp modified branches found at axils and stem tips.

Plant form

Terrestrial plants fall into two obvious categories. Woody plants include trees, shrubs, and woody vines. They typically persist over many growing seasons as above-ground woody structures. Herbaceous plants (sometimes called *herbs*) consist of succulent green tissue which does not persist through an entire year (typically dying back in winter) in Michigan. Plants which form rosettes of leaves close to the ground are an exception. *Subshrubs* are mostly herbaceous plants with a small amount of persistent woody tissue; the upper parts die back as with herbs.

Aquatic plants may be submerged entirely, floating, emergent (lifted somewhat above the water), or merely growing in very wet soil at water's edge.

The stem growth habit of a species may be distinctive. Plants are upright (*erect*) or *prostrate*, growing completely flat on or along the ground. They may also be *decumbent* (stems trail along the ground, but tips are erect), *spreading* (stems are held more or less horizontal), or *ascending* (stems angle upward). Plants dependent on other plants or objects for support are usually considered to be *vines*. These may *climb* (lean on trees or shrubs, anchored with thorns, tendrils, or branches) or *twine* (wrap around stems of the other plant).

The pattern of maturation and flowering is a useful character in identification.Woody species and many herbs are *perennial*, in that the plant lives for more than two growing seasons and (usually) flowers each season. Herbaceous perennials in Michigan die back to a root system, *crown* (a thickened stem with buds), or a flat *rosette* of leaves. Plants which are not perennial exhibit several possible flowering patterns. *Annuals* germinate from seed, mature, flower, fruit, and die in one growing season; most often this is during spring, summer, and fall. However, *winter annuals* germinate in the fall, form a rosette, and bolt (send up a flower stalk) with the onset of warm weather and the correct daylength, often completing their entire life cycle by late spring. *Biennials* germinate in the spring or summer of one growing season, overwinter (usually as a rosette) and then flower and fruit in the second growing season.

Reading a family description

The description of each plant family contains information in a compact format. This representative, though fictional, family description has been annotated to assist in interpretation of the format. Note that the numbers of floral parts are sometimes shown in ranges; in these cases, the number of that part varies, with numbers in parentheses representing unusual conditions. The flowering season is indicated at the end of the description when all Michigan representatives of the family flower in the same season.

Small, monoecious (**look for two sexes of flowers on one plant**), aquatic herbs with opposite, entire leaves. Flowers regular, unisexual (**staminate or pistillate**), small, 1 or 2 per leaf axil (**either condition is likely**); sepals 0; petals 0 (**sepals and petals are absent**); stamens 1–3(5) (**usually 1 to 3 stamens, but can have up to 5**); pistil 1, styles 2 (**a single ovary with two styles**); ovary superior (**a longisection will show that the ovary sits on top of the receptacle**), 4-celled (**a cross-section of the ovary will show 4 compartments**). Fruit splitting into 4 nutlets. Summer.

ILLUSTRATIONS

KEY TO GROUPS

The simplest way to identify a new plant is to go directly to the key for the family to which it belongs. Plant families are listed in the Index under Latin and common names. If the family is not known, this brief key will lead to one of four Group Keys, which will help you find the family for an unfamiliar plant.

These keys to the Plants of Michigan are for identification of gymnosperms (Pinophyta) and flowering plants (Magnoliophyta) only. Non-vascular plants and the ferns (Polypodiophyta) and "fern allies" (Lycopodiophyta, or clubmosses, spikemosses, and quillworts; and Equisetophyta, or horsetails) are not included. Ferns are recognized by their (usually) finely dissected leaves, circinate leaf development ("fiddleheads") and spore clusters (sori) on the backs of leaves or in specialized fronds. Club mosses and horsetails do not have true leaves. See Billington (1952) or Lellinger (1985) for more information.

1a. Trees, shrubs, or woody vines—Group 1, Woody Plants
1b. Herbaceous plants, with little or no woody tissue—2

2a. Plants aquatic, submerged or floating; flowers or fruits may not be readily visible—Group 2, Unusual Plants
2b. Plants terrestrial or with leaves and flowers emergent if growing in shallow water—3

3a. Plant consists of flowers and supporting stems only, with no vegetative parts visible—Group 2, Unusual Plants
3b. Plant consists of both flowers and vegetative parts—4

4a. Leaves are missing or scale-like; plants may be colorless, or yellow to brown, or purplish with no green tissue obvious during the growing season—Group 2, Unusual Plants
4b. Leaves present; plants are green in most parts—5

5a. Leaves linear to sword-shaped, with parallel venation; petioles are absent and leaf bases sheath the stem—6
5b. Leaves narrowly lanceolate to broad in outline, with net or parallel venation; petioles present or absent—7

6a. Floral parts in 3s or 6s—Group 3, Monocots
6b. Floral parts in 4s or 5s—Group 4, Herbaceous Dicots

7a. Leaves simple, with parallel venation; floral parts in 3s or 6s—Group 3, Monocots

7b. Leaves simple or compound, with net venation; floral parts in 4s or 5s—
    Group 4, Herbaceous Dicots

The distinction between monocots (Class Liliopsida) and dicots (Class Magnoliopsida) is not as clear-cut as has often been assumed. The most reliable characters are not good field characters, namely, the number of cotyledons in the seed (1 vs. 2), storage materials in the sieve tube plastids (protein vs. starch), or the arrangement of vascular bundles in the stem (scattered vs. in a single ring) (Zomlefer, 1994). While the floral and leaf features given in couplet 7 *usually* work, there are a number of exceptions among both the monocots and dicots in Michigan flora. The keys to Groups 3 and 4 each include some plants which do not fit neatly into these simple distinctions. For example, **Trillium** spp. (Liliaceae) have net-veined leaves, but floral parts in 3s. **Arisaema** spp. (Araceae) are our only monocots with compound leaves. **Floerkea proserpinacoides** (Limnanthaceae) and members of the Aristolochiaceae are the only Michigan herbaceous dicots with all whorls of floral parts in 3s. The keys have been designed to lead the reader to the correct family where these exceptions exist.

GROUP 1: WOODY PLANTS

Plants with persistent woody stems, including trees, shrubs, and woody vines, may be identified by using this key. Separate keys are provided for plants in flower without leaves and for plants with leaves. All woody plants in Michigan are dicots, except for **Smilax** spp. (Liliaceae), two of which are woody vines. For illustrations and additional information about many woody plants in Michigan consult Barnes and Wagner (1981).

1a. Trees or shrubs bearing flowers (mostly in the early spring) but no leaves—2
1b. Trees or shrubs bearing leaves, with or without flowers—30

Winter Twigs: This key is not designed for identifying woody plants in the "winter condition," i.e., mostly without either leaves or flowers. Harlow's (1959) *Twig Key to Trees and Shrubs* is one such key which will work for many of the Michigan species.

2a. Flowers yellow, appearing in autumn, often after the leaves have fallen—***Hamamelis virginiana***, in HAMAMELIDACEAE, p. 190
2b. Flowers of various colors, appearing in spring, before the leaves have opened—3

3a. Flowers in catkins, without brightly colored petals or petal-like bracts—4
3b. Flowers not in catkins, either with or without brightly colored petals or petal-like bracts—16

4a. Eight or more scales on each leaf bud; buds reddish brown, slender; bark smooth, gray—5
4b. Seven or fewer scales on each leaf bud; buds variously shaped; bark smooth or not, of various colors—6

5a. Large tree; trunk cylindrical or nearly so; buds lance-shaped, over 15 mm long—***Fagus grandifolia***, in FAGACEAE, p. 140
5b. Small tree or large shrub; trunk prominently fluted with projecting longitudinal ridges; buds ovoid, about 5 mm long—***Carpinus caroliniana***, in BETULACEAE, p. 139

6a. Twigs bearing numerous dwarf branches thickly covered with crowded leaf scars—***Betula*** spp., in BETULACEAE, p. 138
6b. Twigs without dwarf branches—7

7a. Three bundle scars (visible as small dots or protrusions) in each leaf scar—8
7b. Four or more bundle scars in each leaf scar—13

8a. Pith (tissue at the center of the stem) angular in cross section—9
8b. Pith circular in cross section—10

9a. Trees; lowermost bud scale directly over the leaf scar; pith 4–5 angled—
   Populus spp., in SALICACEAE, p. 133
9b. Trees or shrubs; lowermost bud scale not directly over the leaf scar; pith
   triangular—***Alnus*** spp., in BETULACEAE, p. 138

10a. Pith chambered (divided into separate cavities by transverse partitions)—***Juglans*** spp., in JUGLANDACEAE, p. 137
10b. Pith entire, not chambered—11

11a. Buds covered with only one scale—***Salix*** spp., in SALICACEAE, p. 133
11b. Buds with more than one bud scale—12

12a. Twigs with tiny, round, yellow oil spots; buds tiny, rounded; shrubs—
   MYRICACEAE, p. 137
12b. Twigs lack yellow oil spots; buds variously shaped; shrubs or trees—
   BETULACEAE, p. 138

13a. Buds clustered near the tips of the twigs—***Quercus*** spp., in FA-
   GACEAE, p. 139
13b. Buds not clustered at the tips of the twigs—14

14a. Pith (tissue at center of the stem) circular in cross section—***Morus*** spp.,
   in MORACEAE, p. 142
14b. Pith five-angled in cross section—15

15a. Buds with three visible bud scales—***Castanea dentata***, in FAGACEAE,
   p. 140
15b. Terminal bud large, with four or more visible bud scales—***Carya*** spp.,
   in JUGLANDACEAE, p. 137

16a. Flowers conspicuous, brightly colored, at least 8 mm wide—17
16b. Flowers inconspicuous, seldom brightly colored and then less than 8
   mm wide—21

17a. Flowers irregular, pink (white in some cultivars), about 1 cm wide—
   Cercis canadensis, in LEGUMINOSAE, p. 204
17b. Flowers regular, white or red—18

18a. Flowers dull purple, about 3 cm wide, with 6 petals—***Asimina triloba***,
   in ANNONACEAE, p. 172
18b. Flowers white, usually less than 2 cm wide, with 5 conspicuous
   petals—19

19a. Ovary 1, superior, in the center of the flower—***Prunus*** spp., in ROSACEAE, p. 191
19b. Ovary inferior, appearing as a swelling below the calyx at the apex of the pedicel—20

20a. Flowers in racemes or solitary—***Amelanchier*** spp., in ROSACEAE
20b. Flowers in flattened or rounded branching clusters—***Aronia prunifolia***, in ROSACEAE, p. 196

21a. Leaf scars and buds opposite—22
21b. Leaf scars and buds alternate—23

22a. Many bundle scars (visible as small dots or protrusions) in each leaf scar, often forming a central ring or oval—***Fraxinus*** spp., in OLEACEAE, p. 261
22b. Three to seven bundle scars in each leaf scar, distinct and not forming a central pattern—ACERACEAE, p. 228

23a. Branches thorny or prickly—24
23b. Branches not thorny or prickly—25

24a. Thorns (often branched) solitary and scattered along the branches—***Gleditsia triacanthos***, in LEGUMINOSAE, p. 206
24b. Spines in pairs at stem nodes—***Zanthoxylum americanum***, in RUTACEAE, p. 218

25a. Bark and leaves spicy-aromatic when crushed; perianth yellow, composed of 6 tepals—26
25b. Bark and leaves not spicy-aromatic when crushed; perianth various—27

26a. Flowers in sessile lateral clusters—***Lindera benzoin***, in LAURACEAE, p. 172
26b. Flowers in peduncled terminal clusters—***Sassafras albidum***, in LAURACEAE, p. 172

27a. Buds tiny, almost hidden by the leaf scar—28
27b. Buds clearly larger than the adjacent leaf scar—29

28a. Flowers in panicles—***Gymnocladus dioica***, in LEGUMINOSAE, p. 206
28b. Flowers in racemes—***Gleditsia triacanthos***, in LEGUMINOSAE, p. 206
29a. Buds scaly, often pointed—ULMACEAE, p. 141
29b. Buds "naked" (no bud scales), covered with hairs—***Dirca palustris***, in THYMELEACEAE, p. 240

30a. Leaves narrow, needle-like or very small and scale-like, mostly ever-green—31
30b. Leaves broader, flat or rolled, but not needle-like or scale-like—36

31a. Leaves densely white- or gray-pubescent; low bushy or matted shrubs with yellow flowers—***Hudsonia tomentosa***, in CISTACEAE, p. 235
31b. Leaves green—32

32a. Leaves scale-like, appressed—33
32b. Leaves needle-like, pointed and not appressed—34

33a. Leaves opposite or whorled—CUPRESSACEAE, p. 94
33b. Leaves alternate—***Tamarix parviflora***, in TAMARICACEAE, p. 235

34a. Leaves opposite or whorled—CUPRESSACEAE, p. 94
34b. Leaves alternate, spirally arranged (appearing alternate), or in clusters (fascicles)—35

35a. Shrubs; all leaves spirally arranged—***Taxus canadensis***, in TAXA-CEAE, p. 93
35b. Trees; leaves spirally arranged or in fascicles—PINACEAE, p. 93

36a. Twining or climbing vines—37
36b. Trees or shrubs, stems not climbing or twining—47

37a. Stems thorny, especially near the base—38
37b. Stems not thorny—40

38a. Stems climbing by tendrils arising at the base of the leaves—***Smilax*** spp., in LILIACEAE, p. 117
38b. Tendrils absent—39

39a. Leaves simple—***Lycium barbarum***, in SOLANACEAE, p. 285
39b. Leaves compound—***Rosa*** spp., in ROSACEAE, p. 191

40a. Leaves compound—41
40b. Leaves simple—44

41a. Leaves pinnately compound, with 5–11 leaflets; petals united, tubular, reddish-orange—***Campsis radicans***, in BIGNONIACEAE, p. 295
41a. Leaves palmately compound, with 3 or 5 leaflets; petals separate, greenish or white—42
42a. Leaves with 5 leaflets—***Parthenocissus*** spp., in VITACEAE, p. 231
42b. Leaves with 3 leaflets—43

43a. Plant climbing by tendril-like leaf stalks—***Clematis*** spp., in RANUN-CULACEAE, p. 165

43b. Plant climbing by hold-fast roots, giving the stem a "fuzzy" appearance—***Toxicodendron radicans***, in ANACARDIACEAE, p. 225

44a. Leaves opposite—***Lonicera*** spp., in CAPRIFOLIACEAE, p. 300

44b. Leaves alternate—45

45a. Plants climbing by tendrils—***Vitis*** spp., in VITACEAE, p. 231

45b. Plants twining—46

46a. Leaf blade ovate-oblong, pinnately-veined, the petiole attached to base of leaf blade—***Celastrus*** spp., in CELASTRACEAE, p. 226

46b. Leaf blade broad, 5–7 angled, palmately-veined, the petiole attached to lower surface (peltate) near the edge—***Menispermum canadense***, in MENISPERMACEAE, p. 171

47a. Leaves opposite—48

47b. Leaves alternate—67

48a. Leaves compound—49

48b. Leaves simple—53

49a. Leaves palmately compound, with 5 or 7 leaflets—HIPPOCASTANACEAE, p. 229

49b. Leaves pinnately compound, with 3–11 leaflets—50

50a. Trees—51

50b. Shrubs—52

51a. Leaflets 3–5, the margins coarsely toothed; fruit a double samara, two wings present—***Acer negundo***, in ACERACEAE, p. 228

51b. Leaflets 5–11, often 7, the margins entire or finely toothed; fruit a samara, one wing present—***Fraxinus*** spp., in OLEACEAE, p. 261

52a. Leaflets 3—***Staphylea trifolia***, in STAPHYLEACEAE, p. 227

52b. Leaflets 5–11—***Sambucus*** spp., in CAPRIFOLIACEAE, p. 300

53a. Leaves palmately veined, or at least with a pair of prominent lateral veins from the base leading to lobes—54

53b. Leaves pinnately veined—55

54a. Leaves with stipules; flowers white, in cymes; fruit a drupe—***Viburnum*** spp., in CAPRIFOLIACEAE, p. 300

54b. Leaves without stipules; flowers purple, red, or yellowish, in racemes or panicles; fruit a double samara—ACERACEAE, p. 228

55a. Leaves serrate—56
55b. Leaves entire—58

56a. Branches terminating in thorns; flowers minute, greenish—***Rhamnus cathartica***, in RHAMNACEAE, p. 230
56b. Branches not terminating in thorns; flowers various—57

57a. Corolla flat; fruit a capsule opening to expose seeds covered with orange-red arils; bark of mature twigs green or with corky wings—***Euonymus*** spp., in CELASTRACEAE, p. 226
57b. Corolla tubular; fruit a drupe or capsule, the seeds lacking arils; bark of mature twigs brown, reddish, or gray, lacking corky wings—CAPRIFOLIACEAE, p. 300

58a. Trees—59
58b. Shrubs—60

59a. Lateral veins of the leaf blade curved forward and running almost parallel to the leaf margin; petals separate from each other; fruit a small drupe—CORNACEAE, p. 253
59b. Lateral veins of the leaf blade spreading and not paralleling the leaf margin; petals united; fruit a long capsule—***Catalpa speciosa***, in BIGNONIACEAE, p. 294

60a. Leaves silvery beneath, densely covered with scales; flowers small, yellowish—ELAEAGNACEAE, p. 240
60b. Leaves green beneath, or somewhat hairy and light-colored; flowers various—61

61a. Aquatic shrub or woody-based herb with lanceolate leaf blades, and stems bending over and into the water; flowers pink—***Decodon verticillata***, in LYTHRACEAE, p. 241
61b. Not truly aquatic, although frequently in wet places—62

62a. Leaves evergreen, as shown by their presence on the older stems; petals united—***Kalmia*** spp., in ERICACEAE, p. 256
62b. Leaves deciduous each year; petals united or separate—63

63a. Leaves dotted with translucent dots; flowers bright yellow—GUTTIFERAE, p. 233
63b. Leaves not dotted with translucent dots; flowers various—64

64a. Lateral veins of the leaf blade curved forward and running almost parallel to the leaf margin; petals separate from each other—CORNACEAE

64b. Lateral veins of the leaf blade spreading and not paralleling the leaf margin; petals united—65

65a. Leaves with stipules; flowers small, white, in dense spherical heads—***Cephalanthus occidentalis***, in RUBIACEAE, p. 298

65b. Leaves without stipules; flowers not in spherical heads—66

66a. Stamens 2; fruit a capsule—***Syringa vulgaris***, in OLEACEAE, p. 261

66b. Stamens 4 or 5; fruit a berry or drupe—CAPRIFOLIACEAE, p. 300

67a. Leaves compound—68

67b. Leaves simple—83

68a. Stems prickly or thorny—69

68b. Stems without prickles or thorns—72

69a. Leaves evenly pinnate (lacking a terminal leaflet)—LEGUMINOSAE

69b. Leaves odd-pinnate (with a terminal leaflet) or trifoliolate—70

70a. Leaflets serrate—ROSACEAE, p. 191

70b. Leaflets entire—71

71a. Leaves aromatic, dotted with translucent glands; flowers greenish-white, regular; fruit 2–5 small follicles—***Zanthoxylum americanum***, in RUTACEAE, p. 218

71b. Leaves neither aromatic nor gland-dotted; flowers white to pale rose, irregular, the upper petal the largest; fruit a legume—***Robinia*** spp., in LEGUMINOSAE, p. 203

72a. Leaflets 3—73

72b. Leaflets 5 to many—76

73a. Leaflets entire or serrulate (finely toothed)—74

73b. Leaflets conspicuously toothed—75

74a. Tall shrub or small tree; fruit a round samara in a terminal cyme—***Ptelea trifoliata***, in RUTACEAE, p. 218

74b. Shrub; fruit a whitish drupe in axillary clusters—***Toxicodendron radicans***, in ANACARDIACEAE, p. 225

75a. Stipules present; flowers white—***Rubus pubescens***, in ROSACEAE

75b. Stipules none; flowers greenish or greenish-yellow—ANACARDIACEAE, p. 224

76a. Leaflets 6–25 mm long—LEGUMINOSAE, p. 203
76b. Leaflets 30 mm long, or more—77

77a. Leaflets entire or with 1–4 teeth near the base—78
77b. Leaflets toothed throughout—80

78a. Leaflets entire, except for 1–4 large grandular teeth near their base; fruit a samara—***Ailanthus altissima***, in SIMAROUBACEAE, p. 219
78b. Leaflets entire; fruit a drupe or a legume—79

79a. Leaves twice-pinnate; stipules present; fruit a legume—***Gymnocladus dioica***, in LEGUMINOSAE, p. 206
79b. Leaves once-pinnate; stipules absent; fruit a drupe—ANACAR-DIACEAE, p. 224

80a. Juice milky—ANACARDIACEAE, p. 224
80b. Juice not milky—81

81a. Stipules present; fruit an orange to red drupe—***Sorbus*** spp., in ROSACEAE, p. 191
81b. Stipules absent; fruit a nut or blue berry—82

82a. Tree; flowers greenish, axillary or in catkins; fruit a nut—JUGLAN-DACEAE, p. 137
82b. Shrub; flowers yellow, in a dense raceme; fruit a blue berry—***Mahonia aquifolium***, in BERBERIDACEAE, p. 170

83a. Leaves 3–10 mm long, spreading, completely rolled into a tube—***Empetrum nigrum***, in EMPETRACEAE, p. 224
83b. Leaves mostly flat, not rolled into a tube—84

84a. Leaves evergreen, as shown by their presence on the older parts of the stem; petals united—ERICACEAE, p. 256
84b. Leaves deciduous; petals united or separate—85

85a. Stems or branches thorny; leaf margins may be spiny-margined- 86
85b. Stems or branches without thorns; leaf margins not spiny—93

86a. Leaves conspicuously palmately veined—87
86b. Leaves pinnately veined, or sometimes with smaller lateral veins arising from the end of the petiole—88

87a. Leaves 5 cm wide or less—***Ribes*** spp., in GROSSULARIACEAE
87b. Leaves 15 cm wide or more—***Oplopanax horridus***, in ARALIACEAE

88a. Leaves entire—89
88b. Leaves toothed or somewhat lobed—91

89a. Tree—*Maclura pomifera*, in MORACEAE, p. 142
89b. Shrub—90

90a. Petals separate, yellow—*Berberis thunbergii*, in BERBERIDACEAE,
     p. 170
90b. Petals united into a tube, purplish—*Lycium barbarum*, in
     SOLANACEAE, p. 285

91a. Leaf margins spiny-toothed; spines often three-pointed—*Berberis vulgaris*, in BERBERIDACEAE, p. 170
91b. Leaf margins not spiny-toothed; spines on twigs not branched—92

92a. Thorns project laterally along the branches—ROSACEAE, p. 191
92b. Thorns terminating the branches—*Rhamnus cathartica*, in RHAMNACEAE, p. 230

93a. Leaves palmately veined, or with one or more pairs of lateral veins
     from the base of the leaf—94
93b. Leaves pinnately veined—107

94a. Leaves entire—95
94b. Leaves toothed or lobed—96

95a. Twigs and leaves spicy-aromatic when crushed; leaf blades ovate; flowers yellow, regular—*Sassafras albidum*, in LAURACEAE, p. 172
95b. Twigs and leaves not spicy-aromatic when crushed; leaf blades heart-shaped; flowers pink (rarely white), irregular—*Cercis canadensis*, in LEGUMINOSAE, p. 204

96a. Leaves toothed—97
96b. Leaves palmately lobed—101

97a. Lateral veins straight and parallel, running to the teeth of the leaf blade
     margins—*Betula* spp., in BETULACEAE, p. 138
97b. Lateral veins curved or branched, and not running straight to the teeth
     of the leaf blade margins—98

98a. Leaves heart-shaped, asymmetrical, oblique at the base—*Tilia americana*, in TILIACEAE, p. 232
98b. Leaves symmetrical or nearly so at the base—99

99a. Shrubs, often less than 1 m high; leaf blades ovate to elliptic—***Ceanothus*** spp., in RHAMNACEAE, p. 231
99b. Trees; leaf blades broadly ovate to deltoid—100

100a. Twigs and leaves with milky juice; petioles round—***Morus*** spp., in MORACEAE, p. 142
100b. Twigs and leaves lack milky juice; petioles often laterally compressed—***Populus*** spp., in SALICACEAE, p. 133

101a. Leaf lobes entire—102
101b. Leaf lobes serrate—103

102a. Leaves green beneath; twigs and leaves spicy-fragrant when crushed—***Sassafras albidum***, in LAURACEAE, p. 172
102b. Leaves densely white-tomentose beneath; twigs and leaves not spicy-fragrant when crushed—***Populus alba***, in SALICACEAE, p. 134

103a. Leaves with milky juice—***Morus*** spp., in MORACEAE, p. 142
103b. Leaves without milky juice—104

104a. Trees; bark mottled and falling readily; axillary buds partially covered by the base of the petiole—***Platanus occidentalis***, in PLATANACEAE, p. 190
104b. Shrubs; bark sometimes falling readily; axillary buds visible—105

105a. Stem covered with brown bristles—***Rubus*** spp., in ROSACEAE
105b. Stem not bristly—106

106a. Stipules present; sides of the petiole strongly decurrent on the stem; bundle scars (visible as small dots or protrusions) five, crowded or nearly in contact within the leaf scars (depression left in bark where petiole detached at leaf fall)—***Physocarpus opulifolius***, in ROSACEAE, p. 194
106b. Stipules absent; sides of petiole slightly decurrent or not at all; bundle scars three, distinctly separate—***Ribes*** spp., in GROSSULARIACEAE

107a. Leaves spicy-aromatic when crushed—108
107b. Leaves not spicy-aromatic when crushed—109

108a. Leaves broadly obovate, entire—***Lindera benzoin***, in LAURACEAE
108b. Leaves linear-lanceolate or oblanceolate, conspicuously pinnately lobed, toothed, or entire—MYRICACEAE, p. 137

109a. Leaves entire—110
109b. Leaves toothed or lobed—119

110a. Base of the petiole covering the axillary buds; twigs appear to be jointed—***Dirca palustris***, in THYMELAEACEAE, p. 240

110b. Base of petiole not covering the bud, and twigs lack jointed appearance—111

111a. Leaves bristle-tipped; fruit an acorn; pith (tissue at center of stem) five-angled in cross section—***Quercus imbricaria***, in FAGACEAE

111b. Leaves not bristle-tipped; fruit various; pith not five-angled in cross section—112

112a. Pith with prominent cross-partitions—113

112b. Pith not partitioned—114

113a. Tall shrubs or small trees; axillary buds naked, reddish brown; leaves 15 cm long or longer—***Asimina triloba***, in ANNONACEAE, p. 172

113b. Trees; axillary buds covered with scales, brownish; leaves up to 15 cm long—***Nyssa sylvatica***, in NYSSACEAE, p. 241

114a. Leaf blades dotted beneath with yellowish resinous dots—***Gaylussacia baccata***, in ERICACEAE, p. 258

114b. Leaf blades not dotted beneath with yellowish resinous dots—115

115a. Lateral veins curved forward and almost parallel to the margin of the leaf—***Cornus alternifolia***, in CORNACEAE, p. 253

115b. Lateral veins spreading—116

116a. Leaves lanceolate or linear, much longer than wide—***Lycium barbarum*** in SOLANACEAE, p. 285

116b. Leaves oblong or ovate to elliptical—117

117a. Leaves with purple petioles, which are at least one-fourth as long as the leaf blade—***Nemopanthus mucronatus***, in AQUIFOLIACEAE

117b. Leaves with short green petioles or sessile—118

118a. Buds scaly—***Vaccinium*** spp., in ERICACEAE, p. 256

118b. Buds naked—***Rhamnus frangula***, in RHAMNACEAE, p. 230

119a. Leaves lobed—120

119b. Leaves serrate, doubly serrate, or more coarsely toothed—121

120a. Leaves with four large entire lobes; stem marked with a ring at each node—***Liriodendron tulipfera***, in MAGNOLIACEAE, p. 171

120b. Leaves with many lobes; stem not ringed—***Quercus*** spp., in FAGACEAE, p. 139

121a. Lateral leaf veins straight and parallel, mostly ending in the teeth of the leaf—122

121b. Lateral leaf veins not straight and parallel—129

122a. Leaves asymmetrical, oblique at the base—123

122b. Leaves symmetrical or nearly so at the base—124

123a. Leaves serrate, the teeth small and sharp—***Ulmus*** spp., in UL-MACEAE, p. 141

123b. Leaves toothed, the teeth large and rounded—***Hamamelis virginiana***, in HAMAMELIDACEAE, p. 190

124a. Pith (tissue at center of stem) triangular in cross section—***Alnus*** spp., in BETULACEAE, p. 138

124b. Pith cylindrical in cross section—125

125a. Leaves, or many of them, crowded on short spur-like branches—126

125b. Leaves scattered, not on short spur-like branches—128

126a. Bark of the trunks separating in thin papery or leathery sheets—***Betula*** spp., BETULACEAE, p. 138

126b. Bark of the trunk not papery or leathery—ROSACEAE, p. 191

127a. Leaves doubly serrate—BETULACEAE, p. 138

127b. Leaves serrate—135

128a. Leaves finely serrate—***Amelanchier*** spp., in ROSACEAE, p. 191

128b. Leaves coarsely serrate—FAGACEAE, p. 139

129a. Leaves asymmetrical, oblique at the base—130

129b. Leaves symmetrical or nearly so at the base—131

130a. Leaf blades heart-shaped, about as broad as long—***Tilia americana***, in TILIACEAE, p. 232

130b. Leaf blades oval or ovate, much longer than wide—***Celtis*** spp., in UL-MACEAE, p. 141

131a. Leaves coarsely or doubly serrate—***Corylus*** spp., in BETULACEAE, p. 138

131b. Leaves simply serrate—132

132a. Leaves with glands on the petiole, base of the leaf blade, or the mid-vein—133

132b. Leaves without glands—135

133a. Leaves with small dark glands on the upper side of the midvein—*Aronia prunifolia*, in ROSACEAE, p. 196
133b. Leaves with glands on the petiole or at the base of the leaf blade—134

134a. Trees or shrubs with slender leaves, frequently conspicuous broad stipules, and axillary buds protected by a single external bud scale—*Salix* spp., in SALICACEAE, p. 133
134b. Trees and shrubs with leaves lanceolate or broader, stipules minute and falling early in the season, and axillary buds covered with more than one bud scale—*Prunus* spp., in ROSACEAE, p. 191

135a. Stipules present or with stipular scars indicating where stipules have been detached—136
135b. Stipules or stipular scars absent—139

136a. Three bundle scars (visible as small dots or protrusions) in each leaf scar (depression in bark left by the petiole at leaf fall)—137
136b. One bundle scar in each leaf scar—138

137a. Trees or shrubs with slender leaves, frequently with large conspicuous stipules, and axillary buds covered by a single external bud scale—*Salix* spp., in SALICACEAE, p. 133
137b. Trees and shrubs with oblong or ovate leaves, small stipules which fall early, and axillary buds with more than one external scale—*Amelanchier* spp., in ROSACEAE, p. 191

138a. Axillary buds superposed (a second one just above the first); petals present; fruit a red or yellow drupe—*Ilex verticillata*, in AQUIFOLIACEAE, p. 226
138b. Axillary buds not superposed; petals absent; fruit a black drupe—*Rhamnus alnifolia*, in RHAMNACEAE, p. 230

139a. Leaves with purple petioles, at least one-fourth as long as the blade—*Nemopanthus mucronata*, in AQUIFOLIACEAE, p. 226
139b. Leaves short-petioled or sessile—140

140a. Stems erect and straight, unbranched or with very few ascending branches—*Spiraea* spp., in ROSACEAE, p. 191
140b. Stems more or less crooked and freely branched, making a spreading shrub—*Vaccinium* spp., in ERICACEAE, p. 256

GROUP 2: UNUSUAL PLANTS

This key is especially designed for identification of the families of aquatic or parasitic flowering plants. Other plants which are unusual in appearance and therefore difficult to place in either the monocots (Group 3) or herbaceous dicots (Group 4) are included here. For example, apparently leafless plants, or those in which the flowering and vegetative stages are completely separated in time will be found here. Green terrestrial plants with recognizable flowers will be found in Group 3 or 4. Refer to Group 1 for all woody plants. Algae are aquatic plants, unicellular or multicellular, and placed in many different divisions. They are not identified here, except for one genus of very large algae (**Chara** spp.) which may be confused with several common aquatic flowering plants.

1a. Plants aquatic, entirely submerged or floating—2
1b. Terrestrial plants, without floating or submerged leaves—38

2a. Aquatic plants, with all or most of the leaves submerged, or leafless—3
2b. Aquatic plants, with the leaves or the whole plant floating on or near the surface—29

Identification of submerged aquatic plants can be extremely difficult, often complicated by the absence of flowers or fruit. An extensive key using vegetative characters of submerged aquatic plants in Michigan is available in Voss (1967).

3a. Submerged aquatics, without noticeable leaves—4
3b. Submerged aquatics, with the leaves linear, lanceolate, or dissected—5

4a. Flowers showy, yellow or purple; corolla 2-lipped—***Utricularia*** spp., in LENTIBULARIACEAE, p. 295
4b. Flowers small and inconspicuous, sessile, purplish or greenish; corolla not 2-lipped—***Myriophyllum tenellum***, in HALORAGACEAE, p. 245

5a. Leaves linear or lanceolate, not lobed or dissected—6
5b. Leaves more or less lobed or dissected—22

6a. Leaves all basal—7
6b. Stem leaves present—13

7a. Leaves flat, often ribbon-like—8
7b. Leaves filiform or tubular—12

8a. Flowers unisexual, one sex basal among the leaves, the other solitary on an elongated pedicel—9
8b. Flowers unisexual or perfect, all in clusters—10

9a. Leaves ribbon-like, 20 cm or longer; male flowers basal, female flower pedicellate—***Vallisneria americana***, in HYDROCHARITACEAE, p. 98
9b. Leaves shorter, to 5 cm long; female flowers basal, male flower pedicellate—***Littorella uniflora***, in PLANTAGINACEAE, p. 297

10a. Flowers white, in panicles or racemes—***Sagittaria*** spp., in ALISMATACEAE, p. 97
10b. Flowers whitish-gray or greenish, in heads—11

11a. Flowers greenish, in a spike of globose heads—SPARGANIACEAE, p. 95
11b. Flowers minute, whitish or gray, in a small head terminating an unbranched, leafless stalk—***Eriocaulon septangulare***, in ERIOCAULACEAE, p. 115

12a. Leaves tubular, hollow, partitioned lengthwise to form two adjacent tubes; flowers very pale blue, 1 cm long or more—***Lobelia dortmanna***, in CAMPANULACEAE, p. 306
12b. Leaves minute, filiform; flowers yellow or purple—***Utricularia*** spp., in LENTIBULARIACEAE, p. 295

13a. Leaves alternate—14
13b. Leaves opposite or whorled—16

14a. Leaves filiform to broadly elliptic or obovate, the blades with an evident midvein; flowers in spikes, greenish—POTAMOGETONACEAE, p. 95
14b. Leaves filiform to linear, midvein not evident; flowers solitary, petals yellow—15

15a. Leaves filiform; flower terminal; sepals 5 and petals 5—***Ranunculus reptans***, in RANUNCULACEAE, p. 165
15b. Leaves flat, linear; flower axillary; tepals 6—***Heteranthera dubia***, in PONTEDERIACEAE, p. 116

16a. Leaves abruptly widened at the base, the margin toothed, often minutely so—NAJADACEAE, p. 96
16b. Leaves not widened at the base, the margin entire—17

17a. Leaves opposite—18
17b. Leaves whorled—20

18a. Leaves thread-like, 3–8 cm long—***Zannichellia palustris***, in ZANNICHELLIACEAE, p. 96
18b. Leaves linear to spathulate, up to 1.5 cm long—19

19a. Apex of submerged leaves split, forming two teeth; petals absent—
CALLITRICHACEAE, p. 224

19b. Apex of submerged leaves entire, not toothed; petals present—***Elatine minima***, in ELATINACEAE, p. 235

20a. Whorled leaves cylindrical—***Chara*** spp., an alga (Division Charophyta)

20b. Whorled leaves flat—21

21a. Stems flexuous, not erect; leaves mostly in whorls of 3—***Elodea*** spp., in HYDROCHARITACEAE, p. 98

21b. Stems erect; leaves in whorls of 4 or more—***Hippuris vulgaris***, in HIP-PURIDACEAE, p. 245

22a. Leaves or branches with numerous small bladders attached, each bladder 1–3 mm long—***Utricularia*** spp., in LENTIBULARIACEAE, p. 295

22b. Leaves or branches without bladders—23

23a. Leaves alternate—24

23b. Leaves opposite or whorled—27

24a. Leaves palmately dissected—***Ranunculus*** spp., in RANUNCU-LACEAE, p. 165

24b. Leaves pinnately divided—25

25a. Leaves twice-pinnate—***Armoracia aquatica***, in CRUCIFERAE, p. 179

25b. Leaves once-pinnate—26

26a. Leaves divided into ovate segments, the terminal one enlarged—***Nasturtium officinale***, in CRUCIFERAE, p. 174

26b. Leaves divided into linear segments, the terminal one not enlarged—***Proserpinaca palustris***, in HALORAGACEAE, p. 244

27a. Leaves pinnately compound—***Myriophyllum*** spp., in HALOR-AGACEAE, p. 244

27b. Leaves palmately compound—28

28a. Leaves whorled, consistently dichotomously forked—CERATOPHYL-LACEAE, p. 164

28b. Leaves opposite or appearing whorled, not dichotomously forked—***Megalodonta beckii***, in COMPOSITAE, p. 329

29a. Plants minute, flattened, consisting of a rounded or ovate thallus (no distinction into stem and leaf); the whole plant floating on or near the surface—LEMNACEAE, p. 114

29b. Plant attached to the soil, with differentiated stem and leaves—30

30a. Leaves circular to elliptic, petiole attached to center of blade or blade with a deep basal slit, with net veins—31

30b. Leaves linear to elliptic; petiole attached to margin of blade, basal slit absent; net or parallel veins—32

31a. Leaf blades entire or with a deep basal slit; flowers large, solitary on long peduncles held on or above the water surface; stamens 12 to many—NYMPHAEACEAE, p. 165

31b. Leaf blades shallowly lobed or crenate; flowers small, several in an umbel; stamens 5—*Hydrocotyle umbellata*, in UMBELLIFERAE

32a. Leaves basal, with parallel veins—*Sagittaria cuneata*, in ALISMAT-ACEAE, p. 97

32b. Stem leaves present, with net or parallel veins—33

33a. Leaves opposite or whorled—34

33b. Leaves alternate—35

34a. Leaves less than 2 cm long—CALLITRICHACEAE, p. 224

34b. Leaves more than 2 cm long—POTAMOGETONACEAE, p. 95

35a. Leaves net-veined, with a single midvein; lanceolate or elliptical in out-line—*Polygonum amphibium*, in POLYGONACEAE, p. 150

35b. Leaves parallel-veined—36

36a. Leaves not over 20 cm long—POTAMOGETONACEAE, p. 95

36b. Leaves over 20 cm long—37

37a. Leaf base forms a sheath around the stem; flowers in 1 or more bracted spikelets—GRAMINEAE, p. 98

37b. Leaf sheath absent; flowers in a spike of spherical heads—SPARGANI-ACEAE, p. 95

38a. Brown, yellow, or white plants, without green color—39

38b. Plants with normal green color, at least in some parts—44

39a. Plants growing on or twining around stems or branches of other plants—40

39b. Plants emerging from the ground, stems not twining—41

40a. Small brown plants, growing as parasites on branches of black spruce (rarely on other gymnosperms)—*Arceuthobium pusillum*, in VIS-CACEAE, p. 144

40b. Stems resembling yellow or white threads, twining around plant stems—CUSCUTACEAE, p. 269

41a. Plants emerging in late winter, appearing to consist only of yellowish flowers in a partly-underground reddish spathe—*Symplocarpus foetidus*, in ARACEAE, p. 113

41b. Plants with erect stems, appearing later in the year—42

42a. Corolla regular; stamens 6–12—MONOTROPACEAE, p. 256

42b. Corolla irregular; stamens 1 or 4 or not evident—43

43a. Sepals and petals each 3—*Corallorhiza* spp., in ORCHIDACEAE

43b. Sepals 5; corolla of united petals, 2-lipped—OROBANCHACEAE

44a. Stem flattened, thick and fleshy, leafless, spiny—CACTACEAE, p. 239

44b. Stem not spiny—45

45a. Flowers replaced by bulblets—46

45b. Flowers present, not replaced by bulblets—47

46a. All leaves simple, linear, entire; bulblets in a terminal umbel—*Allium* spp., in LILIACEAE, p. 117

46b. Most leaves pinnately compound, upper leaves simple and much reduced; bulblets in upper leaf axils and sometimes in terminal umbels—*Cicuta bulbifera*, in UMBELLIFERAE, p. 251

47a. Leaves absent when plant is flowering, flower peduncle may be scaly—48

47b. Leaves and flowers present—54

48a. Flowers small, greenish or brownish, without obvious colored petals—49

48b. Flowers with conspicuous white or colored petals—50

49a. Perianth of 6 small chaffy tepals; flowers in cymes—*Juncus* spp., in JUNCACEAE, p. 116

49b. Perianth absent, each flower in the axil of a single chaffy bract within a terminal spikelet—*Eleocharis* spp., in CYPERACEAE, p. 108

50a. Inflorescence a composite head, with several or many small flowers closely aggregated into a dense head surrounded by a calyx-like involucre of small bracts—51

50b. Flowers separate, variously clustered, but never crowded into involucral heads—52

51a. Flowers yellow—*Tussilago farfara*, in COMPOSITAE, p. 328

51b. Flowers whitish—*Petasites* spp., in COMPOSITAE, p. 318

52a. Flowers regular, tepals 6, white—***Allium tricoccum***, in LILIACEAE, p. 123
52b. Flowers irregular, sepals 3 or 5, corolla pinkish, purple or yellow—53

53a. Sepals 3; corolla pinkish—***Arethusa bulbosa***, in ORCHIDACEAE, p. 132
53b. Sepals 5; corolla yellow or purple—***Utricularia*** spp., in LENTIBU-LARIACEAE, p. 295

54a. Leaves pitcher-shaped, open at the top—***Sarracenia purpurea***, in SAR-RACENIACEAE, p. 185
54b. Leaves flat, variously shaped—55

55a. Leaves reduced to small, often brownish scales—56
55b. Leaves various, but not reduced to scales—61

56a. Stem erect, unbranched or with only 1 or 2 branches—57
56b. Stem freely branched—58

57a. Plants of moist sandy meadows, open woods, and bog margins; scales opposite; corolla regular, 4-lobed—***Bartonia virginica***, in GEN-TIANACEAE, p. 262
57b. Plants in mud or water; scales few, alternate; corolla irregular, 2-lipped—***Utricularia*** spp., in LENTIBULARIACEAE, p. 295

58a. Leaves numerous and close, concealing the stem—59
58b. Leaves spreading, not concealing the stem—60

59a. Plants succulent, growing in highway median areas; leaves opposite, tiny papery scales—***Salicornia europaea***, in CHENOPODIACEAE, p. 152
59b. Plants shrubby, growing on sand dunes and in adjacent forests; leaves alternate, not papery—***Hudsonia tomentosa***, in CISTACEAE, p. 235

60a. Leaf scales subtending filiform branches, mostly alternate; petals 6, greenish—***Asparagus officinalis***, in LILIACEAE, p. 117
60b. Leaf scales distinctly opposite; petals 5, yellow—***Hypericum gentianoides***, in GUTTIFERAE, p. 234

61a. Plant with a rosette of small basal leaves bearing large glandular hairs on the upper surface; cauline leaves absent—DROSERACEAE, p. 186
61b. Basal rosette of leaves absent; cauline leaves succulent—62

62a. Sepals 2—***Portulaca*** spp., in PORTULACACEAE, p. 156
62b. Sepals 4 or 5—CRASSULACEAE, p. 186

GROUP 3: MONOCOTS

Monocots (Class Liliopsida) have one cotyledon (seed leaf) in the seed and scattered vascular bundles within the stem. Most monocots have parallel-veined, often linear, leaves and flower parts in 3s or 6s. However, there are several exceptions to this rule. It is important to look at all the characters displayed in determining whether one is examining a monocot. *Arisaema* spp. (Jack-in-the-pulpit) are the only monocots in our flora which have compound leaves. The *Araceae* have a specialized inflorescence consisting of a fleshy spike (spadix) and a large enclosing bract (spathe). Leaves of most *Araceae* and all *Dioscoreaceae*, *Trillium* spp., and *Smilax* spp. are both broad and net-veined. Trees and shrubs in Michigan are either dicots or gymnosperms (see Group 1, Woody Plants). Keys to floating and/or submerged aquatic monocots are in Group 2, Unusual Plants. Several dicots which could be confused with monocots are included in this key.

1a. Twining or climbing plants; leaves broad and net-veined—2
1b. Non-climbing herbs; leaves narrow or broad—4

2a. Stem twining, tendrils absent; flowers in spikes or panicles—***Dioscorea villosa***, in DIOSCOREACEAE, p. 125
2b. Stem climbing or scrambling, tendrils present; flowers in umbels or racemes—3

3a. Leaf blades lobed; petals 5 or 6, united—CUCURBITACEAE, p. 305
3b. Leaf blades entire; tepals 6, separate—***Smilax*** spp., in LILIACEAE, p. 117

4a. Flowers in a head and plants with milky juice—***Tragopogon*** spp., in COMPOSITAE, p. 309
4b. Flowers various, but if in a head, then plants without milky juice—5

5a. Margin of the leaf blade spiny—***Eryngium yuccifolium***, in UMBELLIFERAE, p. 247
5b. Margin of the leaf blade not spiny—6

6a. Flowers in a dense, fleshy spike (*spadix*), surrounded or subtended by a green or colored bract (*spathe*), the combination resembling a single flower; perianth minute or absent; leaves broad, often net-veined—ARACEAE, p. 113
6b. Flowers not in a spathe-spadix arrangement; leaves often narrow, linear to lanceolate, seldom broad and net-veined—7

7a. Flowers in a solitary head terminating an unbranched, leafless stalk—8
7b. Flowers variously arranged; if in a head, the head not terminal and solitary—9

8a. Perianth whitish or grayish, woolly—***Eriocaulon septangulare***, in ERI-
OCAULACEAE, p. 115
8b. Petals yellow, glabrous—XYRIDACEAE, p. 115

9a. Flowers greenish, yellowish, or brownish, never brightly colored, the perianth absent, or inconspicuous, dry, or chaffy in texture; individual
flowers small, but sometimes grouped into conspicuous clusters—10
9b. Flowers in which all or part of the perianth is petaloid, conspicuous,
white or colored—16

10a. Flowers arise in, and are mostly hidden by, the axils of dry, membranous, or chaffy scales which are arranged into spikelets of uniform size
and structure; spikelets variously arranged; fruit a grain or an achene—
11
10b. Flowers not subtended individually by dry, membranous, or chaffy
scales or not hidden by such scales; fruit various—12

11a. Leaf sheaths often split on the side opposite the blade; leaves usually 2-
ranked, i.e., in two longitudinal rows with the third leaf above the first;
stems usually rounded or flat, never triangular, almost always hollow;
stem nodes solid, conspicuous—GRAMINEAE, p. 98
11b. Leaf sheaths closed into a continuous tube; leaves usually 3-ranked;
stems frequently triangular, usually solid; stem nodes neither solid nor
conspicuous—CYPERACEAE, p. 108

12a. Flowers unisexual, male and female flowers differing in appearance—
13
12b. Flowers perfect, uniform in appearance—14

13a. Inflorescence a dense spike, the pistillate flowers at the base and staminate flowers at the apex of the spike—TYPHACEAE, p. 94
13b. Inflorescence a spike of globose heads; the lower heads pistillate, the
upper staminate—SPARGANIACEAE, p. 95

14a. Inflorescence a fleshy spike, appearing to be lateral near the middle of
a flattened stem—***Acorus calamus***, in ARACEAE, p. 114
14b. Inflorescence a cyme, raceme, or spike-like raceme—15

15a. Inflorescence a cyme of solitary or (often) clustered flowers; ovary 1—
JUNCACEAE, p. 116
15b. Inflorescence a raceme or spike-like raceme; ovaries 3 or 6—
JUNCAGINACEAE, p. 96

16a. Flowers regular, with the petals of approximately the same size and shape—17

16b. Flowers irregular, with the petals of each flower not of the same size or shape—26

17a. Ovaries 2 or more, separate or barely united with each other at the base—18

17b. Ovary 1 in each flower—20

18a. Leaves compound or pinnately divided; ovaries 2 or 3—***Floerkea proserpinacoides***, in LIMNANTHACEAE, p. 224

18b. Leaves simple, entire; ovaries 3 or more—19

19a. Petals and sepals both pink; inflorescence an umbel—***Butomus umbellatus***, in BUTOMACEAE, p. 97

19b. Petals white, sepals green; inflorescence a panicle or raceme of 3-flowered whorls—ALISMATACEAE, p. 97

20a. Perianth differentiated into greenish sepals and 2 or 3 colored petals—21

20b. Perianth not differentiated, all divisions essentially alike (tepals)—22

21a. Leaves broad, net-veined; flower solitary, terminal—***Trillium*** spp., in LILIACEAE, p. 117

21b. Leaves narrow, with parallel veins; flowers in an umbel—***Tradescantia*** spp., in COMMELINACEAE, p. 115

22a. Ovary superior, appearing within the perianth—23

22b. Ovary inferior, appearing below the perianth—24

23a. Tepals 6; stamens 3; mud flats or in shallow water—***Heteranthera dubia***, in PONTEDERIACEAE, p. 116

23b. Tepals 4 or 6; stamens 4 or 6—LILIACEAE, p. 117

24a. Leaves broad, net-veined; tepals reddish- or purplish-brown; stamens 6 or 12—ARISTOLOCHIACEAE, p. 145

24b. Leaves narrow, with parallel veins; tepals blue, white, or yellow; stamens 3 or 6—25

25a. Stamens 6; tepals bright yellow—***Hypoxis hirsuta***, in AMARYLLIDACEAE, p. 125

25b. Stamens 3 (concealed by prominent petaloid style branches in *Iris* spp.); tepals blue, white, or yellow—IRIDACEAE, p. 125

26a. Flowers not blue; ovary inferior; floral structure complex; stamens attached to other parts of the flower and not resembling ordinary stamens
in form or structure—ORCHIDACEAE, p. 127

26b. Flowers blue; ovary superior; stamens distinct from the other parts of
the flower—27

27a. Base of leaf blade cordate; perianth of 6 blue tepals—***Pontederia cordata***, in PONTEDERIACEAE, p. 116

27b. Base of leaf blade tapering to leaf sheath; sepals green—***Commelina
communis***, in COMMELINACEAE, p. 115

GROUP 4: DICOTS

With rare exceptions, Dicots (Class Magnoliopsida) have two cotyledons (seed leaves) and a ring of vascular bundles within the stem. Most dicots have net-veined leaves and flower parts in 4s or 5s ("4-merous" or "5-merous"). However, some dicots exhibit characters usually found in monocots, namely, flower parts in 3s or 6s and/or narrow leaves with parallel veins. It is important to look at all of the characters displayed to decide whether one is examining a dicot. Several monocots which could be confused with dicots are included in this key. Plants with flowers which consist of a large fleshy spike of tiny sessile flowers (spadix) surrounded by a large bract (spathe) are members of the Araceae (monocots). All plants with compound leaves in Michigan are dicots except for **Arisaema** spp. (Jack-in-the-pulpit). Trees and shrubs in Michigan are either dicots or gymnosperms. To identify trees, shrubs, or woody vines, see Group 1, Woody Plants. Families of aquatic plants and parasitic plants (usually not green) are found in Group 2, Unusual Plants.

1a. Leaves all or principally basal; flower stalk either completely leafless or with only bracts or scales—2
1b. Leaves present on the stem, either one or more in number, sometimes limited to a single opposite pair—41

2a. Leaves compound or dissected—3
2b. Leaves simple and entire, lobed, or toothed but not dissected—15

3a. Leaves 2 or 3 times compound or dissected—4
3b. Leaves once-compound or dissected—7

4a. Flowers in racemes—***Dicentra*** spp., in FUMARIACEAE, p. 173
4b. Flowers in umbels—5

5a. Sepals 10–15 mm long, white to pink or purplish; petals absent—
Anemonella thalictroides, in RANUNCULACEAE, p. 170
5b. Sepals and petals less than 5 mm long; petals white or greenish—6

6a. Leaflets 5–15 cm long—***Aralia nudicaulis***, in ARALIACEAE, p. 246
6b. Leaflets less than 2 cm long—***Erigenia bulbosa***, in UMBELLIFERAE, p. 251

7a. Leaflets 2—***Jeffersonia diphylla***, in BERBERIDACEAE, p. 171
7b. Leaflets 3 or more—8

8a. Leaflets entire or very finely toothed—9
8b. Leaflets coarsely toothed or lobed—12

9a. Inflorescence consisting of a greenish or purplish bract (spathe) wholly or partly enclosing a fleshy spike (spadix) of flowers—***Arisaema*** spp., in ARACEAE, p. 113

9b. Inflorescence not a spathe-spadix combination; flowers not subtended by large bracts—10

10a. Flowers irregular, in dense head-like umbels or racemes—LEGUMI-NOSAE, p. 203

10b. Flowers regular, solitary or in loose clusters—11

11a. Leaflets obcordate, not over 2 cm long—OXALIDACEAE, p. 216

11b. Leaflets ovate-elliptic, 3–6 cm long—***Menyanthes trifoliata***, in MENYANTHACEAE, p. 262

12a. Flowers with one perianth whorl (calyx), the sepals petaloid—RA-NUNCULACEAE, p. 165

12b. Flowers with colored or white petals and green or colored sepals—13

13a. Petals and sepals each 4—***Dentaria*** spp., in CRUCIFERAE, p. 174

13b. Petals and sepals each 5—14

14a. Petals white or yellow; stamens many; ovaries more than 1—ROSACEAE, p. 191

14b. Petals pink; stamens 5; ovary 1—***Erodium cicutarium***, in GERANI-ACEAE, p. 217

15a. Leaves orbicular and peltate; aquatic or mud plants—***Hydrocotyle umbellata***, in UMBELLIFERAE, p. 248

15b. Leaves of various shapes but not peltate; aquatic or land plants—16

16a. Inflorescence consisting of a purplish bract (spathe) wholly or partly enclosing a fleshy spike (spadix) of flowers, at or partly beneath the surface of the ground—***Symplocarpus foetidus***, in ARACEAE, p. 113

16b. Inflorescence not a spathe-spadix combination, terminating an aerial stem—17

17a. Leaves pitcher-shaped, open at the top—***Sarracenia purpurea***, in SAR-RACENIACEAE, p. 185

17b. Leaves flat, variously shaped—18

18a. Inflorescence a dense head of small flowers subtended by an involucre of small, green bracts; sepals absent or present as bristles or scales; petals 5, united and tubular or often prolonged into a flat, strap-like structure; stipules absent (see Figs. 41–45)—COMPOSITAE, p. 309

18b. Inflorescence various; if an involucrate head, then the petals are not tubular or prolonged if united and the sepals are green; stipules present or not—19

19a. Flower solitary on a scape—20
19b. Flowers several to many, variously arranged—30

20a. Flowers obviously irregular, one or more petals prolonged into a spur—21
20b. Flowers regular or nearly so, spurred corolla absent—22

21a. Leaves sessile or nearly so; stamens 2—LENTIBULARIACEAE, p. 295
21b. Leaves petioled; stamens 5—VIOLACEAE, p. 236

22a. Flowers yellow—*Ranunculus* spp., in RANUNCULACEAE, p. 165
22b. Flowers not yellow—23

23a. Ovary 1—24
23b. Ovaries numerous—29

24a. Leaves lobed or divided—25
24b. Leaves entire or toothed—26

25a. One leaf, palmately lobed; stamens numerous—*Sanguinaria canadensis*, in PAPAVERACEAE, p. 173
25b. More than one leaf, blades deeply divided into 2 lobes; stamens 8—*Jeffersonia diphylla*, in BERBERIDACEAE, p. 171

26a. Flowers dull red; leaves hairy—*Asarum canadense*, in ARISTOLOCHIACEAE, p. 145
26b. Flowers white or grayish; leaves glabrous—27

27a. Flowers unisexual; perianth (rarely) 3- or (usually) 4-merous—*Littorella uniflora*, in PLANTAGINACEAE, p. 297
27b. Flowers perfect; perianth 5-merous—28

28a. Fertile stamens 5 (additional staminodes often present); petals white with green or yellow veins—*Parnassia* spp., in SAXIFRAGACEAE
28b. Stamens 10; petals white, the veins not distinctly colored—*Moneses uniflora*, in PYROLACEAE, p. 255

29a. Leaves lobed or divided—RANUNCULACEAE, p. 165
29b. Leaves crenate or toothed—*Dalibarda repens*, in ROSACEAE, p. 198

30a. Flowers in dense spikes, spike-like racemes, or a small head—31
30b. Flowers in open, loose clusters—34

31a. Peduncle leafless below the spike—32
31b. Peduncle with several bracts—33

32a. Flowers minute, whitish or gray, 2-merous, in a small head; tubular
     sheath surrounds the base of the peduncle—***Eriocaulon septangulare***,
     in ERIOCAULACEAE, p. 115
32b. Flowers larger, whitish to greenish or brownish, 4-merous, in spikes;
     peduncle lacks a basal sheath—***Plantago*** spp., in PLANTAGINACEAE

33a. Leaves entire—***Goodyera*** spp., in ORCHIDACEAE, p. 127
33b. Leaves toothed—***Besseya bullii***, in SCROPHULARIACEAE, p. 288

34a. Flowers in umbels—***Primula*** spp., in PRIMULACEAE, p. 259
34b. Flowers in racemes, panicles, or flat-topped clusters—35

35a. Leaves tubular, hollow, partitioned lengthwise to form two adjacent
     tubes; plant is an emergent aquatic—***Lobelia dortmanna***, in CAMPAN-
     ULACEAE, p. 306
35b. Leaves flat; plants of various habits—36

36a. Leaves bearing large glandular hairs on the upper surface; bog plant
     with flowers in racemes—DROSERACEAE, p. 186
36b. Leaves pubescent or smooth, but not with long glandular hairs; plants
     of various habits—37

37a. Sepals 6; petals absent; flowers minute, green; leaves frequently lobed
     at the base—***Rumex acetosella***, in POLYGONACEAE, p. 147
37b. Sepals and petals both present—38

38a. Sepals and petals each 4—***Draba*** spp., in CRUCIFERAE, p. 174
38b. Sepals and petals each 5—39

39a. Petals united, corolla irregular—LENTIBULARIACEAE, p. 295
39b. Petals separate, corolla regular—40

40a. Styles 2; stamens 5 or 10—SAXIFRAGACEAE, p. 187
40b. Style 1; stamens 10—PYROLACEAE, p. 254

41a. Cauline (stem) leaves all or chiefly opposite or whorled (the bracts of
     the flower clusters may be alternate)—42
41b. Cauline leaves all or chiefly alternate—125

42a. A vine, the stem twining or climbing—43
42b. An herb, the stem neither twining or climbing—45

43a. Leaves compound, trifoliolate; sepals colored, over 2 cm wide—
   Clematis spp., in RANUNCULACEAE, p. 165
43b. Leaves simple; sepals green, much less than 2 cm wide—44

44a. Leaves entire; petals pink to dark purple—*Vincetoxicum* spp., in AS-
   CLEPIADACEAE, p. 265
44b. Leaves lobed or merely serrate; petals absent—*Humulus lupulus*, in
   CANNABACEAE, p. 143

45a. One pair of cauline leaves present—46
45b. More than one pair of cauline leaves—51

46a. Leaves entire—47
46b. Leaves toothed or lobed—48

47a. Leaves broadly kidney-shaped; flowers solitary; sepals 3, united—
   Asarum canadense, in ARISTOLOCHIACEAE, p. 145
47b. Leaves linear or lanceolate; flowers in racemes; sepals 2, separate—
   Claytonia spp., in PORTULACACEAE, p. 156

48a. Flowers in racemes; petals deeply fringed—*Mitella diphylla*, in SAX-
   IFRAGACEAE, p. 188
48b. Flowers solitary or few in a cluster; petals entire or nearly so—49

49a. Petals 6 or more—*Podophyllum peltatum*, in BERBERIDACEAE
49b. Petals 4 or 5—50

50a. Petals 4—*Dentaria diphylla*, in CRUCIFERAE, p. 179
50b. Petals 5—*Geranium* spp., in GERANIACEAE, p. 217

51a. Inflorescence a dense head of small flowers subtended by an involucre
   of small, green bracts; sepals absent or present as bristles or scales;
   stipules absent—52
51b. Inflorescence various; if an involucrate head, then the petals are not
   prolonged if united and the sepals are green; stipules present or not—
   53

52a. Petals 5, united and tubular or often prolonged into a flat, strap-like
   structure (see Figs. 42–43); stamens 5—COMPOSITAE, p. 309
52b. Petals 4, united but not prolonged; stamens 4—*Knautia arvensis*, in
   DIPSACACEAE, p. 305

53a. Inflorescence a cyathium, consisting of several petaloid bracts, enclosing a few inconspicuous flowers which lack petals (see Fig. 26). Usually several flowers consisting only of a single stamen and one consisting of a 3-lobed ovary with three styles are present; juice milky—*Euphorbia* spp., in EUPHORBIACEAE, p. 220

53b. Inflorescence not a cyathium; if flowers are surrounded by petaloid bracts, the flowers perfect; juice milky or not—54

54a. Flowers small and inconspicuous, the perianth none or greenish or chaffy, not petaloid in appearance—55

54b. Flowers generally larger, with a white or colored petaloid perianth—75

55a. Leaves compound or deeply pinnately divided—56

55b. Leaves entire or toothed—58

56a. Aquatic or mud plants; leaves deeply pinnately divided; flowers in emergent terminal spikes or axillary—HALORAGACEAE, p. 244

56b. Terrestrial plants; leaves palmately compound—57

57a. Flowers in axillary clusters—***Cannabis sativa***, in CANNABACEAE, p. 143

57b. Flowers in an umbel—***Panax*** spp., in ARALIACEAE, p. 246

58a. Leaves all whorled—59

58b. Leaves all or mostly opposite—62

59a. Aquatic or mud plants—60

59b. Terrestrial plants—61

60a. Stems erect, emergent; leaves small, entire; flowers solitary in leaf axils—***Hippuris vulgaris***, in HIPPURIDACEAE, p. 245

60b. Stems flexuous, mostly submerged; leaves scale-like, toothed, or finely pinnately divided; flowers in emergent terminal spikes or axillary—HALORAGACEAE, p. 244

61a. Stems prostrate, often in sandy open areas; flowers axillary—***Mollugo verticillata***, in MOLLUGINACEAE, p. 156

61b. Stems erect, often in woodlands; flowers in a terminal, bracted cluster—***Cornus canadensis***, in CORNACEAE, p. 253

62a. Flowers in terminal or axillary spikes, racemes, panicles, or other clusters—63

62b. Flowers solitary or few in the axils of the leaves—69

63a. Leaves scalelike, without an expanded blade—***Bartonia virginica***, in
    GENTIANACEAE, p. 262
63b. Leaves with an expanded blade—64

64a. Flower pedicels longer than the perianth—65
64b. Flowers sessile or nearly so—66

65a. Flowers 5-merous; leaves entire—CARYOPHYLLACEAE, p. 156
65b. Flowers 2-merous; leaves toothed—***Circaea*** spp., in ONAGRACEAE

66a. Flowers subtended by sharp-pointed bracts; plant white-woolly—
    Froelichia gracilis, in AMARANTHACEAE, p. 154
66b. Flowers not subtended by bracts; plant green, the stems glabrous or pu-
    bescent—67

67a. Stipules absent; stems covered with appressed or spreading hairs—
    Lechea spp., in CISTACEAE, p. 235
67a. Stipules present; stems glabrous or covered with stiff, stinging hairs—
    68

68a. Plants of sandy areas; stems glabrous—CARYOPHYLLACEAE, p. 156
68b. Plants of wet areas; stems glabrous or covered with stiff, stinging
    hairs—URTICACEAE, p. 143

69a. Plants of dry areas; stem erect or sometimes prostrate; leaves linear to
    oblong—CARYOPHYLLACEAE, p. 156
69b. Plants of wet or muddy areas; stems and leaves various—70

70a. Leaves sessile—71
70b. Leaves with an evident petiole—74

71a. Small mud plants up to 5 cm high; flowers 2-merous—72
71b. Plants 10 cm or more high; flowers 4- or 5-merous, or sometimes 3- or
    6-merous, but never 2-merous—73

72a. Stipules present; petals present—***Elatine minima***, in ELATINACEAE
72b. Stipules absent; petals absent—CALLITRICHACEAE, p. 224

73a. Flowers sessile or nearly so—LYTHRACEAE, p. 240
73b. Flowers on long pedicels—***Stellaria borealis***, in CARYOPHYL-
    LACEAE, p. 158

74a. Leaves round, ovate, or kidney-shaped, rounded at the base, crenate or
    lobed—***Chrysosplenium americanum***, in SAXIFRAGACEAE, p. 187

74b. Leaves lanceolate to ovate, entire—***Ludwigia palustris***, in ONA-GRACEAE, p. 243

75a. Plants with several to many small flowers closely aggregated into a dense head surrounded or subtended by an involucre of green, brown, or white bracts—76
75b. Flowers solitary or variously clustered, but not in involucrate heads—80

76a. Involucre of 4 conspicuous white bracts, much larger than the small flower-cluster—***Cornus canadensis***, in CORNACEAE, p. 253
76b. Involucral bracts green or brown—77

77a. Corolla 4-lobed; stamens 4—78
77b. Corolla 5-lobed or 5 free petals; stamens 2, 4, or 10—79

78a. Stems prickly—***Dipsacus*** spp., in DIPSACEAE, p. 305
78b. Stems not prickly—***Plantago arenaria***, in PLANTAGINACEAE, p. 297

79a. Stems often square; corolla 5-lobed, often 2-lipped; stamens 2 or 4—LABIATAE, p. 277
79b. Stems round; petals 5, free, corolla not 2-lipped; stamens 10—***Petrorhagia prolifera***, in CARYOPHYLLACEAE, p. 156

80a. The conspicuous portion of the perianth (usually the petals) composed of separate parts—81
80b. The conspicuous portion of the perianth composed of united parts—101

81a. Cauline leaves compound or deeply divided—82
81b. Leaves entire or toothed, or with one or two small lobes near the base only—88

82a. Cauline leaves pinnately lobed or compound—83
82b. Cauline leaves palmately lobed or compound—85

83a. Flowers pink; stamens 5; fruit not bristly—***Erodium cicutarium***, in GERANIACEAE, p. 217
83b. Flowers yellow; stamens 10 or many; fruit bristly—84

84a. Stems erect; stamens many—***Stylophorum diphyllum***, in PAPAVER-ACEAE, p. 173
84b. Stems prostrate; stamens 10—***Tribulus terrestris***, in ZYGOPHYL-LACEAE, p. 218

85a. Ovaries several in each flower; stamens many—RANUNCULACEAE
85b. Ovary 1; stamens 5, 6, or 10—86

86a. Stamens 6—***Dentaria*** spp., in CRUCIFERAE, p. 179
86b. Stamens 5 or 10—87

87a. Flowers greenish or white, about 2 mm broad; stamens 5—***Panax*** spp., in ARALIACEAE, p. 246
87b. Flowers pink, reddish-purple, or white, 5 mm or more broad; stamens usually 10—***Geranium*** spp., in GERANIACEAE, p. 217

88a. Petals 3—89
88b. Petals 2 or 4–7—92

89a. Stem with a single whorl of 3 entire leaves; sepals 3—***Trillium*** spp., in LILIACEAE, p. 121
89b. Stem with several to many sets of opposite or whorled, entire or toothed leaves; sepals 3 or 5—90

90a. Leaves sharply toothed; sepals 3, red to purple—***Impatiens glandulifera***, in BALSAMINACEAE, p. 229
90b. Leaves entire; sepals 5, most or all green—91

91a. Sepals 5, the 2 lateral ones enlarged and petaloid; leaves mostly linear—POLYGALACEAE, p. 219
91b. Sepals 5, all green and alike; leaves needle-like—***Lechea*** spp., in CISTACEAE, p. 236

92a. Petals 2, deeply notched; stamens 2—***Circaea*** spp., in ONAGRACEAE
92b. Petals 4–7; stamens 4 or more—93

93a. Leaves with 3–5 principal veins—***Rhexia virginica***, in MELASTOMATACEAE, p. 241
93b. Leaves with one principal midvein—94

94a. Ovaries 4 or 5; leaves succulent—CRASSULACEAE, p. 186
94b. Ovary 1; leaves not succulent—95

95a. Leaves dotted with translucent dots (seen when the leaf is held to the light)—GUTTIFERAE, p. 233
95b. Leaves not dotted with translucent dots—96

96a. Style 1 or none—97
96b. Styles 2–6—99

97a. Base of flower flat, lacking a hypanthium; sepals 5—*Chimaphila* spp., in PYROLACEAE, p. 254

97b. Base of flower tubular, the sepals and petals attached to a cup-shaped, tubular, or globose hypanthium; sepals 4–6—98

98a. Ovary superior; sepals 4–6, alternate with similar hypanthium lobes—LYTHRACEAE, p. 240

98b. Ovary inferior; sepals 4; hypanthium lobes absent—ONAGRACEAE

99a. Sepals 2—*Portulaca* spp., in PORTULACACEAE, p. 156

99b. Sepals 5—100

100a. Petals white or pinkish; styles 2–5 (rarely 6); stamens mostly 5 or 10, sometimes fewer, the filaments free—CARYOPHYLLACEAE, p. 156

100b. Petals yellow or white with a yellow base; styles 5; stamens 5, the filaments united below—LINACEAE, p. 216

101a. Perianth of one whorl, either petals or sepals absent—102

101b. Perianth consisting of both calyx and corolla—104

102a. Sepals purplish, petals absent; flowers 3–5 (rarely 1), subtended by a greenish or brownish, spreading, 5-lobed involucre of bracts—NYCTAGINACEAE, p. 155

102b. Petals white to pinkish or yellow, sepals absent; flowers solitary or in clusters, but without a subtending involucre of bracts—103

103a. Stamens 3—VALERIANACEAE, p. 304

103b. Stamens 4 or 5—RUBIACEAE, p. 298

104a. Stamens more numerous than the petals—POLYGALACEAE, p. 219

104b. Stamens as many as or fewer than the petals—105

105a. Top of the ovary very deeply 4-lobed; style 1—LABIATAE, p. 277

105b. Ovary not deeply lobed—106

106a. Stamens as many as the petals—107

106b. Stamens fewer than the petals—116

107a. Ovaries 2—108

107b. Ovary 1—109

108a. Corolla reflexed, seldom spreading; corona present, erect; stamens united with each other and the stigmas (Fig. 30); stem unbranched—ASCLEPIADACEAE, p. 265

108b. Corolla not reflexed; corona absent; stamens separate, not united with the stigma; stem branched—APOCYNACEAE, p. 264

109a. Ovary inferior, appearing as a swelling below the calyx at the apex of the pedicel—110
109b. Ovary superior, located in the center of the flower—111

110a. Corolla 3- or 4-lobed—RUBIACEAE, p. 298
110b. Corolla 5-lobed—*Triosteum* spp., in CAPRIFOLIACEAE, p. 300

111a. Leaves toothed—*Phyla lanceolata*, in VERBENACEAE, p. 277
111b. Leaves entire—112

112a. Stamens located opposite each petal—PRIMULACEAE, p. 259
112b. Stamens located between the petals or corolla lobes, or else attached deeply within the tubular corolla so that their position is not easily ascertained—113

113a. Corolla (actually a petaloid calyx) narrowed above the ovary; fruit an achene—NYCTAGINACEAE, p. 155
113b. Corolla not narrowed above the ovary; fruit a capsule—114

114a. Corolla saucer-shaped, funnelform, or bell-shaped—GENTIANACEAE, p. 262
114b. Corolla salverform—115

115a. Ovary 3-celled—*Phlox* spp., in POLEMONIACEAE, p. 270
115b. Ovary 1-celled—GENTIANACEAE, p. 262

116a. Corolla (actually a petaloid calyx) narrowed above the ovary; fruit an achene—NYCTAGINACEAE, p. 155
116b. Corolla not narrowed above the ovary; fruit a capsule—117

117a. Stamens 3—VALERIANACEAE, p. 304
117b. Stamens 2 or 4—118

118a. Stamens 2—119
118b. Stamens 4—120

119a. Flowers in dense heads—*Justicia americana*, in ACANTHACEAE
119b. Flowers solitary or in racemes or spikes—SCROPHULARIACEAE

120a. Corolla distinctly 2-lipped, irregular—121
120b. Corolla not distinctly 2-lipped, its 5 lobes all alike or nearly so—122

121a. Both corolla and calyx 2-lipped; the upper lip with 3 awl-shaped teeth, the lower with 2 short teeth; flowers in slender terminal spikes—***Phryma leptostachya***, in VERBENACEAE, p. 277
121b. Only the corolla 2-lipped; calyx not obviously 2-lipped, its teeth equal or nearly so—SCROPHULARIACEAE, p. 286

122a. Flowers in nodding pairs at the top of a slender stalk—***Linnaea borealis*** var. ***longiflora***, in CAPRIFOLIACEAE, p. 301
122b. Flowers solitary or in clusters; not in nodding pairs—123

123a. Flowers pedicellate, not in spikes—SCROPHULARIACEAE, p. 286
123b. Flowers sessile or nearly so, in spikes—124

124a. Corolla short, 1 cm or less long—VERBENACEAE, p. 276
124b. Corolla 1.5 cm or more long—SCROPHULARIACEAE, p. 286

125a. A vine, the stem twining (without tendrils) or climbing/scrambling (with tendrils)—126
125b. An herb, the stem neither twining or climbing/scrambling; tendrils absent—135

126a. Stem climbing or scrambling, tendrils present—127
126b. Stem twining, tendrils absent—129

127a. Leaves pinnately compound, the terminal leaflet modified into a tendril—LEGUMINOSAE, p. 203
127b. Leaves simple—128

128a. Leaf blades lobed; petals 5 or 6, united—CUCURBITACEAE, p. 305
128b. Leaf blades entire; tepals 6, separate—***Smilax*** spp., in LILIACEAE

129a. Leaves pinnately dissected or compound—130
129b. Leaves entire or lobed—131

130a. Leaves once-compound; stipules present—LEGUMINOSAE, p. 203
130b. Leaves two or more times compound; stipules absent—***Adlumia fungosa***, in FUMARIACEAE, p. 173

131a. Leaves broad, palmately lobed, peltate—***Menispermum canadense***, in MENISPERMACEAE, p. 171
131b. Leaves various, petiole attached at the margin—132

132a. Plant has sheathing stipules surrounding the stem at the base of each leaf (an ocrea)—POLYGONACEAE, p. 145

132b. Stipules not encircling the stem, or absent—133

133a. Leaves with many prominent, roughly parallel veins; tepals separate—
Dioscorea villosa, in DIOSCOREACEAE, p. 125
133b. Leaves with a single, prominent midvein; petals united—134

134a. Corolla saucer-shaped; flower about 1 cm wide—*Solanum
dulcamara*, in SOLANACEAE, p. 285
134b. Corolla funnelform; flower 2–8 cm wide—CONVOLVULACEAE

135a. Plant has sheathing stipules surrounding the stem at the base of each
leaf (an ocrea, Fig. 16)—POLYGONACEAE, p. 145
135b. Stipules not encircling the stem, or absent—136

136a. Inflorescence a dense head of small flowers subtended by an involucre
of small, green bracts; sepals absent or present as bristles or scales;
petals 5, united and tubular or often prolonged into a flat, strap-like
structure; stipules absent (Figs. 41–45)—COMPOSITAE, p. 309
136b. Inflorescence various; if an involucrate head, then the petals are not
tubular or prolonged if united and the sepals are green; stipules present or not—137

137a. Inflorescence a cyathium, consisting of several petaloid bracts, enclosing a few inconspicuous flowers which lack petals (Fig. 26; box,
p. 221). Usually several flowers consisting only of a single stamen
and one consisting of a 3-lobed ovary with three styles are present;
juice milky—*Euphorbia* spp., in EUPHORBIACEAE, p. 220
137b. Inflorescence not a cyathium; if flowers are surrounded by petaloid
bracts, the flowers perfect; juice milky or not—138

138a. Leaves sword-shaped, finely parallel-veined, with bristly margins;
flowers in dense heads—*Eryngium yuccifolium*, in UMBELLIF-
ERAE, p. 247
138b. Leaves not sword-shaped—139

139a. Flowers small and inconspicuous, the perianth none or greenish or
chaffy, not petaloid in appearance—140
139b. Flowers large or small, but with a white or colored petaloid perianth—166

140a. Aquatic or mud plants; leaves scale-like, toothed, or finely pinnately
divided; flowers in emergent terminal spikes or axillary—HALOR-
AGACEAE, p. 244
140b. Terrestrial plants; leaves and flowers variously arranged—141

141a. Leaves compound, deeply lobed, or divided—142
141b. Leaves simple, not deeply lobed or divided—152

142a. Leaves once-compound, lobed, or divided—143
142b. Leaves 2 or 3 times compound—149

143a. Leaves once-pinnately compound or divided—144
143b. Leaves once-palmately compound or lobed—147

144a. Sepals and petals 3—*Floerkea proserpinacoides*, LIMNAN-
THACEAE, p. 224
144b. Sepals and petals 4 or 6—145

145a. Leaves compound—*Sanguisorba canadensis*, in ROSACEAE, p. 202
145b. Leaves pinnately divided—146

146a. Sepals and petals 4; stamens 2 or 6—*Lepidium* spp., in CRU-
CIFERAE, p. 174
146b. Sepals and petals 6; stamens 10 or more—*Reseda lutea*, in
RESEDACEAE, p. 185

147a. Flower solitary, terminal—*Hydrastis canadensis*, in RANUNCU-
LACEAE, p. 168
147b. Flowers many, in axillary or terminal clusters—148

148a. Flowers in dense terminal umbels or heads—UMBELLIFERAE,
p. 247
148b. Flowers in axillary spikes or panicles—*Cannabis sativa*, in
CANNABACEAE, p. 143

149a. Stamens 5; flowers in umbels—150
149b. Stamens 6 or many; flowers in racemes, panicles, or cymes—151

150a. Styles 5—*Aralia* spp., in ARALIACEAE, p. 246
150b. Styles 2—UMBELLIFERAE, p. 247

151a. Stamens 6; flowers in cymes—*Caulophyllum thalictroides*, in
BERBERIDACEAE, p. 170
151b. Stamens many; flowers in racemes or panicles—RANUNCULACEAE

152a. Flowers minute, subtended by palmately cleft axillary bracts—*Acalypha rhomboidea*, in EUPHORBIACEAE, p. 223
152b. Flowers various, not subtended by palmately cleft axillary bracts—
153

153a. Flowers axillary, solitary or in few-flowered clusters—154
153b. Flowers often many, in terminal, or terminal and axillary, clusters—
160

154a. Leaves linear—CHENOPODIACEAE, p. 152
154b. Leaves broader than linear—155

155a. Flower-clusters with bracts as long as or longer than the flowers—156
155b. Flowers without conspicuous bracts—157

156a. Leaf apex rounded or obtuse—*Amaranthus* spp., in AMARAN-
THACEAE, p. 154
156b. Leaf apex acuminate—*Parietaria pensylvanica*, in URTICACEAE

157a. Stems glabrous—158
157b. Stems pubescent—159

158a. Plants usually in dry, sandy areas; stems round in cross section; fruit a
orange or red fleshy drupe—*Geocaulon lividum*, in SANTALACEAE
158b. Plants of wet areas; stems 4-angled in cross section; fruit a capsule—
Ludwigia polycarpa, in ONAGRACEAE, p. 243

159a. Stems covered with stiff, stinging hairs; flowers in axillary cymes—
Laportea canadensis, in URTICACEAE, p. 143
159b. Stems pubescent but lacking stiff hairs; flowers nodding in the axils
of the leaves—*Hybanthus concolor*, in VIOLACEAE, p. 236

160a. Individual flowers on distinct pedicels—161
160b. Individual flowers crowded in dense clusters, sessile—163

161a. Leaves finely serrate; flowers in a cyme; ovaries 5—*Penthorum sedoides*, in PENTHORACEAE, p. 187
161b. Leaves entire or toothed, sometimes lobed; flowers in a raceme, panicle, or axillary cluster; ovary 1—162

162a. Lower leaves toothed or pinnately lobed; flowers in a raceme—*Lepidium* spp., in CRUCIFERAE, p. 174
162b. All leaves entire; flowers in a panicle or axillary cluster—CIS-
TACEAE, p. 235

163a. Stem and leaves covered with stellate hairs—*Croton glandulosus*, in
EUPHORBIACEAE, p. 223
163b. Stem and leaves glabrous or if pubescent, the hairs not stellate—164

164a. Leaves entire, cordate—*Saururus cernuus*, in SAURURACEA, p. 133

164b. Leaves entire or (often) toothed, not cordate—165

165a. Flowers subtended by sharp-pointed bracts—***Amaranthus*** spp., in AMARANTHACEAE, p. 154
165b. Flowers without subtending bracts, or sometimes with bracts which are not sharp-pointed—CHENOPODIACEAE, p. 152

166a. Plant sap or juice milky or colored—167
166b. Plant sap or juice watery, not colored—171

167a. Corolla irregular, 1- or 2-lipped; stamens often protruding—***Lobelia*** spp., in CAMPANULACEAE, p. 306
167b. Corolla regular—168

168a. Petals reflexed; corona present (Fig. 30); ovaries 2—***Asclepias hirtella***, in ASCLEPIADACEAE, p. 266
168b. Petals not reflexed; corona not present; ovary 1—169

169a. Petals separate—PAPAVERACEAE, p. 172
169b. Petals united—170

170a. Stamens attached to the corolla tube—CONVOLVULACEAE, p. 268
170b. Stamens attached at the very base of the corolla or top of the ovary—CAMPANULACEAE, p. 306

171a. Flowers closely aggregated into a dense head subtended by an involucre of small bracts; stipules present; leaves trifoliolate or rarely with five leaflets—LEGUMINOSAE, p. 203
171b. Flowers solitary or in clusters, but not in involucrate heads; stipules present or absent; leaves various—172

172a. Leaves covered with stellate hairs; calyx of 5 sepals, the outer 2 much narrower than the outer 3—***Helianthemum*** spp., in CISTACEAE, p. 236
172b. Leaves glabrous or hairy; if stellate hairs present, calyx not as above— 173

173a. Flowers irregular, one or more of the sepals or petals differing in size, shape, or orientation—174
173b. Flowers regular—192

174a. Stamens 2 or 4—SCROPHULARIACEAE, p. 286
174b. Stamens 5 or more—175

175a. Stamens 5—176
175b. Stamens more than 5—183

176a. Petals separate—177
176b. Petals united—180

177a. Flowers in panicles or umbels; perianth parts lack a spur—178
177b. Flowers solitary, or in few-flowered clusters; a spur formed from a
      petal or sepal—179

178a. Flowers greenish or purplish, in a panicle—***Heuchera*** spp., in SAX-
      IFRAGACEAE, p. 187
178b. Flowers white or purplish, in compound umbels—UMBELLIFERAE,
      p. 247

179a. Flowers red-orange to pale yellow, in summer—BALSAMINACEAE,
      p. 229
179b. Flowers blue, yellow, white, or greenish, in spring—VIOLACEAE, p.
      236

180a. Anthers united; stamens often protruding from the irregular corolla—
      CAMPANULACEAE, p. 306
180b. Anthers separate; corolla almost regular—181

181a. Corolla saucer-shaped; some or all stamen filaments hairy—***Verbas-
      cum*** spp., in SCROPHULARIACEAE, p. 286
181b. Corolla funnelform; stamen filaments not hairy—182

182a. Corolla dull yellow and purple—***Hyoscyamus niger***, in
      SOLANACEAE, p. 285
182b. Corolla blue—***Echium vulgare***, in BORAGINACEAE, p. 272

183a. Stamens 6–9—184
183b. Stamens 10 or more—187

184a. Leaves compound or dissected—185
184b. Leaves simple—186

185a. Leaves pinnately compound or dissected; stamens 6, the filaments
      joined in two groups of three—FUMARIACEAE, p. 173
185b. Leaves palmately compound or trifoliolate; stamens 6–9, the filaments
      all long, separate—CAPPARACEAE, p. 174

186a. Flowers solitary on a scaly basal branch; perianth S-shaped; stamens
      6—***Aristolochia serpentaria***, in ARISTOLOCHIACEAE, p. 145

186b. Flowers in spikes, heads, or racemes; perianth not S-shaped; stamens (6) 7 or 8—POLYGALACEAE, p. 219

187a. Leaves simple; styles 2—SAXIFRAGACEAE, p. 187
187b. Leaves simple or compound; style 1 or the 3 stigmas sessile—188

188a. Leaves compound—189
188b. Leaves simple, often deeply divided—190

189a. Leaflets 3; stipules absent; petals 4—*Polanisia dodecandra*, in CAP-PARACEAE, p. 174
189b. Leaflets 3 or more; stipules usually present; petals 5—LEGUMI-NOSAE, p. 203

190a. Leaves entire—*Crotalaria sagittatis*, in LEGUMINOSAE, p. 208
190b. Leaves deeply divided—191

191a. Leaves palmately divided; sepals blue; petals blue, 2, prolonged backward into a single spur—*Consolida ambigua*, in RANUNCU-LACEAE, p. 166
191b. Leaves pinnately divided; sepals green; petals yellow, often 6, not prolonged into spurs—*Reseda lutea*, in RESEDACEAE, p. 185

192a. Perianth of one whorl (usually only the calyx)—193
192b. Perianth consisting of both calyx and corolla—198

193a. Leaves entire—194
193b. Leaves toothed, lobed, or compound—196

194a. Tepals 6, greenish yellow—*Smilax* spp., in LILIACEAE, p. 117
194b. Sepals whitish—195

195a. Stamens 5; fruit a green or yellow drupe—*Comandra umbellata*, in SANTALACEAE, p. 144
195b. Stamens 10; fruit a dark purple berry—*Phytolacca americana*, in PHYTOLACCACEAE, p. 155

196a. Stamens 5—UMBELLIFERAE, p. 247
196b. Stamens 6 or more—197

197a. Stamens 6—*Caulophyllum thalictroides,* in BERBERIDACEAE, p. 171
197b. Stamens more than 6—RANUNCULACEAE, p. 165

198a. Corolla composed of united petals—199
198b. Corolla composed of separate petals—214

199a. Ovaries 2; flowers orange-red, in umbels—*Asclepias tuberosa*, in AS-
    CLEPIADACEAE, p. 266
199b. Ovary 1; flowers various—200

200a. Stamens 6 to many—201
200b. Stamens 5 or fewer—202

201a. Leaves compound with 3 obcordate leaflets; stamens 10, united
    below—OXALIDACEAE, p. 216
201b. Leaves simple, often lobed; stamens many, united by their filaments
    into a tube—MALVACEAE, p. 232

202a. Stamens 2 or 4—SCROPHULARIACEAE, p. 286
202b. Stamens 5—203

203a. Top of the ovary deeply 4-lobed; flowers often in helicoid cymes—
    BORAGINACEAE, p. 272
203b. Top of the ovary not deeply lobed—204

204a. Flowers solitary, either terminal or axillary—205
204b. Flowers in terminal or axillary clusters—207

205a. Calyx concealed by 2 bracts—*Calystegia* spp., in CONVOLVU-
    LACEAE, p. 268
205b. Calyx not concealed by bracts—206

206a. Ovary inferior, appearing as a small swelling below the calyx at the
    base of the flower; stigmas as 3–5 lobes at tip of style—CAMPANU-
    LACEAE, p. 306
206b. Ovary superior; stigma 1—SOLANACEAE, p. 284

207a. Some or all stamen filaments hairy—208
207b. Stamen filaments not hairy—209

208a. Leaves lobed or divided; flowers tubular, blue to purple or white—
    HYDROPHYLLACEAE, p. 271
208b. Leaves not lobed; flowers saucer-shaped, yellow or white—*Verbas-
    cum* spp., in SCROPHULARIACEAE, p. 286

209a. Anthers clustered and often touching, longer than the filaments—210
209b. Anthers separate from each other, usually shorter than the filaments—
    211

210a. Ovary inferior, appearing as a small swelling below the calyx at the base of the flower; stigmas as 3–5 lobes at tip of style—CAMPANU-LACEAE, p. 306
210b. Ovary superior; stigma 1—SOLANACEAE, p. 284

211a. Leaves compound or very deeply lobed—212
211b. Leaves simple or with shallow lobes—213

212a. Leaf-segments linear or oblong, irregular; stamens longer than the corolla tube, the filaments hairy—HYDROPHYLLACEAE, p. 271
212b. Leaves pinnately compound or divided into narrow segments; stamens not longer than the corolla tube, the filaments glabrous—POLEMO-NIACEAE, p. 270

213a. Flowers in a terminal spike or raceme; fruit a capsule—PRIMU-LACEAE, p. 259
213b. Flowers in axillary clusters; fruit a berry—SOLANACEAE, p. 284

214a. Petals 3—215
214b. Petals 4 or more—216

215a. Leaves deeply lobed or compound; petals white—***Floerkea proser-pinacoides***, in LIMNANTHACEAE, p. 224
215b. Leaves simple, not lobed; petals small, reddish—***Lechea*** spp., in CIS-TACEAE, p. 235

216a. Petals 4—217
216b. Petals 5 or more—221

217a. Leaves compound—218
217b. Leaves simple, sometimes deeply lobed or dissected—219

218a. Stamens 6 or more; leaflets entire—CAPPARACEAE, p. 174
218b. Stamens 4; leaflets toothed—***Sanguisorba canadensis***, in ROSACEAE, p. 202

219a. Stamens 2 (rarely) or 6, 4 long and 2 short; leaves often deeply lobed or dissected—CRUCIFERAE, p. 174
219b. Stamens 4 or 8—220

220a. Ovary 1, inferior; leaves not succulent—ONAGRACEAE, p. 242
220b. Ovaries 4, superior; leaves succulent—CRASSULACEAE, p. 186

221a. Petals 5—222

221b. Petals 6 or more—240

222a. Sepals 2— *Portulaca* spp., in PORTULACACEAE, p. 156
222b. Sepals more than 2—223

223a. Stamens 4 or 5—224
223b. Stamens 6 or more—231

224a. Flower solitary, terminating the stem; petals white with green or yellow veins—*Parnassia* spp., in SAXIFRAGACEAE, p. 187
224b. Flowers more than 1, either axillary (1 or more per axil) or in terminal
  clusters—225

225a. Flowers axillary, solitary or several, greenish, nodding—*Hybanthus
  concolor*, in VIOLACEAE, p. 236
225b. Flowers in terminal clusters—226

226a. Flowers in umbels—227
226b. Flowers in spike-like racemes, panicles, or cymes—229

227a. Flowers pink; style 1—*Erodium cicutarium*, in GERANIACEAE, p.
  217
227b. Flowers yellow or white; styles 2 or 5—228

228a. Styles 2—UMBELLIFERAE, p. 247
228b. Styles 5—*Aralia* spp., in ARALIACEAE, p. 246

229a. Leaves compound; flowers in slender spike-like racemes—*Agrimonia*
  spp., in ROSACEAE, p. 191
229b. Leaves simple; flowers in cymes or panicles—230

230a. Flowers blue or yellow, in cymes—LINACEAE, p. 216
230b. Flowers greenish, white, or pink, in panicles—*Heuchera* spp., in
  SAXIFRAGACEAE, p. 187

231a. Stamens 6–10—232
231b. Stamens more than 10—238

232a. Leaves simple, entire, toothed, or deeply palmately lobed—233
232b. Leaves compound—235

233a. Leaves deeply palmately lobed—*Geranium* spp., in GERANIACEAE,
  p. 217
233b. Leaves simple, entire or toothed—234

234a. Ovaries and styles 5—CRASSULACEAE, p. 186
234b. Ovary 1, styles 2 —SAXIFRAGACEAE, p. 187

235a. Leaflets entire—236
235b. Leaflets toothed—237

236a. Leaflets 3, obcordate—OXALIDACEAE, p. 216
236b. Leaflets 3 or more, not obcordate—LEGUMINOSAE, p. 203

237a. Petals yellow—*Agrimonia* spp., in ROSACEAE, p. 191
237b. Petals white or pink—*Dictamnus albus*, in RUTACEAE, p. 218

238a. Stamens united by their filaments into a tube—MALVACEAE, p. 232
238b. Stamens separate from each other—239

239a. Leaves with stipules—ROSACEAE, p. 191
239b. Leaves without stipules—RANUNCULACEAE, p. 165

240a. Leaves entire; petals 6, pink—*Lythrum alatum*, in LYTHRACEAE, p. 241
240b. Leaves divided or compound; petals 6 or more, white, yellowish, yellowish-green, or purplish—241

241a. Flowers in a slender terminal raceme—*Reseda lutea*, in RESEDACEAE, p. 185
241b. Flowers solitary or clustered, but not in a slender raceme—242

242a. Stamens 6; petals 6, yellowish-green or purplish; leaflets with a few large terminal lobes—*Caulophyllum thalictroides*, in BERBERIDACEAE, p. 171
242b. Stamens numerous; petals 6–10, white; leaflets toothed nearly to the base—*Actaea* spp., in RANUNCULACEAE, p. 165

THE PLANTS OF MICHIGAN

Division PINOPHYTA—The Gymnosperms

TAXACEAE, The Yew Family

Dioecious evergreen shrubs, with spirally arranged, linear leaves. Male cones small; seeds solitary, on axillary branches and covered with fleshy red tissue, resembling a berry.

One species in Michigan, a low shrub with ascending branches (to 2 m high)—**Ground-hemlock,** *Taxus canadensis*

PINACEAE, The Pine Family

Monoecious trees, usually evergreen, with needle-like leaves either spirally arranged (appearing to be alternate) or in clusters (fascicles). Male cones small, not woody; female cones larger and often woody, the seeds in pairs on spirally arranged scales.

1a. Leaves grouped in fascicles—2
1b. Leaves alternate, not in fascicles—5

2a. Leaves mostly in fascicles of ten or more, on short lateral wart-like branches, deciduous each autumn; often in bogs (to 20 m high)—**Tamarack,** *Larix laricina*
2b. Leaves in fascicles of two or five (*Pinus* spp., **Pine**)—3

3a. Leaves in fascicles of five (to 70 m high)—**White Pine,** *Pinus strobus*
3b. Leaves in fascicles of two—4

Pinus strobus, the white pine, is the Michigan state tree. At one time significant areas of the Upper Peninsula and the northern Lower Peninsula were covered with immense specimens (150–200 feet high) of this species. From 1870–1890 Michigan led the nation in lumber production, much of which was white pine. Most had been logged by the early 1900s, at which time the fungus **Cronartium ribicola**, the causal agent of white pine blister rust, was introduced from Europe. Although white pine was subsequently planted on state lands, this disease made production of lumber-quality trees infeasible. A remnant of the original majestic forest survives in Hartwick Pines State Park, near Grayling.

4a. Leaves 8–15 cm long (to 40 m high)—**Red Pine,** *Pinus resinosa*
4b. Leaves 2–4 cm long (to 20 m high)—**Jack Pine,** *Pinus banksiana*

5a. Leaves four-sided (*Picea* spp., **Spruce**)—6
5b. Leaves flattened—7

6a. Young twigs pubescent, cones ovoid; often in bogs (to 10–25 m high)—
   Black Spruce, *Picea mariana*
6b. Young twigs glabrous, cones cylindrical; NM (to 25 m high)—**White
   Spruce**, *Picea glauca*

7a. Leaves short-stalked, 15 mm long or less (to 30 m high)—**Hemlock**,
   Tsuga canadensis
7b. Leaves sessile, 15–30 mm long; NM (to 25 m high)—**Balsam Fir**, *Abies
   balsamea*

CUPRESSACEAE, The Cypress Family

Monoecious or dioecious trees or shrubs, usually evergreen, with needle-like or scale-like leaves either opposite or whorled. Male cones
small; female cones either dehiscent and woody with brown scales or
indehiscent and berry-like with bluish scales.

1a. Leafy twigs soft and flattened; female cones woody with brown scales (to
   20 m high)—**White-cedar** or **Arbor Vitae**, *Thuja occidentalis*
1b. Leafy twigs not distinctly flattened; female cones berry-like with bluish
   scales (*Juniperus* spp., **Juniper**)—2

2a. Leaves in whorls of three; often a spreading or ascending shrub, growing
   in dense mats (0.5–2 m high)—**Ground Juniper**, *Juniperus communis*
   var. *depressa*
2b. Leaves opposite—3

3a. Erect tree; SLP (to 20 m high)—**Red-cedar**, *Juniperus virginiana*
3b. Prostrate or spreading shrub; NM, espp. Great Lakes shores (to 10 cm
   high)—**Creeping Juniper**, *Juniperus horizontalis*

Division MAGNOLIOPHYTA—The Flowering Plants

Class LILIOPSIDA—The Monocotyledons

TYPHACEAE, The Cat-tail Family

Monoecious perennial herbs with long, alternate, linear, erect leaves.
Flowers unisexual, in terminal spikes, the male flowers above the female; perianth of few to many bristles or narrow scales; stamens often

3, the filaments separate or united; pistil 1, style 1, ovary superior, 1-celled. Fruit an achene. Summer.

1a. Staminate and pistillate portions of the flower-spike usually contiguous, the latter 2 cm or more in diameter; leaf blades often 8 mm or more wide; marshes and ditches (1–3 m high)—**Common Cat-tail**, *Typha latifolia*
1b. Staminate and pistillate portions of the spike separated, the latter 2 cm or less in diameter; leaf blades less than 8 mm wide; marshes and ditches (1–1.5 m high)—**Narrow- leaved Cat-tail**, *Typha angustifolia*

Cat-tail plants with an appearance intermediate between **Typha latifolia** and **Typha angustifolia**, and with bluish-green foliage, are examples of the naturally occurring hybrid, **Typha ×glauca.**

SPARGANIACEAE, The Bur-reed Family

Monoecious perennial herbs with alternate, linear, mostly erect leaves. Flowers unisexual, in dense, spherical heads; perianth of 3–6 tepals; stamens 1–8; pistil 1, style 1, ovary superior, 1–(rarely 2) celled. Fruit an achene. Summer.

Seven species occur in Michigan, of which the most common are:

1a. Leaves floating; NM—*Sparganium angustifolium*
1b. Leaves stiff and erect, emergent—2

2a. Stigmas two, inflorescence branched; marshes and shores (0.5–1.2 m high; summer)—**Bur-reed**, *Sparganium eurycarpum*
2b. Stigma one, inflorescence unbranched; shores, shallow lakes, etc. (5–60 cm high; summer)—*Sparganium chlorocarpum*

POTAMOGETONACEAE, The Pondweed Family

Perennial aquatic herbs with alternate or nearly subopposite, submerged or floating leaves. Flowers regular, perfect, in axillary or terminal spikes; perianth of 4 tepals; stamens 4; pistils and styles (1–)4, each ovary superior, 1-celled. Fruit an achene or small drupe.

Thirty species of *Potamogeton* occur in Michigan, among which two of the most conspicuous are:

1a. Floating leaves with cordate leaves—*Potamogeton natans*
1b. Leaves all submerged, stems flattened—*Potamogeton zosteriformis*

> For the identification of the remaining species of **Potamogeton**, consult Voss (1972). Try to gather fruiting material; sterile pondweeds are very difficult to identify.

ZANNICHELLIACEAE, The Horned Pondweed Family

Monoecious perennial, submerged aquatic herbs with opposite, thread-like entire leaves. Flowers unisexual, small, in axillary clusters; perianth 0; stamen 1; pistils and styles often 4 or 5, each ovary superior, 1-celled. Fruit a nutlet.

One species in Michigan, leaves thread-like, 3–8 cm long—***Zannichellia palustris***

NAJADACEAE, The Naiad Family

Monoecious or dioecious annual, submerged aquatic herbs with opposite or whorled leaves with minutely toothed margins. Flowers unisexual, small, axillary, solitary or in few-flowered clusters; perianth 0; stamen 1; pistil 1, style 1, ovary superior, 1-celled. Fruit an achene.

Five species in Michigan, of which the three most widespread are:

1a. Leaves sharply and coarsely toothed—***Najas marina***
1b. Leaves entire or with minute teeth—2

2a. Leaves less than 1 mm wide, apex tapers to slender point—***Najas flexilis***
2b. Leaves usually over 1 mm wide, apex acute, not long-tapered—***Najas guadalupensis***

JUNCAGINACEAE, The Arrow-grass Family

Perennial herbs with linear, cylindrical leaves. Flowers regular, perfect, inconspicuous, in racemes; perianth of 6 tepals; stamens 6; pistils and styles 3 or 6, each ovary superior, 1-celled. Fruit a cluster of follicles.

1a. Stem leaves present, each with a terminal pore; flowers in a loose-bracted raceme; bogs (20–40 cm high; late spring)—***Scheuchzeria palustris***
1b. Leaves all basal, terminal pore absent; flowers numerous in a bractless, spike-like raceme; marshes, shores (late spring and summer)— (***Triglochin*** spp., **Arrow-grass**)—2

2a. Stigmas six; fruit (usually seen at the base of the raceme) ovoid or oblong, rounded at the base (20–80 cm high)—***Triglochin maritimum***
2b. Stigmas three; fruit linear, narrowed at the base (20–40 cm high)—
Triglochin palustre

ALISMATACEAE, The Water-plantain Family

Perennial herbs with basal, broad-bladed leaves and scape-like stems.
Flowers regular, perfect or unisexual (the plants then monoecious), in
panicles or racemes; sepals 3, petals 3, white; stamens 6–many; pistils
and styles 3–many, either in a ring or a spiral, each ovary superior,
1-celled. Fruit an achene. Summer.

1a. Pistils in a ring; perfect flowers in panicles; marshes, ditches (0.2–1 m
tall or more)—**Water-plantain**, *Alisma plantago-aquatica* [incl. *A. trivi-
iale* & *A. subcordatum*—CQ]
1b. Pistils in a head; flowers in a raceme of 3-flowered whorls, the lower
pistillate, the upper staminate; shores, in shallow water (3–120 cm
high)—(*Sagittaria* spp., **Arrowhead**)—3

2a. Leaves ovate to linear, not sagittate at base—3
2b. Leaves broad or narrow, sagittate at base—4

3a. Pistillate (basal) flowers sessile or nearly so (10–80 cm high)—*Sagittaria rigida*
3b. Pistillate flowers with obvious pedicels (3–50 cm high)—*Sagittaria
graminea*

4a. Basal lobes small, short, linear (10–80 cm high)—*Sagittaria rigida*
4b. Basal lobes of the leaf conspicuous, triangular, almost or quite as long as
the terminal portion—5

5a. Beak of the achene very short and erect (10–50 cm high)—*Sagittaria
cuneata*
5b. Beak of the achene sharp, curved at right angles to the body (10–120 cm
high)—**Wapato**, *Sagittaria latifolia*

BUTOMACEAE, The Flowering-rush Family

Perennial emergent aquatic herbs with long, linear, erect leaves. Flowers regular, perfect, in a terminal umbel; sepals 3, pink; petals 3, pink;
stamens 9; pistils and styles 6, united below, each ovary superior,
1-celled. Fruit a cluster of follicles. Summer.

One species in coastal marshes, SE (1–1.5 m high)—**Flowering-rush**, *Butomus umbellatus*

HYDROCHARITACEAE, The Frog's-bit Family

Dioecious perennial submerged aquatic herbs. Flowers regular or not, unisexual, solitary or in a few-flowered axillary cyme enclosed in a bract (spathe); sepals 3; petals 0, 1, or 3, sometimes much-reduced; stamens 2 (*Vallisneria*) or 9 (*Elodea*), 3 staminodes present in pistillate flowers; pistil 1, style(s) 1 or 3, ovary inferior, 1-celled. Fruit a capsule or indehiscent. Summer.

1a. Leaves ribbon-like, all basal, 20 cm long or more—**Tape-grass**, *Vallisneria americana*
1b. Leaves on the stem, mostly in whorls of three, 2 cm long or less (*Elodea* spp., **Water-weed**)—2

2a. Leaves mostly 2 mm wide—*Elodea canadensis*
2b. Leaves mostly 1–1.3 mm wide—*Elodea nuttallii*

GRAMINEAE (POACEAE), The Grass Family

Annual or perennial herbs with linear or narrow sheathing leaves. Flowers regular, perfect (or occasionally unisexual, the plants then monecious), variously arranged in spikelets; sepals 0; petals 0; stamens mostly 3; pistil 1, styles 2 or 3, ovary superior, 1-celled. Fruit a grain. Most flower in late spring and summer.

The grass inflorescence is composed of individual units called *spikelets*. Classification of grasses depends chiefly upon the structure and arrangement of the spikelets; it is important that the plants to be identified bear mature flowers. Spikelets typically consist of a short axis, the *rachilla*, almost or quite concealed by several chaffy *bracts*. The two lower bracts are termed *glumes*, and have no flowers in their axils. The conspicuous bracts above the glumes are the *lemmas*. In the axil of each lemma, and often concealed by it, is a smaller bract, the *palea*. Between the lemma and the palea is a single flower which often consists of three stamens and one pistil. The number of flowers in a spikelet is therefore normally equal to the number of lemmas. The spikelets are grouped in racemes, spikes, or panicles of various sizes. Some species have a distinctive appendage, the *ligule*, at the junction of the leaf sheath and the blade.

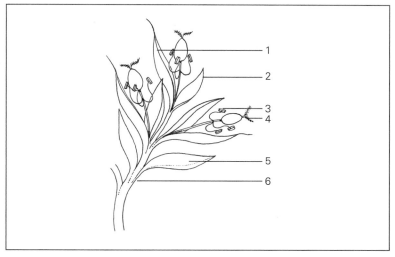

Figure 9: Schematic grass spikelet: 1, lemma; 2, palea;
3, stamen; 4, ovary; 5, glume; 6, rachilla

Of the large number (over 230) of grasses in Michigan, only the most common and distinctive are included here. For a more complete treatment, refer to Dore and McNeill (1980) or Voss (1972).

1a. Spikelets resembling a bur, subtended by an ovoid thorny involucre; sandy areas (20–80 cm high)—**Sandbur,** *Cenchrus longispinus*
1b. Spikelets without a thorny involucre—2

2a. Spikelets unisexual, either borne in separate inflorescences or separated within the same inflorescence—3
2b. Spikelets all perfect or any unisexual spikelets scattered among perfect spikelets—4

3a. Staminate spikelets borne in a terminal raceme, pistillate spikelets clustered in "ears" in leaf axils below; often cultivated (to 2.5 m high)—**Corn,** *Zea mays*
3b. Lower branches of the panicle spreading, bearing staminate flowers, the upper branches erect, with pistillate flowers; in shallow water (2–4 m high)—**Wild-rice,** *Zizania aquatica*

4a. Spikelets arranged in two rows to form a definite spike—5
4b. Spikelets in panicles or spikes, the spikelets never in definite rows—20

5a. Spikelet rows both occur on one side of the axis—6
5b. Spikelets alternating on opposite sides of the axis—12

6a. Spikelets in a terminal digitate cluster of spikes or a single dense spike; weedy areas—7
6b. Spikelets in a panicle or terminal raceme of short spikes—10

7a. Spike dense, unbranched, stems often prostrate; SLP—**Hard or Fairgrounds Grass,** *Sclerochloa dura*
7b. Spikes in a terminal digitate cluster—8

8a. Spikes over 2 mm wide; glumes keeled; SLP (30–60 cm high)—**Goosegrass** or **Yard Grass,** *Eleusine indica*
8b. Spikes less than 2 mm wide; glumes rounded (*Digitaria* spp., **Crabgrass**)—9

9a. Leaf blades and sheaths usually glabrous (sheath summit may be ciliate)—*Digitaria ischaemum*
9b. Leaf blades and sheaths hairy, especially near the plant base; SLP (30–60 cm high)—*Digitaria sanguinalis*

10a. Inflorescence a stout coarse terminal panicle (30–70 cm high)—**Barnyard Grass,** *Echinochloa crusgalli*
10b. Inflorescence a terminal raceme of spreading or ascending spikes—11

11a. Spikelets in 20 or more spreading spikes, each shorter than 1.5 cm; dry prairies, SLP (30–100 cm high)—**Grama Grass** or **Side-oats Grama,** *Bouteloua curtipendula*
11b. Spikelets in 20 or fewer ascending (or appressed to the axis) spikes, most longer than 4 cm; marshes (1–2 m high)—**Cord Grass,** *Spartina pectinata*

12a. Spikelets in pairs or triads at each joint—13
12b. Spikelets single at each joint, forming a loose, open or interrupted spike—17

13a. Glumes absent or unequal and filiform; spikelets spreading at maturity; woods (0.6–1.5 m high)—**Bottlebrush Grass,** *Hystrix patula* [*Elymus hystrix*—CQ]
13b. Glumes present and of equal length; spikelets ascending, forming a dense spike—14

14a. Spikelets in pairs at each joint (*Elymus* spp., **Wild Rye**)—15
14b. Spikelets in a triad at each joint (*Hordeum* spp., **Barley**)—16

15a. Spikes curved or nodding; awns of lemmas curved outward (0.8–1.5 m high)—***Elymus canadensis***

15b. Spikes erect; awns of lemmas straight (0.5–1.2 m high)—***Elymus virginicus***

16a. Awns of the lemmas 2–5 (rarely 6) cm long, the axis of the spike disintegrating with age (30–70 cm high)—**Squirrel-tail**, ***Hordeum jubatum***

16b. Awns of the lemmas 6–16 cm long, the axis remaining intact; weedy areas, often cultivated (60–120 cm high)—**Barley**, ***Hordeum vulgare***

17a. Spikelets with their edges toward the axis of the spike; weedy areas, often cultivated (30–100 cm high)—**Ryegrass**, ***Lolium perenne***

17b. Spikelets with their sides toward the axis of the spike—18

18a. Lemma with a spiny keel, prolonged into a long awn; weedy areas, often cultivated (0.5–1.2 m high)—**Rye**, ***Secale cereale***

18b. Lemma keeled, but the keel not spiny; awn present or not—19

19a. Annual; glumes over 3 mm broad, ovate; weedy areas, often cultivated (0.5–1.2 m high)—**Wheat**, ***Triticum* ×*aestivum***

19b. Perennial; glumes less than 2.5 mm broad, lanceolate (0.5–1 m high)—**Quack Grass**, ***Agropyron repens*** [***Elytrigia repens***—CQ]

20a. Spikelets contain one fertile flower—21

20b. Spikelets contain two or more fertile flowers—48

21a. Spikelets grouped into a dense, solitary, cylindrical spike—22

21b. Spikelets arranged in panicles or in panicles or racemes of spikes—28

22a. Spikelets with one or more long bristles arising from the base of the spikelet; weedy areas (***Setaria*** spp., **Foxtail Grass**)—23

22b. Spikelets without bristles, awnless or with short awns not more than 3 mm long—26

23a. Bristles five or more at the base of each spikelet, orange to golden (0.3–1.3 m high)—**Yellow Foxtail**, ***Setaria glauca***

23b. Bristles one to three at the base of each spikelet, green or purple—24

24a. Spikelets about 3 mm long; fertile lemma smooth, yellowish (to 2.5 m high)—**Foxtail Millet**, ***Setaria italica***

24b. Spikelets about 1.5–3 mm long; fertile lemma roughened, pale green—25

25a. Inflorescence erect or nodding at the tip; spikelets less than 2.5 mm long (20 cm–2.5 m high)—**Green Foxtail,** *Setaria viridis*

25b. Inflorescence nodding from the base; spikelets longer than 2.5 mm; SLP (0.5–2 m high)—**Giant Foxtail,** *Setaria faberi*

26a. Spike-like panicle thickened in the middle, more than 1 cm thick; Great Lakes dunes (0.5–1 m high)—**Beach Grass,** *Ammophila breviligulata*

26b. Spike little or not at all thickened in the middle, less than 1 cm thick—27

27a. Glumes awned; stem erect, unbranched; fields (0.5–1 m high)—**Timothy,** *Phleum pratense*

27b. Glumes unawned; stem branched at the base; wet areas (20–50 cm high)—*Alopecurus aequalis*

28a. Spikelets numerous, in two to six long, symmetrical, widely divergent spikes; prairies, road and rail rights-of-way (1–3 m high)—**Big Bluestem** or **Turkeyfoot,** *Andropogon gerardii*

The stately Big Bluestem (***Andropogon gerardii***) and Little Bluestem (***Andropogon scoparius***) are important components of the prairie plant community, especially the tallgrass prairie. Much of the Michigan prairie has been lost to agriculture and development, but "presettlement" vegetation (plant communities present before European settlement) included many pockets of prairie, including tallgrass prairie (Comer et al. 1995). Railway rights-of-way which are infrequently mowed, tilled, or sprayed and which are burned occasionally have provided a safe harbor for prairie plants (Albert 1995). Prairie remnants still remain, especially in Berrien County, and prairie restoration projects exist at many locations.

28b. Spikelets in panicles, racemes, or loose spikes—29

29a. Lemmas with awns 2 mm or more long—30

29b. Lemmas not awned, or with short inconspicuous awns—33

30a. Spikelets in solitary raceme-like spikes; sandy areas, rights-of-way (0.5–1.2 m high)—**Little Bluestem,** *Andropogon scoparius* [*Schizachyrium scoparium*—CQ]

30b. Spikelets in branching panicles—31

31a. Awns less than 5 mm long; woods and weedy areas, SLP (20–90 cm high)—**Nimblewill,** *Muhlenbergia schreberi*

31b. Awns more than 9 mm long—32

32a. Awns straight, not twisted at the base; rich woods (0.5–1 m high)—*Brachyelytrum erectum*

32b. Awns bent, twisted at the base; prairies and dry open woods (1–2.5 m high)—**Indian Grass**, *Sorghastrum nutans*

33a. Spikelet laterally flattened; lemma surface covered with coarse, sometimes stiff hairs; wet areas (to 1.5 m high)—**Cut-grass**, *Leersia oryzoides*

33b. Spikelet plump; lemma surface glabrous or with fine hairs along the veins and margins—34

34a. Margin of lemma thickened, inrolled, clasping the palea (*Panicum* spp., **Panic Grass**)—35

> The genus ***Panicum***, characterized by an open panicle with small spikelets terminating slender, often wavy branches, is the largest genus of grasses in Michigan, including 31 species. Most species of ***Panicum*** (separated by spikelet features) are difficult to distinguish. Four of our common species are included here. For more detailed information see Voss (1972) or Stephenson (1984).

34b. Margin of lemma neither thickened nor inrolled, not clasping the palea—38

35a. Spikelets glabrous—36
35b. Spikelets pubescent along margins—37

36a. Leaves copiously hairy and leaf sheaths pubescent; panicle about half as long as the entire plant (to 70 cm high)—**Witch Grass**, *Panicum capillare*

36b. Leaves glabrous (or pilose basally) and leaf sheaths glabrous (or at most marginally ciliate); panicle expansive, but shorter (to 2 m high)—**Switch Grass**, *Panicum virgatum*

37a. Leaf blades more than 15 mm wide; spikelets more than 3 mm long; woods (0.4–1 m high)—*Panicum latifolium*

37b. Leaf blades less than 15 mm wide; spikelets less than 3 mm long—*Panicum implicatum* [*P. lanuginosum var. implicatum*—CQ]

38a. Panicle slender, the branches erect or ascending—39
38b. Panicle open, the branches long, ascending or spreading—42

39a. Leaves more than 1 cm wide; spikelets laterally compressed; marshes (0.7–1.5 m high)—**Reed Canary-grass**, *Phalaris arundinacea*

39b. Leaves less than 1 cm wide; spikelets compressed or not (*Muhlenbergia* spp., **Muhly**)—40

40a. Glumes one-fourth as long as the third scale, or sometimes one of them absent; woods and weedy areas, SLP (20–90 cm high)—**Nimblewill**, *Muhlenbergia schreberi*

40b. Glumes at least half as long as the lemma; mostly in wet areas—41

41a. Glumes longer than the body of the lemma (30–120 cm high)—**Marsh Wild-timothy**, *Muhlenbergia glomerata*

41b. Glumes shorter than the body of the lemma (30–90 cm high)—**Wood-grass**, *Muhlenbergia mexicana*

42a. Lemma surrounded by long hairs—43

42b. Lemma without basal long hairs—44

43a. Rachilla extended as a bristle; wet areas (0.5–1.5 m high)—**Blue-joint or Reedgrass**, *Calamagrostis canadensis*

43b. Rachilla not extended as a bristle; Great Lakes dunes (to 2 m high)—*Calamovilfa longifolia*

44a. Long hairs present on the outside of the leaf sheath summit; sandy areas (30–100 cm high)—**Dropseed**, *Sporobolus cryptandrus*

44b. Leaf sheath summit not long-hairy (*Agrostis* spp., **Bentgrass**)—45

45a. Primary panicle branches dividing only beyond the middle (30–90 cm high)—**Ticklegrass**, *Agrostis hyemalis*

45b. Primary panicle branches dividing and bearing flowers below their middle—46

46a. Plants stoloniferous, lower nodes often bent, rooting; often in wet areas—**Creeping Bent**, *Agrostis stolonifera*

46b. Plants erect, stolons absent—47

47a. Spikelets (and thus the inflorescence) red or purple (to 1 m high)—**Redtop**, *Agrostis gigantea*

47b. Spikelets pale green; woods (0.5–1 m high)—**Autumn Bent**, *Agrostis perennans*

48a. Glumes longer than the lemmas—49

48b. Glumes shorter than the lemmas—50

49a. Back of lemma pilose; tuft of white hairs at the summit of the sheath; open areas (20–60 cm high)—**Poverty Grass**, *Danthonia spicata*

49b. Back of lemma glabrous; tuft of white hairs not present; open areas, often cultivated (to 1 m high)—**Oats**, *Avena sativa*

50a. Rachilla covered with conspicuous long hairs about equaling the lemmas; tall marsh grass (1–4 m high)—**Reed**, *Phragmites australis*

50b. Rachilla without conspicuous long hairs—51

51a. Spikelets sessile or nearly so, forming crowded clusters or spike-like panicles—52

51b. Spikelets distinctly pedicelled, in a more or less open panicle—53

52a. Spikelets in dense one-sided clusters at the ends of the panicle branches (0.5–1.2 m high)—**Orchard Grass**, *Dactylis glomerata*

52b. Spikelets in an erect spike-like cluster (30–60 cm high)—**June Grass**, *Koeleria macrantha*

53a. Lemmas, exclusive of the awn when present, 6–8 mm long or more—54

53b. Lemmas, exclusive of the awn when present, 6 mm long or less—58

54a. Dense beard of short hairs at the base of the floret; glumes purplish; woods (0.4–1 m high)—**False Melic**, *Schizachne purpurascens*

54b. Beard at the base of the floret absent; glumes purplish or greenish; often in weedy areas (*Bromus* spp., **Brome Grass**)—55

55a. Awn on the lemma 8–10 mm long or more, equaling or exceeding the lemma; (20–70 cm high)—**Downy Chess**, *Bromus tectorum*

55b. Awn on the lemma 6 mm long or less, usually much shorter than the lemma (or absent)—56

56a. Annual; mature lemma margins inrolled (30 cm–1.2 m high)—**Cheat**, *Bromus secalinus*

56b. Perennials; mature lemma margins not inrolled—57

57a. Plant rhizomatous; inflorescence ascending; lemmas purplish (0.5–1 m high)—**Smooth Brome**, *Bromus inermis*

57b. Plant tufted, lacking rhizomes; inflorescence nodding; lemmas greenish; wet areas (0.6–1.2 m high)—**Fringed Brome**, *Bromus ciliatus*

58a. Lemma with seven sharp conspicuous veins from base to apex; wet areas (*Glyceria* spp., **Manna Grass**)—59

58b. Lemma with three to five veins, conspicuous or not—62

59a. Spikelets 10 mm or longer; pedicels ascending, shorter than the spikelets (0.8–1.2 m high)—*Glyceria borealis*

59b. Spikelets shorter than 9 mm; pedicels spreading to undulate, often longer than the spikelets—60

60a. Lemma veins visible, not strongly raised (to 1 m high)—**Rattlesnake Grass**, *Glyceria canadensis*

60b. Lemma veins strongly raised (corrugated)—61

61a. Spikelets up to 4 mm long; glumes obtuse, both less than 1.5 mm long (0.5–1.2 m high)—**Fowl Manna Grass**, *Glyceria striata*

61b. Spikelets over 4 mm long; glumes acute, the larger glume over 1.5 mm long (to 1.5 m high)—*Glyceria grandis*

62a. Glumes and usually the lemmas strongly keeled—63

62b. Back of glumes and lemmas rounded, not keeled—68

63a. Ligule consists of a fringe of hairs; fields and weedy places (*Eragrostis* spp., **Love Grass**)—64

63b. Ligule is a membranous scale (*Poa* spp., **Bluegrass**)—65

64a. Leaf margins and pedicels glandular-warty (10–40 cm high)—**Stink Grass**, *Eragrostis cilianensis*

64b. Leaf margins and pedicels smooth (30–60 cm high)—**Tumble Grass**, *Eragrostis spectabilis*

65a. Annual, with ascending or decumbent flowering stems (5–30 cm high)—**Annual Bluegrass**, *Poa annua*

65b. Perennials, with erect flowering stems—66

66a. Stems strongly flattened (20–70 cm high)—**Canada Bluegrass**, *Poa compressa*

66b. Stems round—67

67a. Lemma five-nerved; ligule 2 mm long or less (30–60 cm high)—**Kentucky Bluegrass**, *Poa pratensis*

67b. Lemma three-nerved; ligule 2.5 mm long or more; damp areas (0.5–1.5 m high)—**Fowl Meadow Grass**, *Poa palustris*

68a. Lemma veins do not converge at the blunt tip; plants of alkali areas, espp. roadsides (10–50 cm high)—**Alkali** or **Reflexed Saltmarsh Grass**, *Puccinellia distans*

Alkali grass, **Puccinellia distans**, is an example of a salt-tolerant species (halophyte) which has become much more widely distributed as a result of human activity. Voss (1972) mapped it from just six counties, all in eastern Michigan. After making a concerted effort to collect this grass in other areas of the state, Garlitz (1992) located it in 35 counties, which extend from the Ohio border to Sault Ste. Marie. It thrives along roadsides that are heavily salted in the winter, in soil too salty for most plants. There are several such "highway halophytes" invading Michigan.

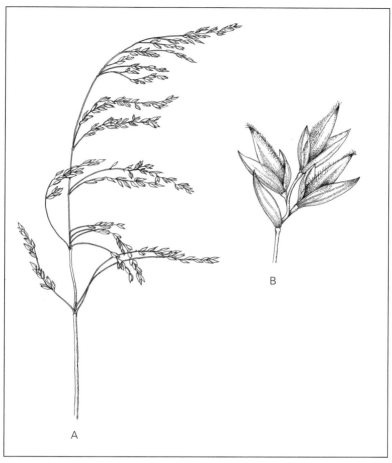

Figure 10: **Poa pratensis:** *A, panicle; B, spikelet (in fruit)*

68b. Lemma veins converge into an acute tip; plants of various habitats
(*Festuca* spp., **Fescue Grasses**)—69

69a. Leaf blades rolled inward; lemmas with conspicuous awns—70
69b. Leaf blades flat; lemmas without awns—71

70a. Perennials with open florets; anthers 3; open, dry areas (0.3–1.0 m
high)—**Red Fescue**, *Festuca rubra* complex (incl. *F. rubra*, *F. ovina*,
F. saximontana [*F. brachyphylla*—CQ])

> The **Festuca rubra** complex is a group of variants with overlapping characteristics, often placed within several different species (see Gleason & Cronquist, 1991).

70b. Tufted annual with closed florets; anther 1; sandy areas, LP (10–60 cm high)—**Six-weeks Fescue**, *Festuca octoflora* [*Vulpia octoflora*—CQ]

71a. Spikelets with five or more florets; borne all along panicle branches; meadows (0.3–1.2 m high)—**Meadow Fescue**, *Festuca pratensis*
71b. Spikelets with two to four flowers; borne beyond the middle of panicle branches; woods (0.6–1.2 m high)—**Nodding Fescue**, *Festuca obtusa* [*F. subverticillata*—CQ]

CYPERACEAE, The Sedge Family

Annual or (often) perennial herbs, often with triangular stems and linear leaves. Flowers regular, often unisexual (the plants then monoecious, rarely dioecious), in small chaffy spikes or spikelets; perianth often of 6 bristles, sometimes 0; stamens often 3; pistil 1, style 1, ovary superior, 1-celled. Fruit an achene, sometimes enclosed in a modified bract (a *perigynium*).

Over 245 species occur in Michigan. Most species grow in wet areas and flower in late spring or early summer. Only the most common are included here. For the remaining species, consult Voss (1972).

1a. All flowers unisexual; spikelets often of two forms, the uppermost often wholly staminate, the lower one or more pistillate; ovary and achene surrounded by a sac, the perigynium. (*Carex* spp., **Sedges**)—2

> **Carex** is represented by over 170 species in Michigan, making it our largest genus. The combination of leaves in three ranks (vertical rows) and an achene enclosed in a sac (perigynium) generally distinguishes a plant as **Carex**, but identification of species is difficult. Mature fruit is necessary; the shape and surface of the perigynium is often diagnostic. Twenty of the more common species are included here. See Voss (1972) for a more detailed treatment.

1b. All flowers perfect; spikelets all alike; ovary and achene not enclosed in a perigynium—21

2a. Achenes flattened; wet areas—3
2b. Achenes three-angled—7

3a. Lateral spikelets peduncled, of two forms—4
3b. Lateral spikelets sessile, all alike—5

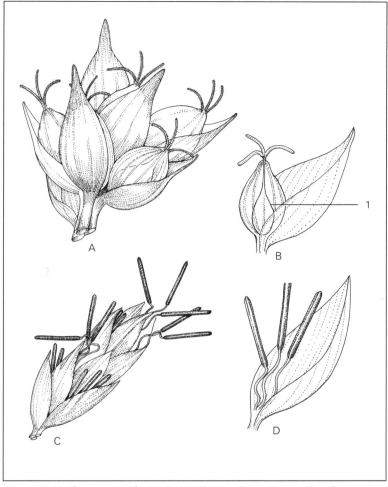

Figure 11: **Carex granularis**: *A, pistillate spikelet; B, pistillate flower, 1, perigynium; C, staminate spikelet, D, staminate flower*

4a. Perigynium golden or orange, somewhat fleshy (3–40 cm high)—***Carex aurea***

4b. Perigynium green (40–160 cm high)—***Carex crinita***

5a. Staminate flowers at the base of the spikelet (20–90 cm high)—***Carex bebbii***

5b. Staminate flowers at the apex of the spikelet (30–100 cm high)—6

6a. Stems sharply angled but soft, over 1.5 mm thick near the inflorescence—*Carex stipata*
6b. Stems not sharply angled but firm, less than 1.5 mm thick near the inflorescence—*Carex vulpinoidea*

7a. Perigynium hairy—8
7b. Perigynium smooth, glabrous—10

8a. Leaf blades 5 mm or more wide; pistillate spikelet globose; perigynium over 1 cm long; woods, mostly SLP (30–90 cm high)—*Carex grayi*
8b. Leaf blades less than 5 mm wide; pistillate spikelet elongate; perigynium less than 5 mm long—9

9a. Pistillate spikelets more than 1 cm long; often in wet areas (30–100 cm high)—*Carex lanuginosa* [*C. pellita*—CQ]
9b. Pistillate spikelets 1 cm or less long; dry, sandy areas (10–50 cm high)—*Carex pensylvanica*

10a. Perigynium tipped with a sharp, straight, two-toothed beak—11
10b. Perigynium tipped with a short untoothed beak or beak absent; chiefly in woods—17

11a. Perigynium less than 1 cm long—12
11b. Perigynium 1 cm or more long—15

12a. Staminate spikelets two to four—13
12b. Staminate spikelet one, terminal—14

13a. Lower leaf sheaths reddened (50–150 cm high)—*Carex lacustris*
13b. Lower leaf sheaths green, not reddened (50–120 cm high)—*Carex rostrata* [*C. utriculata*—CQ]

14a. Lower pistillate spikes nodding (30–100 cm high)—*Carex hystericina*
14b. All pistillate spikes ascending (10–40 cm high)—*Carex viridula*

15a. Pistillate spikelet cylindrical; perigynium beak 6 mm or more long (20–130 cm high)—*Carex lupulina*
15b. Pistillate spikelet ovoid to globose; perigynium beak 4 mm or less long—16

16a. Pistillate spikelet globose, the eight or more perigynia pointing in all directions; mostly SLP (30–90 cm high)—*Carex grayi*

16b. Pistillate spikelet ovoid, the twelve or fewer perigynia mostly spreading to ascending (30–90 cm high)—*Carex intumescens*

17a. Base of plant reddish-tinged; terminal spikelet includes a few perigynia—18
17b. Base of plant reddish or not; terminal spikelet entirely staminate—19

18a. Staminate flowers at the summit of the terminal spikelet (10–30 cm high; spring)—*Carex pedunculata*
18b. Staminate flowers at the base of the terminal spikelet (40–90 cm high)—*Carex gracillima*

19a. Base of perigynia rounded (30–80 cm high)—*Carex granularis*
19b. Base of perigynia tapered, angled—20

20a. Base of plant reddish; leaf blades over 1 cm wide (30–60 cm high)—*Carex plantaginea*
20b. Base of plant brownish; leaf blades less than 1 cm wide; mostly SLP (30–70 cm high)—*Carex blanda*

21a. Scales of the spikelets in two rows along the axis—22
21b. Scales of the spikelets spirally arranged—26

22a. Inflorescence axillary; stem round, jointed (30–100 cm high)—**Three-way Sedge**, *Dulichium arundinaceum*
22b. Inflorescence terminal; stem three-angled, no apparent joints (*Cyperus* spp., **Nut-grass**)—23

23a. Achene lens-shaped; stigmas two (10–40 cm high)—*Cyperus rivularis* [*C. bipartitus*—CQ]
23b. Achene three-sided; stigmas three—24

24a. Spikelets arranged in a dense subglobose head; dry sandy areas (10–50 cm high)—*Cyperus filiculmuis* [*C. lupulinus*—CQ]
24b. Spikelets inserted along an elongated axis—25

25a. Scales yellow, less than 3 mm long; wet or disturbed areas (10–70 cm high)—**Yellow Nut-sedge**, *Cyperus esculentus*
25b. Scales yellow, longer than 3 mm; wet areas (10–100 cm high)—*Cyperus strigosus*

26a. Inflorescence of a single erect, terminal spike; no involucral bracts overtop the spike—27

26b. Inflorescence of one or more spikelets, terminal or lateral; if one, then overtopped by at least one involucral bract—32

27a. Stems with basal leaves; bogs and swamps (20–70 cm high)—**Cotton-grass**, *Eriophorum spissum* [*E. vaginatum*—CQ]
27b. Stems leafless (*Eleocharis* spp., **Spike Rush**)—28

There are 22 species of **Eleocharis** in Michigan. All of these spike-rushes are found in wet areas, especially along shores. Five of the more common species are included here; see Voss (1972) for a more complete treatment.

28a. Spikelet and stem of the same diameter; stem 4-angled; LP (to 1 m high)—*Eleocharis quadrangulata*
28b. Spikelet thicker than the stem; stems round or multi-angled—29

29a. Achenes roughened, three-sided or rounded—30
29b. Achenes smooth, lens-shaped—31

30a. Achenes whitish or gray; stems filiform, round or slightly flattened (3–12 cm high)—*Eleocharis acicularis*
30b. Achenes yellow to brown; stems thicker, four- to eight-angled (15–60 cm high)—*Eleocharis elliptica* [*E. tenuis*—CQ]

31a. Stems soft; plants annual (5–50 cm high)—*Eleocharis obtusa* [*E. ovata*—CQ]
31b. Stems stiff; plants perennial (10–100 cm high)- *Eleocharis smallii* [*E. palustris*—CQ]

32a. Achenes subtended by 15–50 long, silky bristles; bogs and swamps (to 1 m high) (most *Eriophorum* spp., **Cotton-grass**)—33
32b. Achenes subtended by one to eight bristles or not; if so, not as above—34

33a. Scales of spikelet brownish; bristles coppery; spikelets maturing in late summer—**Tawny Cotton-grass**, *Eriophorum virginicum*
33b. Scales of spikelet drab to blackish; bristles white; spikelets maturing in early summer—*Eriophorum viridicarinatum*

34a. Inflorescence lateral, emerging from one side of the stem (*Scirpus* spp. in part, **Bulrush**)—35
34b. Inflorescence terminal—37

The genus **Scirpus** (the bulrushes) includes 19 species in Michigan. These are plants of wet areas or shallow water, often seen along lake shores and in wet ditches. Six of the most common species are included here.

35a. Stems three-angled (0.5–2 m high)—**Three-square**, *Scirpus americanus* [*S. pungens*—CQ]

35b. Stems round (1–3 m high)—36

36a. Stems soft; spikelets ovoid—**Softstem Bulrush**, *Scirpus validus*

36b. Stems firm; spikelets oblong to cylindric—**Hardstem Bulrush**, *Scirpus acutus*

37a. Spikes not conspicuously exceeded by involucral bracts—38

37b. Spikes subtended by involucral bracts which greatly exceed the flower clusters—39

38a. Bristles subtending the achene present; mature inflorescence drooping (to 1.5 m high)—*Scirpus pendulus*

38b. Bristles subtending the achene absent; mature inflorescence erect (to 1 m high)—**Twig-rush**, *Cladium mariscoides*

39a. Spikelets in dense, subglobose heads (to 1.5 m high)—*Scirpus atrovirens*

39b. Spikelets in few-flowered clusters (to 2 m high)—**Wool-grass**, *Scirpus cyperinus*

ARACEAE, The Arum Family

Herbs often with broad, net-veined leaf blades and sheathing petioles. Flowers unisexual (the plants then monoecious or rarely dioecious) or perfect, small, crowded on a fleshy *spadix* to form a conspicuous spike, usually surrounded by a green or colored *spathe*; perianth 0 or 4 or 6 tepals; stamens 2–6; pistil 1, style 1, ovary superior (but often buried in the spadix), 1–3-celled. Fruit a berry or the spadix expanding to form a multiple fruit.

Most members of this family are immediately recognizable when in flower or fruit because of the distinctive inflorescence. A large bract, the *spathe*, encloses the fleshy spike (*spadix*). See Figure 3, page 32.

1a. Spathe sessile, partly underground, often dark red; wet woods and swamps (early spring)—**Skunk-cabbage**, *Symplocarpus foetidus*

1b. Spathe peduncled, all above ground, white to green or purplish (spring and early summer)—2

2a. Leaves compound; moist woods and swamps (spring) (*Arisaema* spp.)—3

2b. Leaves simple (early summer)—4

3a. Leaflets three, spathe pale green or purple, arching over the spadix—
Jack-in-the-pulpit, *Arisaema triphyllum*

3b. Leaflets five to thirteen; spathe green; spadix slender, tapering, longer
than the spathe; SLP—**Green Dragon** or **Dragon-root**, *Arisaema dra-contium*

4a. Leaf blades linear, sword-shaped; spathe similar, resembling an extension of the peduncle; marshes (50–200 cm high; early summer)—**Sweet-
flag**, *Acorus calamus*

4b. Leaf blades broadly ovate to triangular; spathe clearly different from the
peduncle; shallow water—5

5a. Base of leaf blade cordate; spathe white—**Water-arum**, *Calla palustris*

5b. Base of leaf blade more or less sagittate; spathe green; mostly LP—
Arrow-arum, *Peltandra virginica*

LEMNACEAE, The Duckweed Family

Minute monoecious leafless herbs floating on, or just below, quiet waters. Flowers unisexual, exceedingly small and seldom seen; perianth 0;
stamen 1; pistil 1, style 1, ovary superior, 1-celled. Fruit a utricle.

1a. Roots absent (*Wolffia* spp., **Water-meal**)—2

The smallest flowering plants are found in the genus **Wolffia**, an entire plant being less than
1.5 mm long. Our two common species are usually found growing together.

1b. Roots present—3

2a. Plant (and any developing buds) globose—*Wolffia columbiana*

2b. Plant (and any developing buds) longer than broad, the ends more or less
pointed; SLP—*Wolffia punctata*

3a. Roots several from each rounded plant—**Greater Duckweed**, *Spirodela
polyrhiza*

3b. Root single from each rounded plant (*Lemna* spp., **Duckweed**)—4

4a. Thallus rounded; floating on the water—*Lemna minor*

4b. Thallus oblong with a slender stalk attached to the "parent" plant; floating below the surface—**Star Duckweed**, *Lemna trisulca*

XYRIDACEAE, The Yellow-eyed-grass Family

Perennial herbs with basal leaves and erect scapes (floral peduncles). Flowers regular, perfect, in terminal heads or dense spikes; sepals 3; petals 3, yellow; stamens 3, alternating with 3 staminodes; pistil 1, style 1, ovary superior, 1-celled. Fruit a capsule. Summer.

1a. Base of plant bulbous-thickened; scapes ridged, longitudinally twisted; moist sands, SLP (scapes 15–100 cm high)—*Xyris torta*
1b. Base of plant not bulbous-thickened; scapes round, only somewhat longitudinally twisted; wet, peaty areas—2

2a. Leaves less than 2 mm wide; UP and Straits (scapes 5–30 cm high)—*Xyris montana*
2b. Leaves more than 2 mm wide; mostly SLP (scapes 15–70 cm high)—*Xyris difformis*

ERIOCAULACEAE, The Pipewort Family

Monoecious perennial herbs with basal leaves and erect scapes (floral peduncles). Flowers regular or not, unisexual, tiny, in a dense head subtended by bracts; sepals 2; petals 2; stamens 4; pistil 1; style 1; ovary superior, 2-celled. Fruit a capsule. Summer.

One species in Michigan; leaves linear; flower-heads whitish or lead color; shallow water, wet shores (scape 5–20 cm high, to 40 cm in submerged plants)—**Pipewort**, *Eriocaulon septangulare* [*E. aquaticum*—CQ]

COMMELINACEAE, The Spiderwort Family

Herbs with alternate leaves. Flowers regular or not, mostly perfect, in cymes subtended by one or more bracts; sepals 3; petals 3, equal or not; stamens 6 or 3 fertile and 3 staminodes; pistil 1, style 1, ovary superior, (2)3-celled. Fruit a capsule.

1a. Fertile stamens three, sterile stamens three; two blue petals larger than the third (often white) petal; inflorescence subtended by a folded bract; weedy areas (summer)—**Day-flower**, *Commelina communis*
1b. Fertile stamens six; petals all equal, blue (pink or white); subtending bracts flat, not folded; open woods, fields, roadsides (*Tradescantia* spp., **Spiderwort**)—2

2a. Sepals villous; SLP (5–40 cm high; spring)—*Tradescantia virginiana*

2b. Sepals glabrous, or with a tuft of hairs at the apex; LP (40–100 cm high; late spring)—***Tradescantia ohiensis***

PONTEDERIACEAE, The Pickerel-weed Family

Perennial aquatic herbs. Flowers regular or not, perfect, in a spike-like panicle or solitary; perianth of 6 tepals, stamens 3 or 6; pistil 1, style 1, ovary superior, 3-celled. Fruit dry and indehiscent. Summer.

1a. Leaves cordate-sagittate, emergent; flowers blue, in a dense spike-like panicle; stamens 6 (to 100 cm high)—**Pickerel-weed**, *Pontederia cordata*

1b. Leaves linear, submerged (sometimes exposed on mud flats); flowers yellow, solitary; stamens 3—**Water Star-grass** or **Mud-plantain**, *Heteranthera dubia* [*Zosterella dubia*—CQ]

JUNCACEAE, The Rush Family

Annual or (often) perennial herbs, often with linear leaves. Flowers regular, perfect, in cymes of solitary or clustered flowers; sepals 3, green or brown; petals 3, green or brown; stamens 3 or 6; pistil 1, style 1, ovary superior, 1- or 3-celled. Fruit a 3-valved capsule.

1a. Leaf sheaths closed; capsule three-seeded; leaves usually hairy at or near the summit of the sheath; most common in woods (spring) (*Luzula* spp., **Wood Rush**)—2

1b. Leaf sheaths open; capsule many-seeded; leaves glabrous; mostly in open damp areas (summer and autumn) (*Juncus* spp., **Rush**)—3

The genus **Juncus** is represented by 26 species in Michigan. These grass-like plants often inhabit moist areas with mineral soils, especially sandy shores and ditches. Eight of the most common species are included here. As Voss (1972) notes, "well-developed fruit and seeds are necessary for accurate identification " in many **Juncus**. Please refer to his treatment for further assistance.

2a. Flowers solitary at the ends of the inflorescence branches (10–40 cm high)—***Luzula acuminata***

2b. Flowers clustered in short spikes or dense heads (20–50 cm high)— ***Luzula multiflora***

3a. Inflorescence lateral, emerging from one side of the stem; leaves reduced to bladeless sheaths—4

3b. Inflorescence terminal; some leaf blades present—5

4a. Stems clustered in dense clumps (to 1 m high)—***Juncus effusus***
4b. Stems occurring in rows along a long, straight rhizome; espp. wet sandy
shores (40–80 cm high)—***Juncus balticus*** [***J. arcticus***—CQ]

5a. Leaf blades flattened, without cross-partitions (10–80 cm high)—6
5b. Leaf blades round in cross section, with cross-partitions—7

6a. Prolonged area of auricle at the junction of the leaf sheath and blade
whitish, over 1 mm long—**Path Rush**, ***Juncus tenuis***
6b. Prolonged area of auricle at the junction of the leaf sheath and blade absent or yellowish and up to 0.5 mm long—***Juncus dudleyi*** [***J. tenuis*** var.
dudleyi—CQ]

7a. Seeds with a slender, pale appendage at each end (40–100 cm high)—
Juncus canadensis
7b. Seeds with either blunt ends or dark tips—8

8a. Inflorescence of subhemispheric heads; involucral bract shorter than the inflorescence (5–30 cm high)—***Juncus alpinus*** [***J. alpinoarticulatus***—CQ]
8b. Inflorescence of spherical heads; involucral bract extends above the inflorescence—9

9a. Sepals 4 mm long or more, longer than the petals (40–100 cm high)—
Juncus torreyi
9b. Sepals less than 4 mm long, about equaling the petals (10–40 cm
high)—***Juncus nodosus***

LILIACEAE, The Lily Family

Perennial herbs or sometimes woody vines, usually with conspicuous
flowers. Flowers mostly regular, perfect (seldom unisexual, the plants
then monoecious or dioecious); perianth of 6 tepals (4 in *Maianthemum*), sometimes united below; stamens (4)6, free or attached to the
perianth tube; pistil 1, styles 1 or 3 (rarely 0), ovary superior, (2)3-
celled. Fruit a capsule or berry.

1a. Flowers or flower-clusters lateral, axillary or apparently so—2
1b. Flowers or flower-clusters terminal—13

2a. Leaves minute and scale-like, subtending the filiform branches; escaping
to fields and roadsides (70–200 cm high; flowers greenish-yellow,
spring)—**Garden Asparagus**, *Asparagus officinalis*
2b. Leaves broad and flat, not scale-like; mostly in woods—3

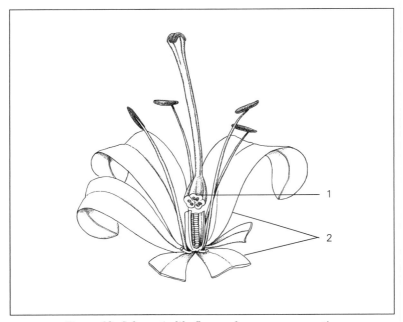

*Figure 12: Schematic lily flower: 1, ovary cross-section
showing 3 carpels; 2, tepal*

3a. Flowers greenish-yellow, unisexual (the plants dioecious), numerous in
rounded umbels; leaves with distinct petioles; vines or twining shrubs
(spring) (***Smilax*** spp., **Carrion-flower** or **Greenbrier**)—4

3b. Flowers of various colors, perfect, in clusters of one to eight; leaves perfoliate, sessile, or clasping; herbs—8

4a. Stems herbaceous or with a woody base; not thorny (***Smilax*** sect. ***Ne-
mexia***, **Carrion-flower**)—5
4b. Stems woody, with prickles (***Smilax*** sect. ***Smilax***, **Greenbrier**)—7

5a. Stems climbing by tendrils; mostly SLP (to 2.5 m high)—***Smilax lasion-
eura*** [*S. herbacea*—CQ]
5b. Stem not climbing; only the upper leaves rarely with tendrils (to 80 cm
high)—6

6a. Pistillate flowers fewer than 25 per umbel—***Smilax ecirrata*** [*S. ecir-
rhata*—CQ]
6b. Pistillate flowers more than 25 per umbel—***Smilax illinoensis***

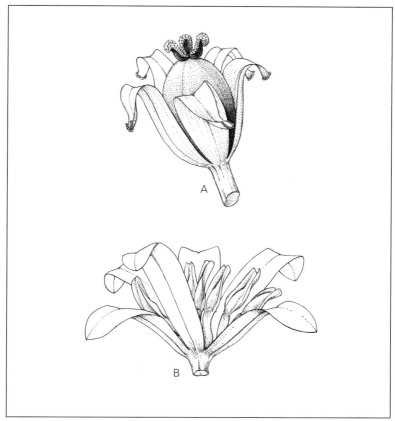

*Figure 13: unisexual flowers of **Smilax** spp.:*
A, pistillate flower; B, staminate flower

7a. Stems (especially young) four-angled; prickles broad with flattened
bases; sandy areas, SLP—**Common Greenbrier**, *Smilax rotundifolia*
7b. Stems round or with more than four angles; prickles needle-like—
Bristly Greenbrier, *Smilax tamnoides* [*S. hispida*—CQ]

8a. Flowers yellow or cream-yellow (spring) (*Uvularia* spp.)—9
8b. Flowers greenish to purple—10

9a. Leaves perfoliate (20–50 cm, rarely to 100 cm high)—**Bellwort**, *Uvularia grandiflora*
9b. Leaves sessile (10–30 cm high)—**Merrybells**, *Uvularia sessilifolia*

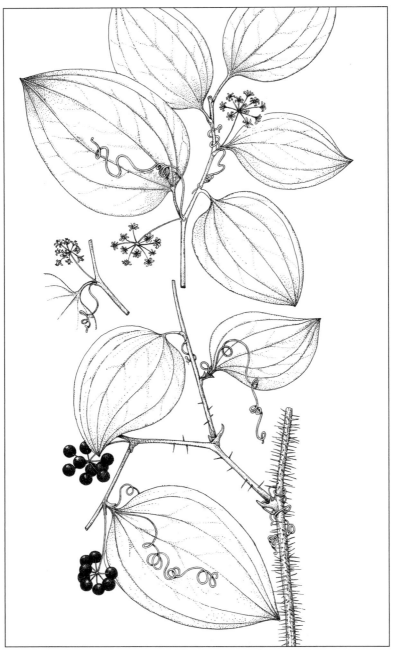

*Figure 14: **Smilax tamnoides***

10a. Perianth segments (tepals) united into a tube (late spring) (***Polygonatum*** spp., **Solomon-seal**)—11

10b. Tepals separate (spring) (***Streptopus*** spp., **Twisted-stalk**)—12

11a. Leaves minutely pubescent beneath (50–90 cm high)—**Small Solomon-seal**, *Polygonatum pubescens*

11b. Leaves smooth beneath; SLP (40–120 cm, or rarely to 250 cm high)—**Great Solomon-seal**, *Polygonatum biflorum*

12a. Leaves distinctly clasping the stem; flowers greenish-white; NM (40–100 cm high)—***Streptopus amplexifolius***

12b. Leaves closely sessile (or slightly clasping); flowers reddish to purple (30–80 cm high)—***Streptocarpus roseus***

13a. Stem with a single whorl of three leaves; woods (spring) (***Trillium*** spp., **Wake-robin** or **Trillium**)—14

Some of Michigan's native plants are protected by Public Act 182 of 1962, commonly called the "Christmas-tree law". Under this law, it is illegal to remove or cut **Trillium spp.** from any area without a bill of sale or written permission from the owner.

13b. Stem with more than one whorl of leaves, or leaves basal, along the stem, or absent—20

14a. Flower sessile, red or brown; leaves usually mottled; SLP (10–30 cm high)—**Toadshade**, *Trillium sessile*

14b. Flower peduncled, white, pink, or red; leaves not mottled—15

15a. Ovary with six distinct wing-like angles (20–40 cm high)—16

15b. Ovary obtusely three-angled or lobed; SLP—19

16a. Stigmas erect or nearly so, slender; petals white, becoming pink with age (20–40 cm high)—**Common Trillium**, *Trillium grandiflorum*

16b. Stigmas strongly recurved or spreading, thickened basally; petals white to maroon—17

17a. Flower above the leaves on a mostly erect peduncle; petals often maroon—**Stinking Benjamin** or **Red Trillium**, *Trillium erectum*

17b. Flower below the leaves on a recurved peduncle; petals often white—18

18a. Filaments one-fourth as long as the anthers or shorter; SLP—**Bent Trillium**, *Trillium flexipes*

18b. Filaments over one-fourth as long to as long as the anthers—**Nodding Trillium**, *Trillium cernuum*

19a. Leaves obtuse; petals obtuse, white (8–15 cm high)—**Snow Trillium**, *Trillium nivale*

19b. Leaves acuminate; petals acute, purple-striped at base (20–40 cm high)—**Painted Trillium**, *Trillium undulatum*

20a. Tepals (perianth segments) 5–12 cm long—21

20b. Tepals shorter than 5 cm—26

21a. Leaves all or chiefly basal, stem leaves bract-like or none; tepals not spotted—22

21b. Leaves chiefly or entirely on the stem; tepals purple-spotted (flowers yellow, orange, or red, in summer) (*Lilium* spp., **Lily**)—24

22a. Leaves numerous, green, linear or sword-shaped; escapes from cultivation (summer)—23

22b. Leaves a single pair, often brown-mottled, oblong or lanceolate (spring)—28

23a. Flowers orange; inflorescence a terminal umbel-like cluster; roadsides (scapes to 1 m high)—**Day-lily**, *Hemerocallis fulva*

23b. Flowers white or creamy; inflorescence a tall panicle; sandy areas, SW (scapes 1–3 m high)—**Yucca**, *Yucca filamentosa*

24a. Flowers erect (30–80 cm high)—**Wood Lily**, *Lilium philadelphicum*

24b. Flowers nodding, tepals recurved—25

25a. Leaves alternate (but crowded), with axillary bulblets; escape from cultivation (60–120 cm high)—**Tiger Lily**, *Lilium lancifolium*

25b. Leaves whorled, without axillary bulblets (60–150 cm high)—**Michigan Lily**, *Lilium michiganense*

26a. Flowers solitary; woods (spring)—27

26b. Flowers in clusters, not solitary—30

27a. Stem with a single pair of basal, often brown-mottled leaves; flower yellow or white (10–20 cm high) (*Erythronium* spp., **Trout-lily** or **Dogtooth-violet**)—28

27b. Stem leafy, leaves green; flowers yellow or cream yellow (*Uvularia* spp., **Bellwort**)—29

28a. Flower yellow—**Yellow Trout-lily**, *Erythronium americanum*

28b. Flower white—**White Trout-lily**, *Erythronium albidum*

29a. Leaves sessile (10–30 cm high)—**Merrybells**, *Uvularia sessilifolia*

29b. Leaves perfoliate (20–50 cm, rarely to 100 cm high)—**Bellwort**, *Uvularia grandiflora*

30a. Stem bearing two whorls of three to nine leaves; moist woods (30–70 cm high; flowers pale yellow, spring and early summer)—**Indian Cucumber-root**, *Medeola virginiana*

30b. Stem leaves not in two whorls, or all leaves basal or absent—31

31a. Flowers in umbels; leaves all or chiefly basal or absent—32

31b. Flowers in racemes or panicles; leaves various—39

32a. Plant with the odor of onions or garlic; umbel of seven or more flowers (or bulblets); fruit a capsule (late spring or summer) (*Allium* spp., **Onion**)—33

32b. Plant not with the odor of onions; umbel of three to six greenish-yellow flowers; fruit a blue berry (scapes 15–40 cm high, late spring)—**Cornlily**, *Clintonia borealis*

33a. Leaves oblong, 2–5 cm wide, present before, but not with, the greenish-white flowers; woods (scapes 12–35 cm high)—**Wild Leek** or **Ramps**, *Allium tricoccum*

33b. Leaves linear, present with the flowers—34

34a. Umbel nodding or horizontal; LP (petals rose or white; scapes 30–60 cm high)—**Nodding Wild Onion**, *Allium cernuum*

34b. Umbel erect—35

35a. Leaves flattened; umbels with some (or all) flowers often replaced by bulblets—36

35b. Leaves cylindrical; umbel with or without bulblets—37

36a. Involucral bracts two or three; SLP (20–60 cm high)—**Wild Garlic**, *Allium canadense*

36b. Involucral bract one; escape from cultivation, LP (30–80 cm high)—**Garlic**, *Allium sativum*

37a. Stems thick, over 5 mm in diameter, inflated below; escape from cultivation—**Onion**, *Allium cepa*

37b. Stems less than 5 mm in diameter, not thickened below—38

38a. Pedicels longer than the flowers; bulblets present; sandy areas, LP (30–100 cm high)—**Field Garlic**, *Allium vineale*

38b. Pedicels equaling or shorter than the flowers; bulblets not present; often a garden escape (10–50 cm high)—**Chives**, *Allium schoenoprasum*

Allium schoenoprasum is an example of a plant in which both native and introduced populations occur in Michigan. The common garden plant is a European variety (*var. schoenoprasum*) which grows in dense clumps and does escape to waste ground while the native circumpolar variety (*var. sibiricum*) is taller, with few flowering stems; it occurs locally on shorelines and cliffs in the NW Upper Peninsula.

39a. Leaves lanceolate to ovate, not more than eight times as long as broad—40

39b. Leaves linear or grass-like, at least twelve times as long as broad—45

40a. Principal leaves all basal (or appearing so), stem leaves none or bract-like—41

40b. Principal leaves on the stem (spring)—42

41a. Inflorescence a spike-like raceme; flowers white, more or less sessile, ascending or spreading; mostly SLP (40–100 cm high; summer)—**Colic-root**, *Aletris farinosa*

41b. Inflorescence a curved raceme; flowers white, on thin curved pedicels, nodding; escape from cultivation (scapes 10–20 cm high; spring)—**Lily-of-the-valley**, *Convallaria majalis*

42a. Tepals and stamens four (5–20 cm high; flowers white)—**Wild Lily-of-the-valley**, *Maianthemum canadense*

42b. Tepals and stamens six (*Smilacina spp.*, **False Solomon-seal**)—43

43a. Inflorescence a panicle; woods (30–80 cm high)—**False Spikenard**, *Smilacina racemosa*

43b. Inflorescence a raceme—44

44a. Leaves one to four, usually three; inflorescence overtops the leaves; swamps and bogs (10–40 cm high)—*Smilacina trifolia*

44b. Leaves six or more; inflorescence overtopped by the uppermost leaves; dunes and woods (20–60 cm high)—*Smilacina stellata*

45a. Tepals united; bright blue; escape from cultivation (scapes 10–20 cm, or sometimes to 40 cm high, spring)—**Grape-hyacinth**, *Muscari botryoides*

45b. Tepals separate; blue, greenish, yellowish, or white—46

46a. Styles three; flowers greenish-white (summer)—47

46b. Style one; flowers blue or white—48

47a. Tepals with two glands near the base; inflorescence a panicle; Great Lakes shores and wet calcareous areas (20–60 cm high)—**White Camas**, *Zigadenus glaucus* [*Z. elegans* var. *glaucus*—CQ]

47b. Perianth segments without glands; inflorescence a dense raceme; wet shores, ditches, bogs (scape 10–50 cm high)—**False Asphodel**, *Tofieldia glutinosa*

48a. Tepals white, with a green medial stripe on the outer surface; escape from cultivation (scapes 10–30 cm high; spring)—**Star-of-Bethlehem**, *Ornithogalum umbellatum*

48b. Tepals blue or rarely white, without an outer green stripe—49

49a. Tepals with three to seven nerves; inflorescence a long, many-flowered raceme; river bottoms, SE (scapes 30–60 cm high; early summer)—**Wild-hyacinth**, *Camassia scilloides*

49b. Tepals with one nerve; inflorescence a short, few-flowered raceme; escape from cultivation, lawns (scapes to 10 cm high; early spring)—**Squill**, *Scilla siberica*

DIOSCOREACEAE, The Yam Family

Dioecious, perennial, twining herbs with broad, net-veined leaves. Flowers regular, unisexual, in panicles or spikes; perianth of 6 tepals, united below; stamens 6, attached to the perianth base; pistil 1, styles 3, ovary inferior, 3-celled. Fruit a 3-winged capsule. Summer.

One species in Michigan; leaves ovate-cordate; mostly SLP (to 5 m high; summer)—**Wild Yam**, *Dioscorea villosa*

AMARYLLIDACEAE, The Amaryllis Family

Perennial herbs with basal, linear leaves. Flowers regular, perfect; perianth of 6 tepals, united below; stamens 6, attached to the perianth base; pistil 1, style 1, ovary inferior, 3-celled. Fruit a capsule. Spring.

One species in Michigan; flowers yellow, 1–2.5 cm wide; SLP (to 60 cm high)—**Star-grass**, *Hypoxis hirsuta*

IRIDACEAE, The Iris Family

Perennial herbs with leaves in two overlapping rows. Flowers regular, perfect; perianth of 6 tepals or as 3 sepals and 3 petals, sometimes united below; stamens 3, attached to the base of the sepals (or outer tepals), the filaments united below or not; pistil 1, style 1, 3-branched, ovary inferior, 3-celled. Fruit a capsule.

1a. Flowers 5 cm wide or larger; perianth of three petals and three sepals; the petaloid style branches both hiding the stamens and overarching the sepals (*Iris* spp., **Iris**)—2

1b. Flowers about 1 cm wide; perianth of six similar tepals; stamens visible and style branches narrow, not petaloid (flowers arise from a large bract (spathe) terminating a two-edged stem 10–50 cm high, spring and early summer)(*Sisyrinchium spp.*, **Blue-eyed-grass**)—7

2a. Flowering stems 20 cm or less high (spring)—3

2b. Flowering stems 40–100 cm high (early summer)—4

3a. Upper surface of the sepals bearded; an escape from cultivation, SLP— *Iris pumila*

3b. Upper surface of the sepals lacking a beard; native, Straits area—**Dwarf Lake Iris**, *Iris lacustris*

4a. Upper surface of the sepals bearded; roadside escapes, often cultivated— **Flag**, *Iris germanica* and hybrids

---

Iris germanica is one of several Eurasian species which have been used by breeders to create many of the horticultural varieties of "tall bearded" irises commonly grown today. The origin of this species is not known; it may also be the product of hybridization.

4b. Upper surface of the sepals lacking a beard; wet areas—5

5a. Flowers yellow; LP—**Yellow Flag**, *Iris pseudacorus*

5b. Flowers blue or violet, sometimes white—6

6a. Bright yellow spot at the base of the sepal blade; LP—**Southern Blue Flag**, *Iris virginica*

6b. Base of the sepal blade either unspotted or with a greenish-yellow spot; NM—**Wild Blue Flag**, *Iris versicolor*

7a. Spathes long-peduncled, arising from leaf-axils; stem appears to be branched—*Sisyrinchium angustifolium*

7b. Spathes sessile, terminal; stem appears unbranched—8

8a. Spathes two on each flowering stem; mostly SLP—*Sisyrinchium albidum*

8b. Spathe single—9

9a. Plant wiry, stems less than 1.5 mm wide and not (or barely) winged; calcareous shores—*Sisyrinchium mucronatum*

9b. Plant more stout, stems 1.5–3 mm wide and winged; moist fields and clearings, mostly NM—*Sisyrinchium montanum*

ORCHIDACEAE, The Orchid Family

Perennial herbs. Flowers irregular, perfect; sepals 3, two may be fused; petals 3, the two lateral petals similar to each other and often resembling the sepals, while the third petal, known as the *lip*, is normally the lowest petal and differs from the others in size, shape, and sometimes color (including being variously inflated or prolonged into a spur); stamen(s) 1(2), adherent to the style, forming the *column*; pistil 1, style 1, ovary inferior, 1-celled. Fruit a capsule which opens by vertical slits.

The more common species are included here. See Case (1987) for a comprehensive treatment of Michigan orchids.

Some of Michigan's native plants are protected by Public Act 182 of 1962, commonly called the "Christmas-tree law". Under this law, it is illegal to remove or cut orchids from any area without a bill of sale or written permission from the owner. A number of Michigan orchids are also rare, threatened, or endangered species and therefore should be collected only under special circumstances.

1a. Lip conspicuously sac-like, inflated—2

1b. Lip neither sac-like nor inflated, often flattened—12

2a. Flowers one to three, lip slipper-shaped—3

2b. Flowers three to many, lip basally sac-like, not slipper-shaped—8

3a. Plant with a single, basal leaf; woods, NM (scape 10–20 cm high; late spring)—**Calypso**, *Calypso bulbosa*

3b. Plant either with a leafy stem or with 2 basal leaves (late spring and early summer) (*Cypripedium* spp., **Lady's-slipper**)—4

4a. Plant with two basal leaves; lip pink, rarely white (scape 20-40 cm high, spring)—**Pink** or **Stemless Lady's-slipper**, *Cypripedium acaule*

4b. Plant with leafy stem; lip yellow to white—5

5a. Lip yellow (20–80 cm high; spring to early summer)—**Yellow Lady's-slipper**, *Cypripedium calceolus*

5b. Lip white, with or without crimson or purple markings—6

6a. Outside of the lip entirely white; bogs and meadows, SLP (15–40 cm high; late spring)—**White Lady's-slipper**, *Cypripedium candidum*

6b. Outside of the lip white, with crimson or purple markings—7

7a. Lip 2.5 cm long or less; the two lower sepals separate, greenish-brown; wooded dunes, NM (10–40 cm high; late spring)—**Ram's-head Lady's-slipper**, *Cypripedium arietinum*

7b. Lip 3 cm long or more; the two lower sepals united, white; bogs and swamps (40–100 cm high; summer)—**Showy Lady's-slipper**, *Cypripedium reginae*

8a. Leaves alternate on the stem; flowers greenish to purple; woods (to 80 cm high; summer)—**Helleborine**, *Epipactis helleborine*

8b. Leaves all basal (or nearly so), prominently net-veined, and frequently blotched with white; flowers whitish; woods, esp. coniferous (summer) (*Goodyera* spp., **Rattlesnake-plantain**)—9

9a. Perianth 6–10 mm long; lip with elongated point; leaves with a white or pale green midvein; NM (scape 20–40 cm high)—*Goodyera oblongifolia*

9b. Perianth 6 mm long or less; lip sac-like; other leaf veins white or pale green—10

10a. Inflorescence a densely flowered raceme, the flowers appearing at all angles around the stem (scape 20–40 cm high)—*Goodyera pubescens*

10b. Inflorescence either one-sided or loosely flowered—11

11a. Inflorescence a strongly one-sided raceme; mostly NM (scape 10–20 cm high)—*Goodyera repens*

11b. Inflorescence a loosely flowered, spiraled raceme (scape 15–35 cm high)—*Goodyera tesselata*

12a. Plants appear to lack chlorophyll, with most vegetative parts brown, purple, or yellow; leaves scale-like—13

12b. Plants with chlorophyll, vegetative parts green—17

13a. Surface of lip with 3 parallel ridges, margin of lip toothed and rolled upward; woods, mostly LP (scape 30–60 cm high, late spring)—**Putty-root**, *Aplectrum hyemale*

13b. Surface of lip with one or two parallel ridges, margin of lip entire, barely toothed, or lobed, not rolled upward (*Corallorhiza* spp., **Coral-root**)—14

14a. Lip with two lateral lobes or teeth—15

14b. Lip entire, or barely toothed—16

All orchids have specialized associations with fungi called *mycorrhizae*. Usually, mycorrhizal fungi are associated with the roots in a mutualistic relationship in which the photosynthetic orchid supplies carbohydrate. The coral-root orchids, **Corallorhiza** *spp.*, contain no chlorophyll and have lost their photosynthetic capability. According to Zelmer and Currah (1995) **Corallorhiza trifida** derives its nutrition from the digestion of fungal hyphae contained in its yellowish coralloid rhizomes. The fungus was also part of an ectomycorrhizal association with **Pinus contorta**. Ectomycorrhizal fungi are known to assist trees and shrubs in uptake of phosphorus, receiving photosynthate in return. In this case, the orchid evidently derives carbohydrate nutrition from the tree via the fungus which is associated with both plants.

15a. Stems yellowish or greenish; sepals and petals one-nerved (petals may be weakly three-nerved); lip usually less than 4.5 mm long, white, spotted or not; woods and swamps (10–30 cm high; spring and early summer)—**Early Coral-root**, *Corallorhiza trifida*

15b. Stems purplish or yellowish; sepals and petals three-nerved; lip over 4.5 mm long, white, usually purple-spotted; woods (20–60 cm high; summer)—**Spotted Coral-root**, *Corallorhiza maculata*

16a. Flower, exclusive of ovary, to 4 mm long; sepals and petals not purple-striped; dry woods, SLP (10–20 cm high; late summer)—**Fall Coral-root**, *Corallorhiza odontorhiza*

16b. Flower over 8 mm long; sepals and petals purple-striped; woods and swamps, NM (20-45 cm high; spring and early summer)—**Striped Coral-root**, *Corallorhiza striata*

17a. Flowers in a spike-like obviously twisted raceme; small, yellowish or greenish-white (10–40 cm high; late summer and autumn) (*Spiranthes spp.*, **Ladies'-tresses**)—18

Nine species of **Spiranthes** occur in Michigan. Identification of the species is difficult and often relies on careful inspection of the tiny lip; see Case and Catling (1982) for a technical key, Case (1987) for a simpler one. Habitat provides a good clue in identification; species known to inhabit dry areas are rarely found in wet areas and vice versa. The three most widely distributed species are included here.

17b. Flowers solitary or in clusters, but never in a twisted raceme—20

18a. Plants in dry sandy areas or fields; flowers often in one row;—**Slender Ladies'-tresses**, *Spiranthes lacera*

18b. Plants in moist or wet areas; flowers in two or three rows—19

19a. Lip constricted near the apex—**Hooded Ladies'-tresses**, *Spiranthes romanzoffiana*

19b. Lip oblong, not constricted—**Nodding Ladies'-tresses**, *Spiranthes cernua*

20a. Base of the lip prolonged into a spur, 2–50 mm long—21
20b. Lip not prolonged into a spur—36

21a. Leaves one to three, all basal—22
21b. Leaves all on the stem (*Habenaria* spp., **Rein Orchid**, in part)—27

22a. Flowers purple or magenta, or with white markings (*Orchis* spp.)—23
22b. Flowers greenish, yellowish, or white (*Habenaria* spp. in part)—24

> The names used here follow Voss (1972) and reflect a conservative approach both to species and to generic limits. Case (1987) used nomenclature which reflects the views of many orchid specialists, in which **Habenaria** and **Orchis** are split into smaller genera. Since many recent orchid texts use these newer names, we have included them in brackets.

23a. Plant with one leaf; bogs, NM (scape 10–20 cm high; early summer)—
     Round-leaved Orchis, *Orchis rotundifolia* [*Amerorchis rotundifolia*—Case]
23b. Plant with two leaves; woods, SLP and western UP (scape 10–25 cm high; late spring)—**Showy Orchis**, *Orchis spectabilis* [*Galearis spectabilis*—Case]

24a. Plant with a single leaf, at least twice as long as broad—25
24b. Plant with two often orbicular leaves, less than twice as long as broad—26

25a. Lip apex rounded, not toothed; coniferous woods and bogs, NM (scape 10–30 cm high)—**Blunt-leaf Orchid**, *Habenaria obtusata* [*Platanthera obtusata*—Case]
25b. Lip with three lobes or teeth at apex; moist woods, ditches (10–40 cm high)—**Club-spur Orchid**, *Habenaria clavellata* [*Platanthera clavellata*—Case]

26a. Flower stalk without bracts below the raceme; flowers yellowish-green; woods (scape 20–40 cm high)—**Hooker's Orchid**, *Habenaria hookeri* [*Platanthera hookeri*—Case]
26b. Flower stalk bearing bracts below the raceme; flowers greenish-white; woods (scape 30–60 cm high)—**Round-leaved Orchid**, *Habenaria orbiculata* [*Platanthera orbiculata*—Case]

27a. Lip fringed (edge deeply cut many times); wet areas, esp. bogs and marshes—28
27b. Lip not fringed—32

28a. Lip fringed and clearly three-lobed—29

28b. Lip fringed but not three-lobed (40–100 cm high; summer)—31

29a. Flowers purple (30–150 cm high)—**Purple Fringed Orchid**,
  Habenaria psycodes [*Platanthera psycodes*—Case]
29b. Flowers white, yellowish, or greenish—30

30a. Spur 2 cm long or more; sepals over 6 mm long; mostly SLP (40–100
  cm high; summer)—**Prairie Fringed Orchid**, *Habenaria leucophaea*
  [*Platanthera leucophaea*—Case]
30b. Spur 2 cm long or less; sepals less than 6 mm long (30–80 cm high;
  summer)—**Ragged Fringed Orchid**, *Habenaria lacera* [*Platanthera*
  lacera—Case]

31a. Flowers orange; SLP—**Orange Fringed Orchid**, *Habenaria ciliaris*
  [*Platanthera ciliaris*—Case]
31b. Flowers white; bogs, LP—**White Fringed Orchid**, *Habenaria ble-*
  phariglottis [*Platanthera blephariglottis*—Case]

32a. Lip with two to three evident teeth at apex—33
32b. Lip without apical teeth—34

33a. Stem leaves three or more; rich woods (20–50 cm high; spring and
  summer)—**Bracted Orchid**, *Habenaria viridis* [*Coeloglossum viride*—
  Case]
33b. Stem leaves one or two; moist woods, ditches (10–40 cm high; summer)—**Club-spur Orchid**, *Habenaria clavellata* [*Platanthera clavel-*
  lata—Case]

34a. Lip oblong, truncate at the apex; wet woods, LP (30–70 cm high; summer)—**Tubercled Orchid**, *Habenaria flava* [*Platanthera flava*—Case]
34b. Lip lanceolate, tapering toward the apex; moist areas—35

35a. Flowers white (to 100 cm high)—**Tall White Bog Orchid**, *Habenaria*
  dilatata [*Platanthera dilatata*—Case]
35b. Flowers greenish-yellow (30–100 cm high; summer)—**Tall Northern**
  Bog Orchid, *Habenaria hyperborea* [*Platanthera hyperborea*—Case]

36a. Flowering plant leafless, with leaves reduced to tubular sheaths, or with
  one linear to ovate leaf—37
36b. Flowering plant with two or more oblong to ovate leaves—43

37a. Flowers one to two, pink to purple—38
37b. Flowers more than two, yellowish or greenish-white—40

38a. Leaf lanceolate to elliptic; lip fringed and bearded with yellow hairs; bogs (10–40 cm high; late spring and early summer)—**Rose Pogonia**, *Pogonia ophioglossoides*

38b. Leaf linear or linear-lanceolate or absent; lip not fringed, bearded or not—39

39a. Lip is the uppermost petal, bearded with yellow hairs; bogs and swamps (30–70 cm high; summer)—**Grass-pink**, *Calopogon pulchellus* [*C. tuberosus*—CQ]

39b. Lip is the lowermost petal, not bearded; bogs and swamps (10–30 cm high; late spring)—**Arethusa**, *Arethusa bulbosa*

40a. One ovate or elliptic leaf attached to the stem beneath the flowers; flowers (including lips) greenish-white (scape 10–30 cm high; summer) (*Malaxis* spp., **Adder's-mouth**)- 41

40b. One elliptic leaf, if present, adjacent to flower scape; scape leaves reduced to tubular sheaths; flowers yellowish, the lips white—42

41a. Lip apex narrowly pointed; mature pedicels less than 3 mm long—**White Adder's-mouth**, *Malaxis monophylla* [*M. monophyllos*—CQ]

41b. Lip apex is two- to three-toothed; mature pedicels more than 3.5 mm long—**Green Adder's-mouth**, *Malaxis unifolia*

42a. Basal elliptic leaf present (or not); flowers apically purple-tinged; surface of lip with 3 parallel ridges, margin of lip toothed and rolled upward; woods, mostly LP (scape 30–60 cm high, late spring)—**Putty-root**, *Aplectrum hyemale*

42b. Basal leaf absent; surface of lip with one or two parallel ridges, margin of lip has two lobes, not rolled upward; woods and swamps (10–30 cm high; spring and early summer)—**Early Coral-root**, *Corallorhiza trifida*

43a. Leaves a whorl of five; bogs, LP (20–40 cm high; late summer)—**Whorled Pogonia**, *Isotria verticillata*

43b. Leaves two, opposite (10–30 cm high; summer)—44

44a. Leaves opposite and sessile near the middle of the stem (*Listera* spp., **Twayblade**)—45

44b. Leaves basal (*Liparis* spp., **Twayblade**)—46

45a. Lip deeply two-cleft; flowers greenish to purplish; bogs and swamps, NM—**Heartleaf Twayblade**, *Listera cordata*

45b. Lip wedge-shaped, with two round shallow lobes; flowers greenish-yellow; wet woods, NM—**Broad-leaved Twayblade**, *Listera convallarioides*

46a. Lip about 10 mm long, purple; woods, LP—**Purple Twayblade**, *Liparis liliifolia*

46b. Lip about 5 mm long, yellowish-green; grassy wet areas—**Green Twayblade**, *Liparis loeselii*

Class MAGNOLIOPSIDA—The Dicotyledons

SAURURACEAE, The Lizard's-tail Family

Perennial herbs with alternate, simple leaves. Flowers perfect, small, in dense spikes; sepals 0; petals 0; stamens 6–8; pistils and styles 3–5, each ovary superior, 1-celled. Fruit a mass of indehiscent carpels.

One species in Michigan, a marsh plant with heart-shaped leaves and slender nodding spikes of white flowers; SLP (50–120 cm high; summer)—**Lizard's-tail**, *Saururus cernuus*

SALICACEAE, The Willow Family

Dioecious trees or shrubs with alternate, simple leaves. Flowers unisexual, in catkins appearing with or often before the leaves; sepals 0; petals 0; stamens 1–80; pistil 1, styles 1–4, ovary superior, 1-celled. Fruit a capsule. Spring.

1a. Leaf blades ovate to deltoid, less than twice as long as broad, on petioles 3 cm long or more (***Populus* spp., Poplar**)—2

1b. Leaf blades lanceolate to linear, more than twice as long as broad, on petioles 2.5 cm long or less (***Salix* spp., Willow**)—7

2a. Petioles strongly flattened laterally—3

2b. Petioles not flattened laterally—6

3a. Leaf blades broadly ovate or nearly circular, without a translucent border—4

3b. Leaf blades broadly triangular or deltoid in shape, with a translucent border—5

4a. Leaf blades coarsely toothed, fewer than 15 teeth on each side; sandy soils—**Bigtooth Aspen**, *Populus grandidentata*

4b. Leaf blades finely crenulate or serrate, more than 15 teeth on each side—**Quaking Aspen**, *Populus tremuloides*

5a. Tree with spreading crown; glands present at the junction of the leaf blade and petiole; LP—**Cottonwood**, *Populus deltoides*

5b. Tree with narrow spire-shaped crown; leaf blade glandless; often planted—**Lombardy Poplar**, *Populus nigra* cv. *italica*

6a. Leaf blade densely white tomentose beneath; often planted, escapes to disturbed areas—**White Poplar**, *Populus alba*

6b. Leaf blade glabrous or nearly so beneath—**Balsam Poplar**, *Populus balsamifera*

7a. Trees—8
7b. Shrubs—16

> **Salix** species are often difficult to distinguish. Hybridization among our species of willows is known, producing plants with intermediate characteristics. Of the 24 species known in Michigan, 19 are included here; see Voss (1985) for a complete treatment, including keys to pistillate and staminate specimens lacking leaves.

8a. Leaf blades entire or only with a few apical teeth; wet areas—9
8b. Leaf blades distinctly serrate nearly to the base—10

9a. Leaf blades pubescent above and beneath, veins raised beneath (2–10 m high)—**Beaked** or **Bebb's Willow**, *Salix bebbiana*

9b. Leaf blades glabrous above, veins not raised beneath (2–7 m high)—**Pussy Willow**, *Salix discolor*

10a. Branches and twigs conspicuously drooping; escapes from cultivation—**Weeping Willow**, *Salix alba* var. *tristis*

10b. Branches and twigs not conspicuously drooping—11

11a. Petioles with glands near the junction of the blade; dunes or wetlands—12

11b. Petioles without glands; shores, wet areas—14

12a. Leaf blades whitened and silky beneath; escapes from cultivation to wet areas (to 25 m high)—**White** or **Yellow Willow**, *Salix alba*

12b. Leaf blades green beneath (to 20 m high)—13

13a. Stipules conspicuous; dunes and wetlands (to 6 m high)—**Shining Willow**, *Salix lucida*

13b. Stipules inconspicuous or absent; escapes from cultivation to wet areas (to 20 m high)—**Crack** or **Brittle Willow**, *Salix fragilis* or *S.* ×*rubens* (a hybrid of *S. fragilis* and *S. alba*)

14a. Leaf blades curved (sickle-shaped), green, not glaucous or densely white pubescent beneath (to 20 m or more)—**Black Willow, *Salix nigra***

14b. Leaf blades straight, glaucous or white pubescent beneath; LP—15

15a. Leaf blades glaucous beneath; LP (3–20 m high)—**Peach-leaved Willow, *Salix amygdaloides***

15b. Leaf blades covered with dense, white silky hairs beneath; SLP (to 4 m high)—**Silky Willow, *Salix sericea***

16a. Leaves opposite on older twigs, often purple-tinged; escapes from cultivation to wet areas (1–2.5 m high)—**Basket Willow, *Salix purpurea***

16b. Leaves all alternate—17

17a. Leaf blades entire or only with a few teeth, apical or widely spaced—18

17b. Leaf blades distinctly serrate nearly to the base—28

18a. Leaf blades glabrous above and beneath—19

18b. Leaf blades with at least one pubescent surface—22

19a. Leaf blades linear to very narrowly lanceolate; wet areas (1.5–3 m high)—**Sandbar Willow, *Salix exigua***

19b. Leaf blades lanceolate to ovate—20

20a. Leaf blades entire, apex rounded; stipules absent; peatlands and moist dunes (40–100 cm high)—**Bog Willow, *Salix pedicellaris***

20b. Leaf blades with some apical teeth, apex acute; stipules present or not—21

21a. Base of leaf blade rounded; stipules present; coastal sand dunes (1–4 m high)—**Blueleaf Willow, *Salix myricoides***

21b. Base of leaf blade acute; stipules absent; open wet areas (2–7 m high)—**Slender** or **Meadow Willow, *Salix petiolaris***

22a. Leaf blades white tomentose or covered with silky hairs beneath; wet areas—23

22b. Leaf blades green, glabrous, or glaucous beneath—25

23a. Leaf blades mostly entire, densely white tomentose beneath (to 1 m high)—**Sage Willow, *Salix candida***

23b. Leaf blades finely toothed, covered with silky hairs beneath—24

24a. Base of leaf blade rounded; SLP (to 4 m high)—**Silky Willow,** *Salix sericea*

24b. Base of leaf blade acute or acuminate (1.5–3 m high)—**Sandbar Willow,** *Salix exigua*

25a. Leaf blades linear to narrowly oblanceolate; open wet areas (2–7 m high)—**Slender** or **Meadow Willow,** *Salix petiolaris*

25b. Leaf blades oblanceolate to mostly elliptic to obovate—26

26a. Plant of dry, often upland habitats (1–3 m high)—**Upland Willow,** *Salix humilis*

26b. Plant of wet sites—27

27a. Leaf blades pubescent above and beneath, veins raised beneath (2–10 m high)—**Beaked** or **Bebb's Willow,** *Salix bebbiana*

27b. Leaf blades glabrous above, veins not raised beneath (2–7 m high)—**Pussy Willow,** *Salix discolor*

28a. Petioles with glands near the junction with the blade—29

28b. Petioles without glands—30

29a. Leaf blades whitened beneath, apex acute to acuminate; bogs and calcareous wet areas (1–4 m high; late spring)—**Autumn Willow,** *Salix serissima*

29b. Leaf blades green or pale (not strongly whitened) beneath, apex long-acuminate; wet areas and dunes (to 6 m high)—**Shining Willow,** *Salix lucida*

30a. Leaf blades curved (sickle-shaped), green, not glaucous or densely white pubescent beneath; shores—**Black Willow,** *Salix nigra*

30b. Leaf blades straight, glaucous or white pubescent beneath; LP—31

31a. Leaf blades white tomentose or covered with silky hairs beneath—32

31b. Leaf blades glaucous or whitened beneath—34

32a. Stipules conspicuous; leaf blades ovate-lanceolate or ovate, white tomentose; coastal sand dunes (to 3 m high)—**Sand-dune Willow,** *Salix cordata*

32b. Stipules minute or absent; leaf blades linear to very narrowly lanceolate, covered with silky hairs beneath—33

33a. Base of leaf blade acute to acuminate; coastal sand dunes, wet areas (1.5–3 m high)—**Sandbar Willow,** *Salix exigua*

33b. Base of leaf blade rounded; wet areas, SLP (to 4 m high)—**Silky Willow**, *Salix sericea*

34a. Base of leaf blades acute; open wet areas (2–7 m high)—**Slender** or **Meadow Willow**, *Salix petiolaris*

34b. Base of leaf blades rounded; wet areas or dunes—35

35a. Stipules absent or to 2 mm long; foliage has balsam fragrance when dried; bogs, wet clearings (to 5 m high)—**Balsam Willow**, *Salix pyrifolia*

35b. Stipules longer than 5 mm long; foliage lacks balsam fragrance—36

36a. Leaves thick, glaucous beneath; coastal sand dunes (1–4 m high)—**Blueleaf Willow**, *Salix myricoides*

36b. Leaves not thickened, pale green or white beneath; wet areas (2–4 m high)—**Diamond Willow**, *Salix eriocephala*

MYRICACEAE, The Bayberry Family

Dioecious shrubs with aromatic odor and alternate, simple leaves. Flowers unisexual, in catkins; sepals 0; petals 0; stamens 2–8, mostly 4; pistil 1, styles 2, ovary superior, 1-celled. Fruit an achene or small drupe.

1a. Leaves pinnately lobed; stipules present; dry sandy areas, mostly NM (to 1.5 m high)—**Sweetfern**, *Comptonia peregrina*

1b. Leaves entire or merely serrate; stipules absent; wet areas, NM (to 1.5 m high)—**Sweet Gale**, *Myrica gale*

JUGLANDACEAE, The Walnut Family

Monoecious trees with alternate, pinnately compound leaves. Flowers unisexual, in catkins (or pistillate flower solitary); perianth of 4 tepals, minute and united, or 0; stamens 3–40; pistil 1, styles 2, ovary inferior, 1-celled above, 2- or 4-celled below. Fruit a nut enclosed in a husk which may or may not be dehiscent. Spring.

1a. Leaflets seven to nine, eleven, or more, terminal leaflets not clearly larger than lower leaflets; pith divided by partitions into chambers; fruit husk indehiscent; mostly LP (*Juglans* spp., **Walnut**)—2

1b. Leaflets five to nine, terminal leaflets larger than lower leaflets; pith not partitioned; fruit husk splitting into four valves; SLP, sometimes planted further north (*Carya* spp., **Hickory**)—4

2a. Leaflets seven to nine, entire, glabrous; often planted—**English Walnut**, *Juglans regia*

2b. Leaflets eleven or more, serrate, pubescent at least below—3

3a. Hairs on leaflets mostly stellate; pith of branches brown; bark with flat smooth ridges; fruit ovoid; mostly LP (to 30 m high)—**Butternut**, *Juglans cinerea*

3b. Hairs on leaflets mostly simple; pith of branches cream in color; bark with rough ridges; fruit globose; SLP (to 40 m high)—**Black Walnut**, *Juglans nigra*

4a. Bark of the trunk essentially smooth, not deeply furrowed or shaggy—5

4b. Bark of the trunk deeply furrowed or shaggy—6

5a. Leaflets seven to nine, somewhat pubescent beneath; buds bright yellow, with four to six non-overlapping scales—**Bitternut Hickory**, *Carya cordiformis*

5b. Leaflets five to seven, glabrous beneath or with pubescence only along veins; buds greenish, with more than six overlapping scales—**Pignut Hickory**, *Carya glabra*

6a. Leaflets pubescent beneath—**Kingnut** or **Shellbark Hickory**, *Carya laciniosa*

6b. Leaflets glabrous beneath or with hairs only near teeth or along larger veins—7

7a. Leaflets glabrous beneath or pubescent only along veins—**Pignut Hickory**, *Carya glabra*

7b. Leaflets with small tufts of hairs on the teeth—**Shagbark Hickory**, *Carya ovata*

BETULACEAE, The Birch Family

Monoecious trees or shrubs with alternate, simple leaves. Flowers unisexual, tiny, in catkins, the pistillate catkins sometimes woody and cone-like; sepals 2 or 4, minute and united, or 0; petals 0; stamens several, often 2 or 4, anthers and filament apices often divided; pistil 1, styles 2, ovary inferior, 1-celled above, 2-celled below. Fruit a nut, samara, or achene. Early spring.

1a. Trees, with white or yellowish bark exfoliating in thin papery plates or layers (*Betula* in part, **Birch**)—2

1b. Trees or shrubs; bark dark, smooth or roughened, not exfoliating—4

2a. Bark yellowish (to 30 m high)—**Yellow Birch**, *Betula alleghaniensis*
2b. Bark white or chalky—3

3a. Leaf blades deltoid, apex long-acuminate, glabrous; escape from cultivation, SLP—**European White Birch**, *Betula pendula*
3b. Leaf blades ovate, apex acute or acuminate, pubescent along veins beneath (to 30 m high)—**Paper** or **White Birch**, *Betula papyrifera*

4a. Tree or shrub, with smooth gray bark; trunk fluted with prominent longitudinal ridges (to 10 m high)—**Hornbeam** or **Blue-beech**, *Carpinus caroliniana*
4b. Trees or shrubs; the bark more or less roughened; trunk not fluted—5

5a. Shrub, with leaves 4 cm long or less; swamps and peatlands (1–4 m high)—**Swamp**, **Bog** or **Dwarf Birch**, *Betula pumila*
5b. Shrubs or trees, with leaves 5 cm long or more—6

6a. Pistillate catkins woody, persistent on the plant throughout the year; often forms thickets in wet areas (*Alnus* spp., **Alder**)—7
6b. Pistillate catkins herbaceous, dropping in late autumn—8

7a. Leaf blades serrate, covered with resinous dots beneath; mature nut broadly winged; UP—**Green** or **Mountain Alder**, *Alnus crispa* [*Alnus viridis* subsp. *crispa* - CQ]
7b. Leaf blades serrate and appearing to be either coarsely toothed or even obscurely lobed, resinous dots absent beneath; mature nut not winged—**Speckled Alder**, *Alnus rugosa* [*Alnus incana* subsp. *rugosa* - CQ]

8a. Tree; fruit a cluster of bladder-like sacs each containing a small achene (to 20 m high)—**Hop-hornbeam** or **Ironwood**, *Ostrya virginiana*
8b. Shrubs; fruit a nut within a close-fitting involucre (1–3 m high) (*Corylus* spp., **Hazel**)—9

9a. Involucre of two broad, separate bracts, less than twice as long as the enclosed nut; mostly SLP—**Hazel**, *Corylus americana*
9b. Involucre of united bracts, prolonged into a bristly beak more than twice as long as the enclosed nut; mostly NM—**Beaked Hazel**, *Corylus cornuta*

FAGACEAE, The Beech Family

Monoecious trees (except *Quercus prinoides*, a shrub), with alternate, simple leaves. Flowers unisexual, the staminate in catkins, the pistillate solitary or in a small cluster; perianth of 3–7, mostly 6, tepals, sometimes united below; stamens 3–20; pistil 1, styles (0)3 or 6, ovary inferior, 3- or 6-celled. Fruit one or more nuts enclosed in a cup or bur. Spring.

1a. Leaf blades serrate with numerous sharp-pointed teeth; fruit one to three nuts enclosed in a prickly bur—2

1b. Leaf blades serrate, lobed, or entire, but never serrate with sharp-pointed teeth; fruit an acorn (solitary nut with "cup" of scales); pith five-angled in the young twigs (*Quercus* spp., **Oak**)—3

Distinguishing among species of **Quercus** is often difficult. The shape and lobing of oak leaves is often quite variable, even on the same tree. Hybrid individuals occur, with most members of each subgenus capable of forming hybrids with other members of the same subgenus. Gleason and Cronquist (1991) listed 60 names which have been described from such hybrids. Please consult Voss (1985) for additional discussion and references.

2a. Bark gray, smooth; buds three to four times as long as wide, sharply pointed; leaf teeth less than 2 mm long; nut triangular; rich woods, except western UP (to 30 m high)—**Beech**, *Fagus grandifolia*

2b. Bark rough; buds relatively thicker; leaf teeth 2 mm or longer; nut rounded; LP (to 30 m high, but see box)—**Chestnut**, *Castanea dentata*

The chestnut, **Castanea dentata**, was once well distributed in parts of the Lower Peninsula, both as a native forest species and as planted individuals, and was a valuable source of nuts and lumber in much of eastern North America. Most mature specimens have been killed by the chestnut blight. The fungus which causes this disease, **Endothia parasitica**, was introduced to North America from Asia, first noted in New York City in 1904. Young plants and sprouts are resistant to the disease, but become infected and die before maturity. Some large trees still surviving at sites in the western Lower Peninsula have been the focus of recent research on hypovirulent strains of the fungus, which may protect chestnut trees against becoming diseased. See Brewer (1982, 1995) for further details.

3a. Leaf blades entire, except for a bristle at the tip; SLP—**Shingle Oak**, *Quercus imbricaria*

3b. Leaf blades toothed or lobed—4

4a. Leaf blades toothed or lobed, the points bristle-tipped (*Quercus* subg. *Erythrobalanus*, **Red** or **Black Oaks**)—5

4b. Leaf blades toothed or lobed, the points frequently rounded and without bristles (*Quercus* subg. *Quercus*, **White Oaks**)—9

5a. Length of the lateral leaf-lobes less than one-third the width of the leaf—6

5b. Length of the lateral leaf-lobes more than one-third the width of the leaf—7

6a. Leaf blades glossy above; cup of the acorn hemispherical or top-shaped, covering about one-half of the acorn; dry sandy areas, LP (to 40 m high)—**Black Oak**, *Quercus velutina*

6b. Leaf blades not glossy above; cup of the acorn saucer-shaped, covering one-third or less of the acorn (to 50 m high)—**Red Oak**, *Quercus rubra*

7a. Upper rim of acorn cup fringed; dry sandy areas, LP (to 40 m high)— **Black Oak**, *Quercus velutina*

7b. Upper rim of acorn not fringed—8

8a. Acorn about 1 cm in diameter, cup saucer-like; moist areas, SLP (to 30 m high)—**Pin Oak**, *Quercus palustris*

8b. Acorn 1.5–2 cm in diameter, cup covers about one-half of the acorn; dry sandy areas (to 25 m high)—**Scarlet Oak**, *Quercus coccinea*

9a. Leaf blades deeply pinnately lobed—10

9b. Leaf blades crenate, dentate, or with a few irregular, shallow lobes; SLP—11

10a. Leaf blade divided nearly to the middle by a pair of deep lateral lobes near the middle of the leaf; acorn more than half covered by a fringed cup—**Bur Oak**, *Quercus macrocarpa*

10b. Leaf blade without a median pair of deeper lobes; cup of acorn saucer-shaped, covering up to one-third of the acorn—**White Oak**, *Quercus alba*

11a. Acorn on a peduncle three to seven cm long, the stalk longer than adjacent leaf petioles; wet areas (to 30 m high)—**Swamp White Oak**, *Quercus bicolor*

11b. Acorn sessile or on peduncles not longer than adjacent petioles—12

12a. Tall tree; leaf blades with eight to fourteen sharp coarse teeth on each side; rich woods (to 25 m high)—**Chinquapin** or **Yellow Oak**, *Quercus muehlenbergii*

12b. Shrub; leaf blades with five to eight teeth on each side; sandy areas, SLP (1–3 m high)—**Dwarf Chinquapin Oak**, *Quercus prinoides*

ULMACEAE, The Elm Family

Trees or shrubs with alternate, simple leaves which often have oblique bases. Flowers regular, perfect or unisexual (the plants then monoecious), small; sepals 4–9, united below; petals 0; stamens 4–9; pistil 1, styles 2, ovary superior, 1-celled. Fruit a samara or a drupe. Spring.

1a. Lateral leaf blade veins straight, run directly to marginal teeth; fruit a
    samara (*Ulmus* spp., **Elm**)—2
1b. Lateral leaf blade veins curved, branching before reaching the margin;
    fruit a drupe (*Celtis* spp., **Hackberry**)—5

2a. Leaf blades less than 7 cm long, the base barely asymmetrical and the
    margin simply serrate; agressively spreads into disturbed areas, LP—
    Siberian Elm, *Ulmus pumila*
2b. Leaf blades longer than 7 cm, the base strongly asymmetrical and the
    margin coarsely and doubly serrate; often in woods along rivers and
    streams—3

3a. Older branches with flat corky wings; leaf blades smooth and glabrous
    above; LP and western UP (to 30 m high)—**Cork Elm**, *Ulmus thomasii*
3b. Branches without corky wings; leaf blades smooth or rough above—4

4a. Axillary buds densely covered with reddish hairs; leaf blades very rough
    above; LP and western UP (to 30 m high)—**Slippery** or **Red Elm**,
    Ulmus rubra
4b. Axillary buds glabrous or nearly so; leaf blades smooth or rough above
    (to 40 m high)—**American Elm**, *Ulmus americana*

The loss of most specimens of **Ulmus americana**, the American Elm, to Dutch Elm Disease, is another example of a fungus from logs imported from Asia which has severely altered populations of our native forests. Some urban specimens have been maintained by yearly treatment with a fungicide active against **Ceratocystis ulmi**, but wild populations have not had this protection.

5a. Leaf blades toothed nearly to the base; fruiting pedicels longer than the
    petioles; drupe purple; often in moist areas, mostly SLP (6–15 m high)—
    Hackberry, *Celtis occidentalis*
5b. Lower third of leaf blade entire; fruiting pedicels about as long at the
    petioles; drupe salmon or orange-red; dry sites, SLP (to 5 m high)—
    Dwarf Hackberry, *Celtis tenuifolia*

MORACEAE, The Mulberry Family

Monoecious or dioecious trees with alternate, simple, leaves and milky juice. Flowers unisexual, small, in catkins or dense heads; sepals 4, united below; petals 0; stamens 4; pistil 1, styles 1 or 2, ovary superior, 1-celled. Fruit a fleshy syncarp formed by connation of all the flowers in an inflorescence, enclosing or completely concealing the achenes. Spring.

1a. Leaf blades entire, pinnately veined; fruit green to yellow, globose, about 10 cm in diameter; often planted as fence rows, SLP (to 20 m high)—**Osage-orange**, *Maclura pomifera*

1b. Leaf blades lobed or toothed, palmately veined; fruit white to purple, short-cylindric, 2–3 cm in diameter (*Morus* spp., **Mulberry**)—2

2a. Leaf blades pubescent beneath; fruit purple; moist woods, rare, SLP (to 20 m high)—**Red Mulberry**, *Morus rubra*

2b. Leaf blades glabrous or pubescent only along veins beneath; fruit white to (mostly) purple; escape to disturbed areas, mostly LP (to 25 m high)—**White Mulberry**, *Morus alba*

CANNABACEAE, The Hemp Family

Dioecious herbs with mostly opposite, simple or compound leaves. Flowers unisexual, small, in axillary clusters; sepals 5, united or just below; petals 0; stamens 5; pistil 1, style 1, ovary superior, 1-celled. Fruit an achene.

1a. Twining perennial; leaves serrate or lobed; achenes hidden by large green bracts; long-persisting after cultivation—**Hops**, *Humulus lupulus*

1b. Erect annual; leaves palmately compound, sometimes alternate above; achenes not hidden (1–3 m high)—**Hemp**, *Cannabis sativa*

While neither is a major agricultural crop in Michigan today, both members of the Cannabaceae were cultivated here in past years. Large tracts of land supported *Humulus lupulus* cultivation in the early 1900s; hops are an ingredient in beer. *Cannabis sativa* was (legally) cultivated as a source of fiber hemp in southwestern Michigan during the 1940s.

URTICACEAE, The Nettle Family

Monoecious or dioecious herbs with alternate or opposite, simple, stipulate (absent in *Parietaria*) leaves. Flowers regular, unisexual, in axillary clusters; sepals 2–5, united or not; petals 0; stamens 4 or 5; pistil 1, style 1, ovary superior, 1-celled. Fruit an achene.

1a. Leaves alternate—2

1b. Leaves opposite; wet areas—3

2a. Leaf blades toothed, 8–20 cm long, with stiff, stinging hairs along the veins; wet, often wooded areas (50–100 cm high)—**Wood Nettle**, *Laportea canadensis*

2b. Leaf blades entire, 2–8 cm long, stinging hairs absent; dry to wet sites, sometimes as a weed, mostly SLP (10–40 cm high)—**Pellitory**, *Parietaria pensylvanica*

3a. Flowers in unbranched spikes; mostly LP (40–100 cm high)—**False Nettle**, *Boehmeria cylindrica*
3b. Flowers in branched cymes or panicles—4

4a. Stems opaque; stinging hairs present on inflorescence branches, often on leaf blades (to 2 m high)—**Stinging Nettle**, *Urtica dioica*
4b. Stems translucent; stinging hairs absent; mostly LP (10–50 cm high) (**Clearweed**, *Pilea* spp.)—5

5a. Achenes pale yellow to tan, the surface with irregular purplish spots— *Pilea pumila*
5b. Achenes dark green to black, the surface warty—*Pilea fontana*

SANTALACEAE, The Sandalwood Family

Low herbs with alternate, simple, entire leaves. Flowers regular, mostly perfect, small, greenish-white, in terminal or axillary cymes; perianth of 5 tepals, united or not; stamens 5; pistil 1, style 1, ovary inferior, 1-celled. Fruit a drupe. Michigan species are hemiparasitic, attached via modified roots to a host plant while also having green leaves.

1a. Whitish flowers in several-flowered terminal cymes; drupe dry, yellow or green; calcareous areas (10–40 cm high; spring and early summer)— **Bastard Toad-flax**, *Comandra umbellata*
1b. Greenish-purple flowers in three-flowered axillary cymes; drupe fleshy, orange or red; often in sandy areas, NM (10–30 cm high; summer)— *Geocaulon lividum*

VISCACEAE, The Mistletoe Family

Dioecious parasitic plants, attached to the branches of trees, with reduced scale-like leaves. Flowers regular, unisexual, small, axillary; perianth of 2–4 tepals, united; stamens 3 or 4, filaments absent; pistil 1, style 1, ovary inferior, 1-celled. Fruit a berry.

One species in Michigan, a dwarf brown plant, occurring mostly on branches of black spruce; NM (5–15 mm high; early spring)—**Dwarf Mistletoe**, *Arceuthobium pusillum*

Arceuthobium pusillum is the only mistletoe found in Michigan. This tiny leafless parasite of conifers (only on spruce in Michigan) may often be spotted by a "witches'-broom" of branches on the infected tree. The Christmas Mistletoe, **Phoradendron flavescens**, is a leafy green parasite native to the southern United States.

ARISTOLOCHIACEAE, The Birthwort Family

Perennial herbs with alternate or basal, simple leaves. Flowers regular or irregular, perfect, reddish- or purplish-brown, at or near the ground; sepals 3, united, tubular or S-shaped; petals 0 or as vestigial remnants; stamens 6 or 12, attached (or in proximity) to the style; pistil 1, style 1, ovary inferior, 6-celled. Fruit a capsule. Spring.

1a. Leaves alternate, on the stem; flowers on a basal scaly branch; calyx S-shaped, stamens 6; rich woods, SLP (10–60 cm high)—**Virginia-snake-root**, *Aristolochia serpentaria*

1b. Leaves a single basal pair, bearing one short-stalked flower between them; calyx tubular below, stamens 12; rich woods—**Wild-ginger**, *Asarum canadense*

POLYGONACEAE, The Buckwheat Family

Herbs with alternate, simple, entire leaves, the stipules surrounding the stem above the base of each leaf to form the sheath-like *ocrea*. Flowers regular, perfect or unisexual (the plants then monoecious or dioecious); perianth of 4–6 tepals, often in two whorls and/or united below; stamens 3–9, filaments sometimes united below; pistil 1, styles 2 or 3, ovary superior, 1-celled. Fruit an achene, often 3-angled.

1a. Plant with large, basal leaves, lacking cauline leaves; stamens 9; achenes winged; occasional escapes from cultivation (summer)—**Rhubarb**, *Rheum rhaponticum* [*R. rhabarbarum*—CQ]

1b. Plant with alternate leaves along the stem; stamens 3–8; achenes not winged—2

2a. Tepals six, not petaloid, the three inner ones enlarging in fruit and surrounding the achenes (*Rumex* spp., **Dock**) (summer)—3

2b. Tepals four or five (rarely six), often petaloid, all of similar size in fruit—9

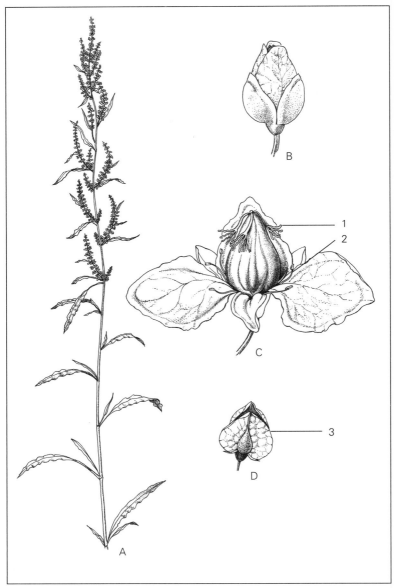

Figure 15: ***Rumex crispus****: A, inflorescence (raceme of spikes);*
B, flower (closed); C, flower (opened); 1, stigma; 2, stamen;
D, fruit; 3, valve

3a. Leaves hastate, the two basal lobes pointing outward; plants dioecious; common weed (10–40 cm high)—**Red** or **Sheep Sorrel**, *Rumex acetosella*

3b. Leaves without basal lobes; plants either monoecious or bearing perfect flowers—4

4a. Lowermost leaves with cordate or subcordate bases; moist, disturbed areas (up to 120 cm high)—**Bitter Dock**, *Rumex obtusifolius*

4b. Leaf bases not cordate or subcordate—5

5a. Leaves with strongly crisped or wavy-curled margins—6

5b. Leaves flat or nearly so, the margin entire—7

6a. Fruiting pedicels with a swollen joint; dry or wet areas (0.5–1.5 m high, ripe fruit in June-July)—**Curly Dock**, *Rumex crispus*

6b. Fruiting pedicels without a swollen joint; wet areas (up to 2.5 m high, ripe fruit in September)—**Great Water Dock**, *Rumex orbiculatus*

7a. Midrib of one valve swollen (forming a grain or tubercle) (to 1 m high)—**Pale Dock**, *Rumex altissimus*

7b. Midribs of all three valves swollen—8

8a. Fruiting pedicels curved, not reflexed, about the same length as the fruiting perianth (to 1 m high)—*Rumex triangulivalvis* [*R. salicifolius*—CQ]

8b. Fruiting pedicels straight, all regularly reflexed, up to three times longer than the fruiting perianth; wet areas, LP (to 1.5 m high)—**Water Dock**, *Rumex verticillatus*

9a. Flowers on slender pedicels that are jointed near the base; flowers white to pink or red, solitary in the axils of closely overlapping bracts, forming a loose, terminal raceme; leaves about 1 mm wide; sandy areas (10–50 cm high; summer and early autumn)—**Jointweed**, *Polygonella articulata*

9b. Pedicels jointed near the summit (if at all); flowers often more crowded or if solitary, are either subsessile in leaf axils or bracts that are not closely overlapping; leaves wider than 1.5 mm—10

10a. Achene exserted from the perianth; plant a glabrous annual with broadly triangular leaves; flowers white; escapes to roadsides, fields, etc. (20–60 cm high; summer)—**Buckwheat**, *Fagopyrum esculentum*

10b. Achene mostly or entirely enclosed in persistent perianth; if leaves are broadly triangular, the stem prickly or the plant a vine or stout perennial (*Polygonum* spp., **Smartweed** or **Knotweed**) (summer and early autumn)—11

> **Polygonum** is a large genus which presents a challenge to the taxonomist. How are species within the genus related? Can groups of species (*sections* or *subgenera*) be recognized? What characters are significant in assessing such relationships? Should the genus be maintained or split into a number of smaller genera? Some accept this genus in a "broad" sense (as here); following the sectional classification used by Gleason and Cronquist (1991), species found in Michigan are members of five sections. Others define **Polygonum** in a "narrow" sense. Ronse Decraene and Akeroyd (1988) concluded that **Polygonum** should be split into several genera based on their study of floral features in 83 species of **Polygonum**; under their scheme, our species would be members of **Polygonum**, **Persicaria**, and **Fallopia**.

11a. Scrambling plants, clinging by sharp recurved prickles on the four-angled stems; leaves sagittate or hastate; wet areas; (1–2 m high) (***Polygonum*** sect. ***Echinocaulon***, (or a section within the genus ***Persicaria***),**Tear-thumb**)—12

11b. Prostrate to erect herbs or twining or trailing vines, recurved prickles absent; leaves various—13

12a. Leaves sagittate, the basal lobes parallel—***Polygonum sagittatum***

12b. Leaves hastate, the basal lobes pointing outward; SLP—***Polygonum arifolium***

13a. Flowers inconspicuous, in small axillary clusters; leaves jointed at the base (***Polygonum*** sect. ***Polygonum***, **Knotweed**)—14

13b. Flowers more or less conspicuous, in obvious spikes or racemes which terminate the stems or branches, or arise from the axils of the upper leaves; leaves not jointed at the base—18

14a. Apex of pedicel recurved; flowers and fruit reflexed; sands, rocks, WM & UP (20–60 cm high)—***Polygonum douglasii***

14b. Apex of pedicel straight; flowers and fruit erect or ascending—15

15a. Leaves sharply folded lengthwise; sandy areas, SLP (10–40 cm high)—***Polygonum tenue***

15b. Leaves flat or nearly so—16

16a. Perianth narrowed above the achene; leaves all similar in size (homophyllous), broadly oblong, oval, or elliptical; disturbed areas (to 50 cm high)—***Polygonum achoreum***

16b. Perianth not narrowed; leaves homophyllous or heterophyllous, mostly narrower—17

17a. Outer three tepals exceed the inner two, especially in fruit; plants erect; sandy shores, mostly NM (30–100 cm high)—***Polygonum ramosissimum***

17b. All tepals approximately the same size; plants prostrate or ascending; roadsides, weedy areas, etc.—**Polygonum aviculare** [*P. aviculare* and *P. arenastrum*—CQ]

> Certain species of **Polygonum** display the characteristic of having leaves of two or more different sizes (*heterophyllous*), in contrast to having all leaves of a similar size (*homophyllous*). This key follows Voss (1985) in referring to **Polygonum aviculare** as a species which may exhibit either of these characteristics. Some authors, including Gleason and Cronquist, 1991, prefer to segregate the homophyllous plants commonly found as a prostrate weed of dooryards and gardens as **Polygonum arenastrum** (Dooryard Knotweed). The name **Polygonum aviculare** is then used to refer only to the sprawling to somewhat erect, heterophyllous plants in this group.

18a. Flowers solitary at the nodes of long, slender spike-like racemes; styles 2, persistent; tepals four; woods, SLP (50–100 cm high)—**Jumpseed** or **Virginia Knotweed**, *Polygonum virginianum*

18b. Flowers clustered at the nodes of shorter spikes or racemes; styles one or three, deciduous; tepals mostly five, or sometimes six—19

19a. Outer tepals winged or keeled in fruit; plant a stout perennial with broad leaves or a twining or trailing vine—20

19b. Outer tepals neither winged not keeled in fruit; habit various (*Polygonum* sect. *Persicaria* or a section within the genus *Persicaria*, **Smartweed**)—24

20a. Stout, erect perennials to 3 m high; escapes from cultivation (*Polygonum* sect. *Pleuropterus* or a section within the genus *Fallopia*)—21

20b. Twining or trailing vines; often in disturbed ground (*Polygonum* sect. *Tiniaria* or a section within the genus *Fallopia*)—22

21a. Base of leaf blade truncate—**Japanese Knotweed** or **Mexican Bamboo**, *Polygonum cuspidatum*

21b. Base of leaf blade cordate—**Giant Knotweed**, *Polygonum sachalinense*

22a. Stem nodes with a ring of bristles at the base; mostly WM & NM—**Fringed False Buckwheat**, *Polygonum cilinode*

22b. Stem nodes glabrous—23

23a. Outer tepals rough and keeled, not winged—**Black-bindweed**, *Polygonum convolvulus*

23b. Outer tepals smooth and strongly winged; LP—**False Buckwheat**, *Polygonum scandens*

24a. Flowers in one or two terminal, thick and many-flowered racemes; flowers pink; often in water, the leaves floating or emergent (to 1 m high)—**Water Smartweed**, *Polygonum amphibium*

> ***Polygonum amphibium*** is extremely variable. At the two extremes of this variation, varieties can be distinguished if desired. Aquatic plants with prostrate stems, floating glabrous leaves, and an ovoid to conic-oblong inflorescence can be referred to ***Polygonum amphibium*** var. ***stipulaceum.*** Terrestrial plants with erect stems, pubescent leaves, and a cylindrical inflorescence can be called ***Polygonum amphibium*** var. ***emersum.*** Intermediates between these two extremes will also be found.

24b. Flowers in few to many axillary and terminal racemes; flowers pink, white, greenish; leaves various but never floating—25

25a. Upper margin of the ocrea at most ciliate, not fringed with bristles; moist areas (to 2 m high)—26
25b. Upper margin of the ocrea fringed with bristles—27

26a. Racemes drooping or nodding at the tip; outer tepals strongly nerved—**Nodding Smartweed**, *Polygonum lapathifolium*
26b. Racemes erect; outer tepals inconspicuously nerved—**Bigseed Smartweed**, *Polygonum pensylvanicum*

27a. Upper margin of ocrea flared outward; racemes nodding; flowers crimson to rose; escaping to weedy areas, SLP (to 2.5 m high)—**Prince's Feather**, *Polygonum orientale*
27b. Upper margin of ocrea not flared outward, appressed to the stem; racemes erect, arching, or nodding; flowers pink to white—28

28a. Tepals covered with tiny with yellowish glandular dots; wet areas—29
28b. Tepals not yellow-dotted—30

29a. Racemes arching, drooping, or nodding at the tip; achene dull-colored (to 0.6 m high)—**Water-pepper**, *Polygonum hydropiper*
29b. Racemes erect or arching; achene smooth and shining (to 1 m high)—*Polygonum punctatum*

30a. Hairs on the peduncles gland-tipped; moist areas, SW (to 1.2 m high)—*Polygonum careyi*
30b. Hairs on the peduncles not gland-tipped or absent—31

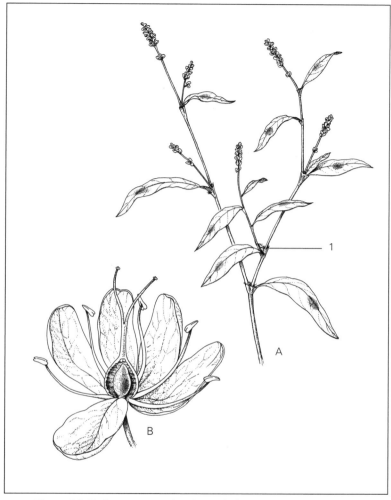

Figure 16: **Polygonum persicaria**: *A, habit with spikes; 1, ocrea;*
B, flower (ovary dissected)

31a. Inflorescence dense, shortly cylindric, 3 cm or less long and more than 5
   mm broad; achenes usually two-sided; dry or moist disturbed areas (to
   0.8 m high)—**Heart's-ease** or **Lady's-thumb**, *Polygonum persicaria*
31b. Inflorescence less dense and elongated, over 3 cm long and less than 5
   mm broad; achenes three-sided; wet areas (to 1 m high)—**Mild Water-**
   pepper, *Polygonum hydropiperoides*

CHENOPODIACEAE, The Goosefoot Family

Herbs with alternate (rarely opposite), simple leaves. Flowers mostly regular, perfect (seldom unisexual, the plants then monoecious), small, mostly in axillary or terminal herbaceous-bracted spikes or panicles; sepals (rarely 0) 1–5, often united below; petals 0; stamens 1–5; pistil 1, styles 2 or 3(–5), ovary superior, 1-celled. Fruit a small nut or utricle, often hidden by the calyx or subtending bracts. The foliage of several species becomes reddened in the autumn. Summer with fruit often maturing in the autumn.

1a. Leaves opposite, reduced to tiny, papery scales; plant succulent; highway medians, SE (10–50 cm high)—**Glasswort**, *Salicornia europaea*
1b. Leaves alternate (esp. above), not reduced to tiny scales; plant succulent or not—2

2a. Leaves linear or nearly so, entire—3
2b. Leaves lanceolate to deltoid, usually toothed or lobed—8

3a. Leaves rather stiff, narrowly linear or thread-like, with spine-like tips; weed in sand or cinders (20–100 cm high)—**Russian-thistle**, *Salsola kali* [*S. tragus*—CQ]
3b. Leaves soft, not spine-tipped—4

4a. Slender spines present in the inflorescence among the flowers; lawns, Alpena (to 30 cm high)—*Chenopodium aristatum*

> The only populations of **Chenopodium aristatum** in North America are in the Alpena area. This species is native to central and eastern Asia, and was probably introduced from Asia via cargo arriving at the port of Alpena. Although the date of introduction is unknown, it was reported to be a common lawn and garden weed in Alpena when the first specimen was collected in 1981 (Gereau & Rabeler, 1984).

4b. Spines not present among the flowers—5

5a. Leaves with a tiny non-green mucronate tip—6
5b. Leaves entirely green—7

6a. Leaves flat; plant widely branched, rather diffuse; sandy areas, esp. along Lake Michigan shore (10–60 cm high)—**Bugseed**, *Corispermum hyssopifolium*
6b. Leaves flat above, rounded below; plant erect to decumbent; highway medians, SE (10–80 cm high)—**Sea-blite**, *Suaeda calceoliformis*

Suaeda calceoliformis and *Salicornea europaea* are two members of a suite of species nick-named "highway halophytes". These plants can tolerate very high amounts of salt in the soil and often grow in areas along or near highways which are heavily salted during the winter. For additional information on these plants see Reznicek (1980).

7a. Upper portion of stem glabrous; leaves glabrous (30–80 cm high); road-sides, railroads, etc.—*Chenopodium subglabrum* [*C. pratericola* and *C. leptophyllum*—CQ]

7b. Upper portion of stem villous; leaves pubescent, margin of blade ciliate; railroads, highway edges, mostly LP (50–100 cm high)—**Summer-cypress**, *Kochia scoparia*

8a. Principal leaf blades with a broad truncate, rounded, or hastate base; disturbed areas—9

8b. Principal leaf blades narrowed to the base—12

9a. Leaf blades broadly ovate, with one to three large sharp projecting teeth on each side (to 1.5 m high)—**Maple-leaved Goosefoot**, *Chenopodium hybridum* [*C. gigantospermum*—CQ]

9b. Leaf blades hastate or triangular-ovate, entire or with many teeth—10

10a. Flowers unisexual, the pistillate flowers (and thus the fruit) enclosed by two small, deltoid bracts; leaf blade hastate, the margin entire or somewhat undulate (to 1 m high)—**Spearscale**, *Atriplex patula*

10b. Flowers perfect, not enclosed by deltoid bracts; leaf blade margin sharply or sinuately toothed (upper leaves may be entire)—11

11a. Flowers in small heads, in the axils or in terminal spikes; the heads becoming red and fleshy at maturity; leaf blades hastate (20–60 cm high)—**Strawberry Blite**, *Chenopodium capitatum*

11b. Flowers in terminal panicles, not becoming red nor fleshy at maturity; leaves triangular-ovate; SLP (10–80 cm high)—**Alkali Blite**, *Chenopodium rubrum*

12a. Foliage covered with glandular hairs and strongly aromatic; roadsides, gravel pits, railroads (20–60 cm high)—**Jerusalem-oak**, *Chenopodium botrys*

12b. Foliage not glandular nor aromatic; sometimes ill-scented—13

13a. Leaves green beneath, soon deciduous; perianth forming a flat wing around the fruit; stem widely and diffusely branched, often becoming a tumbleweed; sandy areas (10–80 cm high)—**Winged Pigweed**, *Cycloloma atriplicifolium*

Chenopodium ambrosioides, Mexican-tea or Wormseed, is a species similar to *C. botrys* and was once grown to produce a vermifuge medicine. A catalog of medicines from Parke, Davis and Company of Detroit (1912) lists an extract of "Chenopodium antihelminticum" (a synonym of *C. ambrosioides*) for $2.00 a pint. Many of the specimens collected in Michigan were from counties surrounding Detroit, possibly escapes from cultivation for this medicinal purpose. The species has rarely been collected in Michigan in the last half century (Voss, 1985).

13b. Leaves white-mealy beneath; perianth not forming a wing around the fruit; stem various—14

14a. Stem erect, not succulent; perianth white-mealy; weedy and disturbed areas (to 1 m high)—**Lambs-quarters**, *Chenopodium album*
14b. Stem prostrate or ascending, succulent; perianth not white-mealy (10–40 cm high); gravelly, often saline areas—**Oak-leaved Goosefoot**, *Chenopodium glaucum*

AMARANTHACEAE, The Amaranth Family

Herbs, with alternate (rarely opposite), simple leaves. Flowers regular, perfect or unisexual (the plants then monoecious or dioecious), inconspicuous, often in axillary or terminal spike-like, scarious-bracted clusters; sepals (rarely 0)1–5, sometimes united, often scarious; petals 0; stamens 1–5, the filaments rarely united and resembling petals; pistil 1, style 1, ovary superior, 1-celled. Fruit an indehiscent utricle, often hidden by the calyx or subtending bracts. Summer.

1a. Leaves opposite, plant white-woolly; sepals united; railroads, SW (20–70 cm high)—**Cottonweed**, *Froelichia gracilis*
1b. Leaves alternate, plant not white-woolly; sepals free (*Amaranthus* spp., **Pigweed** or **Amaranth**)—2

2a. Flower-clusters axillary—3
2b. Flower-clusters in terminal spikes or panicles, sometimes also axillary—4

3a. Plant prostrate or decumbent; sepals five or rarely four; seed about 1.5 mm broad; roadsides, railroads, etc.—*Amaranthus blitoides*
3b. Plant erect or ascending; sepals three; seeds about 1 mm broad (0.3–1.0 m high)—**Tumbleweed**, *Amaranthus albus*

4a. Plants dioecious; pistillate flowers without calyx; swamps or streambanks; SLP (to 2 m high)—**Water-hemp**, *Amaranthus tuberculatus*
4b. Plants monoecious; pistillate flowers with a calyx; weedy plants or escapes of cultivated or disturbed ground—5

5a. Inflorescence stiff, erect—6
5b. Inflorescence lax, nodding—8

6a. Apex of sepals of pistillate flowers rounded to truncate; stems densely
pubescent (to 2 m high)—**Pigweed**, *Amaranthus retroflexus*
6b. Apex of sepals of pistillate flowers acute; stems not densely
pubescent—7

7a. Inflorescence dull greenish; seed obovate in shape, about 1 mm broad;
weedy places (to 2 m high)—*Amaranthus powellii*
7b. Inflorescence bright red, yellow, or green; seed circular in shape, about
1.25 mm broad; an often showy escape from cultivation; SLP—**Prince's
Feather**, *Amaranthus hypochondriacus*

8a. Inflorescence greenish; bracts subulate, sharply awned; apex of sepals of
pistillate flowers acute; shores, pastures, weedy areas, SLP (to 2 m
high)—**Green Amaranth**, *Amaranthus hybridus*
8b. Inflorescence reddish; bracts merely acuminate; apex of sepals of pistillate flowers obtuse or rounded; escape from cultivation—**Purple** or **Red
Amaranth**, *Amaranthus cruentus*

NYCTAGINACEAE, The Four-o'clock Family

Herbs with opposite, simple, entire leaves. Flowers regular, perfect, in
clusters of (1)3–5 flowers surrounded by a calyx-like involucre of
bracts; sepals 5, united, purplish, resembling a corolla; petals 0; stamens 3–5; pistil 1, style 1, ovary superior, 1-celled. Fruit an indehiscent utricle enclosed in the calyx. Summer.

1a. Leaves lanceolate or narrower, sessile (to 1 m high)—**Umbrella-wort**,
Mirabilis hirsuta
1b. Leaves ovate to deltoid, petioled (to 1.5 m high)—**Wild Four-o'clock**,
Mirabilis nyctaginea

PHYTOLACCACEAE, The Pokeweed Family

Perennial herbs with alternate, simple, entire leaves. Flowers regular,
perfect, small, in racemes; sepals 5, whitish; petals 0; stamens 10; pistil 1, styles 10, ovary superior, 10-celled. Fruit a dark-purple berry.
Late summer.

One species in Michigan (1–3 m high)—**Pokeweed**, *Phytolacca americana*

MOLLUGINACEAE, The Carpet-weed Family

Annual prostrate herbs with whorled, simple, entire, oblanceolate leaves. Flowers regular, perfect, small, axillary; sepals 5, green or whitish; petals 0; stamens 3 or 4; pistil 1, styles 3–5, ovary superior, 3–5-celled. Fruit a capsule containing many red seeds. Summer.

One species in Michigan, sometimes forming extensive mats, especially in sandy areas—**Carpetweed**, *Mollugo verticillata*

PORTULACACEAE, The Purslane Family

Herbs with opposite or alternate, simple, entire leaves. Flowers regular, perfect; sepals 2; petals 4–6, usually 5; stamens 5–10, sometimes many; pistil 1, styles 3–9, ovary superior or half-inferior, 1–celled. Fruit a capsule.

1a. Leaves a single pair on each stem, not succulent; flowers pink (10–30 cm high; early spring) (*Claytonia spp.*, **Spring Beauty**)—2
1b. Leaves numerous, succulent; flowers yellow or white to red (summer) (*Portulaca* spp., **Purslane**)—3

2a. Leaves lance-ovate to oblong, not more than eight times as long as wide; petiole is distinct; mostly NM—*Claytonia caroliniana*
2b. Leaves linear or linear-lanceolate, more than eight times as long as wide; petiole is not distinct; mostly SLP—*Claytonia virginica*

3a. Flowers yellow, up to 1 cm wide; leaves flat; mat-forming weed— **Purslane**, *Portulaca oleracea*
3b. Flowers white to red or yellow, 2–5 cm wide, the petals often doubled; leaves round in cross section; occasional escape from cultivation, LP (20–40 cm high)—**Moss-rose**, *Portulaca grandiflora*

CARYOPHYLLACEAE, The Pink Family

Herbs, with opposite or whorled, simple, entire leaves, and stems frequently swollen at the nodes. Flowers regular, perfect or seldom unisexual (the plants then dioecious); sepals (4)5, separate or united; petals (0–)5, often 2-cleft; stamens often 10 in conspicuous flowers, 1–5(–10) in small flowers; pistil 1, styles 2–5(6), ovary superior, 1-celled or rarely 3–5-celled below. Fruit a many-seeded capsule opening by as many or twice as many valves or teeth as styles or an indehiscent, one-seeded utricle.

Of the 60 species of Caryophyllaceae now known in Michigan (Voss, 1985; Rabeler, 1988), only one-quarter of them (15 species) are native to North America. The remaining 45 species have been introduced from Europe or Asia, either accidentally or intentionally. Of the 20 genera, 14 consist solely of introductions.

1a. Stipules present, mostly scarious—2
1b. Stipules absent—7

2a. Leaves elliptic to oblanceolate; petals absent; fruit a 1-seeded indehiscent utricle; sandy areas, SLP (summer)—3
2b. Leaves linear; petals present; fruit a several-seeded dehiscent capsule (10–40 cm high, summer)—4

3a. Plant erect, leaves red-dotted beneath; stipules glabrous (10–40 cm high)—**Whitlow-wort**, *Paronychia canadensis*
3b. Plant forming a prostrate mat, leaves not red-dotted beneath; stipules ciliate—*Herniaria glabra*

4a. Leaves whorled; petals white; styles five; sandy fields, roadsides, etc.—**Spurrey**, *Spergula arvensis*
4b. Leaves opposite; petals pink and/or white; styles three—(*Spergularia* spp., **Sand-spurrey**)—5

5a. Stamens one to five; stipules broadly triangular; saline areas, SLP—*Spergularia marina*
5b. Stamens seven to ten; stipules lanceolate—6

6a. Leaves succulent; saline areas, SE—*Spergularia media*
6b. Leaves not succulent; sandy areas, mostly UP—*Spergularia rubra*

7a. Sepals separate or only briefly basally connate; petals white (sometimes absent)—8
7b. Sepals united for at least one-fourth to one-half their length, the calyx often tubular; petals white to pink or red (rarely absent)—26

8a. Leaves subulate or thread-like; styles as many as valves of the capsule (summer)—9
8b. Leaves spathulate to ovate; styles one-half as many as valves (teeth) of the capsule—10

9a. Leaves opposite; petals shorter than the sepals; styles four, seldom five; lawns, pavement cracks; rocks (2–10 cm high)—**Procumbent Pearlwort**, *Sagina procumbens*

9b. Leaves fascicled in the axils; petals exceed the sepals; styles three; sandy woods and dunes (10–40 cm high)—**Rock Sandwort**, *Arenaria stricta* [*Minuartia michauxii*—Kartesz (1994)]

> The genera **Arenaria** and **Stellaria** are examples of a classical dispute among taxonomists. Some, the "lumpers", recognize the broad, inclusive definition of a particular genus, while others, the "splitters", separate species into other closely related genera. We have indicated both treatments for these two genera. Note how **Stellaria** is divided (or not) into **Stellaria** and **Myosoton**, while some species of **Arenaria** are segregated (or not) into **Moehringia** and **Minuartia**.

10a. Petal apex entire or toothed—11
10b. Petal apex notched or two-cleft, or petals absent—13

11a. Inflorescence an umbel; petal apex toothed; lawns and roadsides, SLP (10-30 cm high; early spring)—**Jagged Chickweed**, *Holosteum umbellatum*
11b. Inflorescence a cyme; petal apex entire (spring through fall)—12

12a. Petals shorter than the sepals; stem pubescence not retrorse; sandy roadsides, lawns, etc. (5–30 cm high; spring and summer)—**Thyme-leaved Sandwort**, *Arenaria serpyllifolia*
12b. Petals exceed the sepals; stem retrorsely pubescent; damp woods, SLP (spring and early summer)—*Arenaria lateriflora* [*Moehringia lateriflora*—Kartesz (1994)]

13a. Capsule ovoid, splitting by valves at maturity; styles often three, rarely four to six (*Stellaria* [incl. *Myosoton*] spp., **Stitchwort** or **Chickweed**)—14
13b. Capsule cylindrical, often curved, opening by terminal teeth at maturity; styles five (*Cerastium* spp., **Mouse-ear Chickweed**)—20

14a. Petals shorter than the sepals, or none—15
14b. Petals as long as the sepals, or longer—16

15a. Leaves ovate, the lower ones petiolate; ubiquitous weed (to 40 cm high; spring through fall)—**Common Chickweed**, *Stellaria media*

> Some early spring collections which key to **Stellaria media**, and which have yellowish green leaves and small, closed flowers may be **Stellaria pallida**, an early spring annual recently found at several sites in the **SLP** (Rabeler, 1988).

15b. Leaves lanceolate to oblong, petioles not distinct; damp forests, NM (to 50 cm high; summer)—**Northern Stitchwort**, *Stellaria calycantha* [*S. borealis*—CQ]

16a. Leaves ovate; styles five; damp shores and meadows (spring through fall)—**Giant Chickweed**, *Myosoton aquaticum* [*Stellaria aquatica*— CQ]

16b. Leaves linear to lanceolate or narrowly elliptic; styles three—17

17a. Inflorescence bracts resembling the leaves, or the flowers axillary and solitary; damp shores, SW and eastern UP (10–20 cm high; summer)— **Fleshy Stitchwort**, *Stellaria crassifolia*

17b. Inflorescence bracts scarious, at most with a green midrib—18

18a. Pedicels and leaves erect; inflorescence few-flowered; sandy shores, eastern UP (15–30 cm high; summer)—**Long-stalked Stitchwort**, *Stellaria longipes*

18b. Pedicels and leaves spreading; inflorescence open, many-flowered—19

19a. Leaves distinctly linear; cymes lateral; sepals and bracts usually glabrous; damp, open areas (15–45 cm high; early summer)—**Long-leaved Stitchwort**, *Stellaria longifolia*

19b. Leaves distinctly broadest near the base; cymes terminal; sepals and bracts ciliate; lawns, roadsides, etc. (30–50 cm high; summer)—**Common Stitchwort**, *Stellaria graminea*

20a. Petals showy, at least twice as long as the sepals—21
20b. Petals shorter than or just exceeding the sepals—22

21a. Stem and leaves white-tomentose; escape from cultivation (spring and early summer)—**Snow-in-summer**, *Cerastium tomentosum*

21b. Stem and leaves green; rocks (UP), grassy areas (SLP) (15–40 cm high; spring and summer)—**Field Chickweed**, *Cerastium arvense*

22a. Uppermost inflorescence bracts with transparent white margins—23
22b. All inflorescence bracts entirely green—25

23a. Flowers 8–10 mm wide; petals deeply notched; stamens ten; capsules curved, mostly over 8 mm long; lawns, gardens, roadsides, etc. (15–50 cm high; spring and summer)—**Common Mouse-ear Chickweed**, *Cerastium fontanum* [*C. vulgatum*—CQ]

23b. Flowers 5–6 mm wide; petals shallowly notched; stamens five (rarely ten); capsules mostly straight and less than 8 mm long; sandy or gravelly roadsides, railroads, etc. (5–25 cm high; spring)—24

24a. Inflorescence bracts with wide scarious margins (upper half of bract may be scarious); mostly LP—**Small Mouse-ear Chickweed**, *Cerastium semidecandrum*

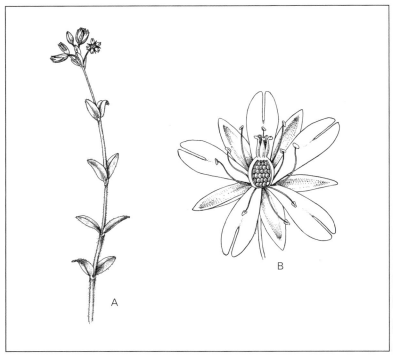

Figure 17: **Cerastium fontanum:** *A, habit; B, flower (ovary dissected)*

24b. Inflorescence bracts with narrow scarious margins, lower may be almost entirely herbaceous; SLP—**Curtis' Mouse-ear Chickweed,**
Cerastium pumilum (Rabeler, 1988)

25a. Pedicels bent near the flower, much longer than the calyx or capsules;
fields and forest edges; SLP and UP (10-45 cm high; spring and early
summer)—**Nodding Chickweed,** *Cerastium nutans*

25b. Pedicels straight, shorter than or equalling the calyx or capsules; lawns,
roadsides, etc., SLP (5-15 cm high; spring)—**Clammy Chickweed,**
Cerastium glomeratum [*Cerastium viscosum*—CQ]

26a. Flowers inconspicuous, axillary, apetalous; leaves linear, subulate; fruit
a utricle; sandy areas (to 15 cm high; summer)—**Knawel,** *Scleranthus
annuus*

26b. Flowers conspicuous, rarely apetalous; leaves various; fruit a capsule—
27

27a. Calyx teeth longer than the calyx tube and the pink petals; fields and roadsides, formerly common in wheat fields (40–100 cm high; late summer)—**Corn-cockle, *Agrostemma githago***

27b. Calyx teeth shorter than the calyx-tube; petals variously colored, not exceeded by calyx teeth—28

28a. Bracts immediately subtending the calyx and/or enclosing much of the inflorescence present; open sandy areas—29

28b. Bracts immediately subtending the calyx absent—35

29a. Area between adjacent sepals green; each sepal has four or more veins (late spring and summer) (***Dianthus* spp., Pink**)—30

29b. Area between adjacent sepals white, membranous; each sepal has one to three veins (summer and early autumn) (***Petrorhagia* spp.**)—34

30a. Flowers sub-sessile, mostly in crowded cymes or heads—31

30b. Flowers solitary and long-stalked—33

31a. Calyx pubescent; flowers few in open, terminal cymes (20–60 cm high)—**Deptford Pink, *Dianthus armeria***

31b. Calyx glabrous; flowers numerous in one or more dense heads—32

32a. Leaves lanceolate to ovate; petals white to dark red (30–60 cm high)—**Sweet William, *Dianthus barbatus***

32b. Leaves linear; petals red; NM (to 95 cm high)—**Cluster-head Pink, *Dianthus carthusianorum***

33a. Petal apex fringed or cleft; subtending bracts one-third or less the length of the calyx tube; mostly NM (10–30 cm high)—**Garden or Grass Pink, *Dianthus plumarius***

33b. Petal apex toothed; subtending bracts about one-half the length of the calyx tube (10–40 cm high)—**Maiden Pink, *Dianthus deltoides***

34a. Flowers in lax cymes or solitary; subtending bracts green, about one-half the length of the calyx; esp. near Lake Michigan (5–40 cm high)—**Saxifrage Pink, *Petrorhagia saxifraga***

34b. Flowers in a head; subtending bracts brown, broadly enclosing nearly the entire calyx of most flowers; near Grand Haven (10–60 cm high)—**Childing Pink, *Petrorhagia prolifera***

35a. Styles two (very rarely three); flowers perfect (summer)—36

35b. Styles three to five or flowers imperfect, staminate—41

36a. Area between adjacent sepals white, membranous (*Gypsophila* spp.,
    Baby's-breath)—37
36b. Area between adjacent sepals green—39

37a. Calyx and pedicels glandular-pubescent; petals pinkish; sandy or cal-
    cium-rich areas, NM (50–100 cm high)—*Gypsophila scorzonerifolia*
37b. Calyx and pedicels glabrous; petals white—38

38a. Calyx less than 2.5 mm long, flowers in large panicles; sandy areas
    (40–100 cm high)—*Gypsophila paniculata*
38b. Calyx 3–5 mm, flowers in cymes; roadsides, NM (20–50 cm high)—
    Gypsophila elegans

> At least three members of the Pink Family (*Gypsophila elegans*, **G. muralis**, **Silene armeria**)
> is sometimes included in "wildflower-in-a-can" seed mixtures, although none are native to North
> America. **Gypsophila elegans** (Baby's-breath) was collected in 1993 from an area reseeded
> after highway construction near Alpena. It is likely that these plants will increase their distribu-
> tion in Michigan.

39a. Calyx ovoid, the tube five-angled or winged; railroads, weedy areas,
    mostly LP (20–60 cm high)—**Cow Herb**, *Vaccaria hispanica*
39b. Calyx tube cylindrical, not angled or winged (*Saponaria* spp., **Soap-
    wort**)—40

40a. Calyx over 15 mm long, glabrous; common along roadsides and rail-
    roads (40–80 cm high)—**Bouncing Bet**, *Saponaria officinalis*
40b. Calyx less than 10 mm long, glandular-pubescent; escape from cultiva-
    tion, Pellston (to 25 cm high)—*Saponaria ocymoides*

41a. Styles five or flowers imperfect, staminate—42
41b. Styles three (most *Silene* spp., **Campion** or **Catchfly**)—44

42a. Flowers unisexual, the petals white or light pink; disturbed areas
    (40–120 cm high; late spring and summer)—**White Campion**, *Silene
    pratensis* [*S. latifolia*—CQ]
42b. Flowers perfect, petals crimson or scarlet; escapes from cultivation to
    fields and woods (summer) (*Lychnis spp.*)—43

43a. Plant white-woolly; petals crimson, the apex not deeply notched;
    mostly LP (40-80 cm high)—**Mullein Pink** or **Rose Campion**, *Lychnis
    coronaria*
43b. Plant green, pubescence less dense; petals scarlet, the apex notched; UP
    (30-60 cm high)—**Maltese-cross**, *Lychnis chalcedonica*

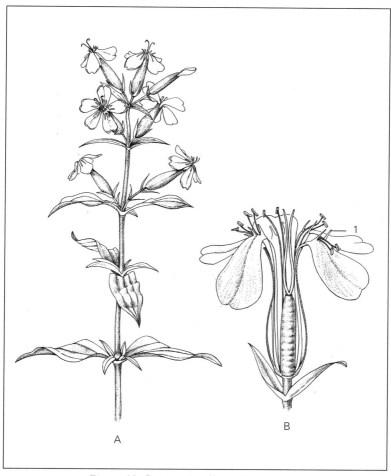

*Figure 18: **Saponaria officinalis**: A, habit;
B, flower (ovary dissected); 1, corona*

44a. Flowers opening in early evening, always wilted during the day; petals
    pinkish above, yellowish beneath; disturbed areas (20–80 cm high;
    summer)—**Night-flowering Catchfly**, *Silene noctiflora*

44b. Flowers open during the day; petals white, pink, or red—45

45a. Calyx pubescent—46
45b. Calyx glabrous—49

46a. Petals crimson, deeply cleft; woods, SLP (20–80 cm high; late spring and summer)—**Fire Pink**, *Silene virginica*

46b. Petals white or pink, not bright red—47

47a. Principal leaves in whorls of four; woods, SW (30–120 cm high; summer)—**Starry Campion**, *Silene stellata*

47b. Principal leaves opposite—48

48a. Inflorescence one-sided (flowers all on the same side of the stem); calyx with ten raised veins; roadsides, fields, etc., mostly LP (30–80 cm high; summer)—*Silene dichotoma*

48b. Inflorescence not one-sided; calyx with thirty raised veins; sandy aeas, SW (20–50 cm high; late spring and summer)—*Silene conica*

49a. Upper internodes of the stem have sticky glandular areas; petals pink (or sometimes absent); calyx with straight veins—50

49b. Upper internodes of the stem lack sticky areas; petals white; calyx with netted veins (summer)—51

50a. Calyx 12 mm long or more, club-shaped; petals pink; escapes from cultivation to fields and roadsides (10–70 cm high; early summer)—**Sweet-William Catchfly**, *Silene armeria*

50b. Calyx 10 mm long or less, tubular; petals white or pink (or absent); dry fields and woods (20–80 cm high; summer)—**Sleepy Catchfly**, *Silene antirrhina*

51a. Calyx much inflated or bladder-like, not closely adhering to the capsule; sandy or calcium-rich areas, esp. NM (20–80 cm high)—**Bladder Campion**, *Silene vulgaris*

51b. Calyx closely adheres to the capsule; railroads, weedy areas (30–80 cm high)—*Silene csereii* [*S. cserei*—CQ]

CERATOPHYLLACEAE, The Hornwort Family

Monoecious submerged aquatic herbs, with whorled, dichotomously forked, finely dissected leaves. Flowers unisexual, inconspicuous; sepals 8–15, minute; petals 0; stamens 10–20; pistil 1, style 1, ovary superior, 1-celled. Fruit an achene.

1a. Margins of leaf segments finely toothed; leaves stiff, not forked more than twice—**Coontail**, *Ceratophyllum demersum*

1b. Margins of leaf segments entire or virtually so; leaves limp, some may be forked three times—**Coontail**, *Ceratophyllum echinatum*

NYMPHAEACEAE, The Water-lily Family

Aquatic herbs, usually with large, alternate, floating leaves which are round or elliptical and palmately veined. Flowers regular, perfect, solitary, on long peduncles; sepals 3–6; petals 3–many; stamens 3–many; pistils 1–40, each with a large discoid stigma, styles 0, each ovary superior, 1 or 6–30-celled. Fruit a berry, achene, or individual nuts embedded in a receptacle.

1a. Leaves peltate (petiole attached to center of lower surface of the blade)—2
1b. Leaves not peltate, with a deep basal sinus—3

2a. Leaf blades round, 30 cm in diameter or more; flowers pale yellow, over 10 cm wide; SLP—**American Lotus**, *Nelumbo lutea*

The American Lotus, **Nelumbo lutea**, is protected by Public Act 182 of 1962, commonly called the "Christmas-tree law". Under this law, it is illegal to remove or cut these plants from any area without a bill of sale or written permission from the owner.

2b. Leaf blades elliptic, 5–15 cm long; flowers purple, up to 3 cm wide— **Water-shield**, *Brasenia schreberi*

3a. Flowers white or tinged with pink; leaf blades elliptic—**Water-lily**, *Nymphaea odorata*
3b. Flowers yellow; leaf blades round *(Nuphar* spp., **Pond-lily**)—4

4a. Petioles round or oval in cross section; leaf blades often above the water; SLP—*Nuphar advena*
4b. Petioles flattened in cross section; leaf blades usually floating—*Nuphar variegata*

RANUNCULACEAE, The Buttercup Family

Herbs or, rarely somewhat woody vines with alternate (rarely opposite or whorled) and/or basal, simple or compound leaves and acrid, watery juice. Flowers mostly regular, perfect (or rarely unisexual, the plants then usually dioecious); sepals 3–20, often 4 or 5 and petaloid; petals 4 or 5(–10) or 0; stamens many; pistils and styles several to many (rarely 1), each ovary superior, 1-celled. Fruit an achene (often tipped with a beak, the persistent style), follicle, berry, or utricle.

1a. Climbing vines, sometimes woody; leaves opposite, mostly trifoliolate; woods and stream banks (late summer) (*Clematis* spp., **Virgin's Bower**)—2

1b. Plant a terrestrial or aquatic herb; leaves alternate or basal, rarely opposite or whorled—3

2a. Sepals white, 2–3 cm wide (2–3 m high)—*Clematis virginiana*

2b. Sepals pink-purple, 5–8 cm wide; UP (to 2 m high)—*Clematis occidentalis*

3a. Aquatic plant with submerged, mostly dissected leaves (late spring and summer)—4

3b. Terrestrial plants, leaves not submerged—7

4a. Petals white—**White Water Crowfoot**, *Ranunculus longirostris*

4b. Petals yellow—5

5a. Leaf blades linear to narrowly oblong; shores, sometimes submerged, mostly NM (summer)—**Creeping Spearwort**, *Ranunculus reptans* [*R. flammula*—CQ]

5b. Leaf blades finely dissected (**Yellow Water Crowfoot**)—6

6a. Petals over 6.5 mm long; submerged leaves divided into hair-like segments—*Ranunculus flabellaris*

6b. Petals less than 6.5 mm long; submerged leaves palmately divided into linear lobes—*Ranunculus gmelinii*

7a. One or more perianth segments (sepals and/or petals) prolonged into a spur—8

7b. Perianth segments without spurs—10

8a. Flowers blue, irregular, the two petals forming one spur; style and follicle one; disturbed areas, LP (30–70 cm high, summer)—**Larkspur**, *Consolida ambigua* [*Delphinium ambiguum*—CQ]

8b. Flowers regular, the five petals each prolonged into a spur; styles and follicles five; (30–200 cm high) (*Aquilegia* spp., **Columbine**)—9

9a. Petal spurs nearly straight; flowers scarlet and yellow; woodlands and clearings (spring)—**Wild Columbine**, *Aquilegia canadensis*

9b. Petal spurs strongly incurved; flowers mostly blue, white, or pink; escape from cultivation (spring, early summer)—**Garden Columbine**, *Aquilegia vulgaris*

10a. Flowers yellow—11
10b. Flowers of various colors, but never yellow—24

11a. Sepals petaloid, yellow; petals absent; leaves crenate or dentate; wet areas (20–60 cm high, early spring)—**Marsh-marigold** or **Cowslip**, *Caltha palustris*
11b. Sepals green or yellowish; petals yellow (*Ranunculus* spp., **Buttercup** or **Crowfoot**)—12

12a. None of the leaves lobed or deeply cleft (base of leaf blades may be cordate)—13
12b. Cauline leaves lobed or deeply cleft; basal leaves may be unlobed—14

13a. Sepals five; leaf blades linear to narrowly oblong; stems prostrate and trailing, rooting at the nodes; shores, mostly NM (summer)—**Creeping Spearwort**, *Ranunculus reptans* [*R. flammula*—CQ]
13b. Sepals three or sometimes four; leaf blades cordate; stems erect, not rooting at the nodes; damp woods, SLP (10–30 cm high)—**Lesser-celandine**, *Ranunculus ficaria*

14a. Stem prostrate or creeping; leaves simple, the blades palmately lobed; wet places (**Yellow Water Crowfoot**)—15
14b. Stem erect or ascending, rarely prostrate; leaves simple or compound (compound if stems prostrate)—16

15a. Petals over 6.5 mm long—*Ranunculus flabellaris*
15b. Petals less than 6.5 mm long—*Ranunculus gmelinii*

16a. Petals shorter than or equalling the sepals, less than 5 mm long; flowers 1 cm broad or smaller—17
16b. Petals longer than the sepals, longer than 5 mm long; flowers 2 cm broad or larger—20

17a. Stem glabrous or with a few short hairs above—18
17b. Stem covered with spreading hairs—19

18a. Most basal leaves not lobed, cauline leaves divided; woods (20–50 cm high; spring)—**Small-flowered Buttercup**, *Ranunculus abortivus*
18b. Basal and cauline leaves similar, divided; wet areas (20–60 cm high; spring and summer)—**Cursed Crowfoot**, *Ranunculus sceleratus*

19a. Fruits tipped with a recurved beak; woods (20–70 cm high; late spring)—**Hooked Crowfoot**, *Ranunculus recurvatus*

19b. Fruits tipped with a straight or slightly curved beak; marshes or wet soil (30–70 cm high; summer)—**Bristly Crowfoot**, *Ranunculus pensylvanicus*

20a. Basal leaves ovate, not lobed; cauline leaves cleft; grasslands, SLP (10–20 cm high; spring)—**Prairie Buttercup**, *Ranunculus rhomboideus*
20b. Basal leaves lobed, cleft, or compound—21

21a. Leaves simple, the terminal lobe of leaf sessile (50–100 cm high; late spring and summer)—**Common** or **Tall Buttercup**, *Ranunculus acris*
21b. Leaves compound, the terminal lobe of the leaf clearly stalked—22

22a. Stems creeping; styles short, obviously curved; disturbed and wet areas, LP (late spring and early summer)—**Creeping Buttercup**, *Ranunculus repens*
22b. Stems erect or ascending; style long and slender, straight or nearly so (spring)—23

23a. The two lateral divisions of the leaf sessile or nearly so; leaflet teeth rounded or obtuse; dry areas (10–30 cm tall; spring)—**Early Buttercup**, *Ranunculus fascicularis*
23b. The two lateral divisions of the leaf on long stalks; leaflet teeth acute; wet areas (15–90 cm high)—**Swamp Buttercup**, *Ranunculus hispidus*

24a. Leaves lobed or divided, but the divisions not separated by definite stalks—25
24b. Leaves clearly compound, all their divisions separated by distinct stalks—33

25a. Leaves all basal—26
25b. Cauline leaves present—28

26a. Leaves trifoliolate, the margins toothed; sepals white; damp mossy areas (scape 5–15 cm high; early spring)—**Goldthread**, *Coptis trifolia*
26b. Leaves simple, their lobes (usually three) entire; sepals pink-purple; rich woods (scape 5–15 cm high; early spring) (*Hepatica* spp., **Hepatica**)—27

27a. Lobes of the leaf obtuse or rounded—**Round-lobed Hepatica**, *Hepatica triloba*
27b. Lobes of the leaf acute—**Sharp-lobed Hepatica**, *Hepatica acutiloba*

28a. Sepals three, usually falling away as soon as the flower opens; petals absent; filaments white, showy; fruit a red berry; rich woods, SLP (20–50 cm high; spring)—**Goldenseal**, *Hydrastis canadensis*

28b. Sepals petal-like, four or more; petals absent; fruit an achene
(*Anemone* spp., **Anemone** or **Windflower**)—29

29a. Cauline leaves sessile or nearly so—30
29b. Cauline leaves on definite petioles—31

30a. Sepals red (rarely cream); achenes densely woolly; Great Lakes shores,
NM (10–60 cm high; late spring)—**Red Anemone**, *Anemone multifida*
30b. Sepals white; achene pubescence thinner (20–80 cm high; late spring
and early summer)—**Canada Anemone**, *Anemone canadensis*

31a. Sepals white; achenes merely pubescent (not woolly), in a subglobose
head; woods (10–20 cm high; spring)—**Wood Anemone**, *Anemone
quinquefolia*
31b. Sepals greenish-white; achenes densely woolly, in an ovoid to cylindrical head (30–100 cm high; summer) (**Thimbleweed**)—32

32a. Head of fruit cylindrical; leaves at the base of the flowering peduncles
4–9—*Anemone cylindrica*
32b. Head of fruit ovoid or oblong; leaves at the base of the flowering peduncles 3—*Anemone virginiana*

33a. Flowers numerous, in panicles or racemes; perianth greenish, flower
color provided by the numerous stamens—34
33b. Flowers solitary or few, in loose clusters; perianth white to purplish;
rich woods, SLP (spring)—40

34a. Inflorescence a branched panicle; fruit an achene (*Thalictrum* spp.,
Meadow-rue)—35
34b. Inflorescence a dense raceme; fruit a follicle or berry—38

35a. Cauline leaves with obvious petioles, not completely expanded when
flowers appear; woods (30–70 cm high; spring)—**Early Meadow-rue**,
Thalictrum dioicum
35b. Cauline leaves sessile or nearly so, completely expanded when flowers
appear (late spring and summer)—36

36a. Leaflets three-lobed, each lobe with additional teeth; shores, UP
(30–100 cm high)—**Northern Meadow-rue**, *Thalictrum venulosum*
36b. Leaflets three-lobed, each lobe without additional teeth; wet areas
(80–200 cm high)—37

37a. Underside of the leaves covered with glands or glandular hairs; leaf
     margin somewhat rolled under; mostly UP and SE—***Thalictrum revo-
     lutum***
37b. Underside of the leaves covered with nonglandular hairs; leaf margin
     flat—**Purple Meadow-rue**, ***Thalictrum dasycarpum***

38a. Racemes slender, 20–80 cm long; fruit a follicle; rare, SLP, but now
     often cultivated (1–2.5 m high; summer)—**Black Snakeroot**, ***Cimi-
     cifuga racemosa***
38b. Racemes short and stout, 3–8 cm long; fruit a berry; rich woods (40–80
     cm high) (***Actaea spp.***, **Baneberry**)—39

39a. Berries white, dark stigma prominent; pedicels significantly thickened
     at maturity (spring)—**White Baneberry** or **Doll's-eyes**, ***Actaea pachy-
     poda***
39b. Berries red (rarely white), dark stigma not prominent; mature pedicels
     slender (spring and early summer)—**Red Baneberry**, ***Actaea rubra***

40a. Sepals white; cauline leaves alternate, basal leaves absent (10–40 cm
     high)—**False Rue-anemone**, ***Isopyrum biternatum***
40b. Sepals white to pink or purplish; cauline leaves whorled, basal leaves
     present (10–30 cm high)—**Rue-anemone**, ***Anemonella thalictroides***

BERBERIDACEAE, The Barberry Family

Shrubs or perennial herbs with alternate, opposite, or basal, simple or
compound leaves. Flowers regular, perfect; sepals 4 or 6; petals 6, 8,
or 9; stamens 6, 8, 12, or 18, opening by two terminal or longitudinal
valves; pistil 1, style 0, ovary superior, 1-celled. Fruit a capsule or
berry. Spring.

1a. Plant woody; leaves simple or pinnately compound; flowers yellow; es-
    capes from cultivation into woods and fields—2
1b. Plant herbaceous; leaves simple, with two leaflets, or ternately com-
    pound; flowers white or greenish-yellow; rich woods—4

2a. Leaves compound; stem not spiny; fruits blue; SW & western UP—**Ore-
    gon-grape**, ***Mahonia aquifolium***
2b. Leaves simple; stem spiny; fruits red—3

3a. Leaf margins entire; spines unbranched; flowers solitary or in few-flow-
    ered clusters (to 2 m high)—**Japanese Barberry**, ***Berberis thunbergii***
3b. Leaf margins spine-toothed; spines often three-pointed; flowers in a
    raceme; LP (to 3 m high)—**Common Barberry**, ***Berberis vulgaris***

> The common barberry, **Berberis vulgaris**, was the subject of organized eradication programs in the early years of the 20th century. From 1918-1930, 18fi million barberry plants were destroyed in North America (Large, 1962). This plant is the alternate host of **Puccinia graminis**, the causal agent of wheat rust, and a major epidemic of the disease was "defeated" by the eradication campaign. A little-enforced law (Act 189 of 1931) still makes it illegal in Michigan to keep "any barberry, mahonia or mahoberberis bushes . . . subject to . . . black stem rust of small grains".

4a. Leaves ternately compound, with numerous leaflets; flowers many, yellowish-green or purplish (30–80 cm high)—5

4b. Leaves deeply lobed or with two large leaflets; flowers solitary, white—6

5a. Sepals yellowish-green; style less than 1 mm long—**Blue Cohosh**, *Caulophyllum thalictroides* var. *thalictroides*

5b. Sepals purplish; style 1 mm or more—**Blue Cohosh**, *Caulophyllum thalictroides* var.*giganteum*

6a. Leaves opposite, cauline, peltate, palmately lobed; flower on a nodding peduncle beneath the leaves; mostly SLP (30–50 cm high)—**May-apple**, *Podophyllum peltatum*

6b. Leaves all basal, either deeply two-lobed or with two leaflets; flower borne on long naked stalk overtopping the leaves; SLP (10–50 cm high)—**Twinleaf**, *Jeffersonia diphylla*

MENISPERMACEAE, The Moonseed Family

Dioecious woody climbing vines with alternate, simple, palmately veined leaves. Flowers regular, unisexual, in cymes or cyme-like panicles; sepals 4–8, often 6 in two whorls; petals 6–9 in two whorls; stamens 12–24; pistils and styles 2–4, often 3, each ovary superior, 1-celled. Fruit a drupe.

One species in Michigan, with five- to seven-angled leaves which are peltate near the edge, and small white flowers; rich, damp woods, mostly SLP (early summer)—**Moonseed**, *Menispermum canadense*

MAGNOLIACEAE, The Magnolia Family

Trees or shrubs with alternate, simple leaves. Flowers regular, perfect, large, solitary, and frequently showy; sepals 3; petals 6; stamens many; pistils and styles many, each ovary superior, 1-celled. Fruit a cone of samaras.

One species in Michigan, a tree with broad four-lobed leaves and large, greenish-yellow flowers; SLP (to 60 m high; late spring)—**Tulip-tree** or **Yellow-poplar**, *Liriodendron tulipifera*

None of the North American species of **Magnolia** are known in Michigan, but several species of mostly Asian origin are cultivated and may become tall shrubs or small trees. They are most easily recognized by the presence of large flowers with a perianth of nine to fifteen white or pink tepals inserted in three to five rows, many stamens, and numerous pistils. Two common examples are **M. ×soulangiana** and **M. stellata**.

ANNONACEAE, The Custard-apple Family

Trees or shrubs with alternate, simple, entire leaves. Flowers regular, perfect; sepals 3, falling early; petals 6, in two whorls; stamens many; pistils and styles 1–15, each ovary superior, 1-celled. Fruit fleshy.

One species in Michigan, a tall shrub or small tree with obovate leaves, large dull-purple flowers, and a large edible yellow fruit; rich, damp woods, SLP (to 10 m high; spring)—**Pawpaw**, *Asimina triloba*

LAURACEAE, The Laurel Family

Dioecious trees or shrubs with spicy-aromatic odor and alternate, simple leaves. Flowers regular, unisexual, small; tepals 6, in two whorls; stamens 9, the anthers opening by terminal valves; pistil 1, style 1, ovary superior, 1-celled. Fruit a colored berry or drupe.

1a. Freely branched shrub; leaves obovate-oblong, entire; flowers yellow, appearing before the leaves; rich, damp woods, SLP (to 5 m high; early spring)—**Spicebush**, *Lindera benzoin*
1b. Tree or tall shrub; at least some of the leaves with two or three (rarely to six) lobes, often resembling mittens; flowers yellow, appearing with the immature leaves; mixed, esp. sandy woods, mostly SLP (to 30 m high; spring)—**Sassafras**, *Sassafras albidum*

PAPAVERACEAE, The Poppy Family

Herbs with milky or colored juice and alternate, opposite, or basal, simple leaves. Flowers regular, perfect; sepals 2; petals 4, 6, or 8–16; stamens many; pistil 1, style 0 or 1, ovary superior, 1-celled. Fruit a capsule dehiscing by valves or a ring of pores.

1a. Leaves palmately lobed, basal; flower with 8 (or more) white petals; rich woods (scape 5–15 cm high; early spring)—**Bloodroot**, *Sanguinaria canadensis*

1b. Leaves pinnately toothed or lobed, basal and cauline; flower with 4 (rarely 6) yellow petals (late spring to summer)—2

2a. Cauline leaves opposite; petals 2–3 cm long; woods, LP (30–50 cm high)—**Celandine** or **Wood Poppy**, *Stylophorum diphyllum*

2b. Cauline leaves alternate; petals less than 1.5 cm long; roadsides, gardens, and woods, SLP and Straits area (30–80 cm high)—**Celandine**, *Chelidonium majus*

Several showy members of the Papaveraceae have been cultivated in Michigan and may occasionally be found. They include the Corn Poppy, **Papaver rhoeas**, the Opium Poppy **Papaver somniferum**, both annual species with red to purple or white petals, the Oriental Poppy (**Papaver orientale**), a perennial species with large red, orange, or rose-colored petals, and the orange-flowered California Poppy **Eschscholzia californica**, an annual often included in wildflower mixes.

FUMARIACEAE, The Fumitory Family

Herbs with watery juice and alternate or basal, compound or dissected leaves. Flowers irregular, perfect, sometimes bilaterally symmetrical; sepals 2, small; petals 4, in two pairs, sometimes united, one or both of the outer pair with a spurred or sac-like base; stamens 6, the filaments joined in two groups of 3; pistil 1, style 1, ovary superior, 1-celled. Fruit a capsule.

1b. One outer petal spurred or sac-like at the base (*Corydalis* spp.)—2
1a. Both outer petals spurred or sac-like at the base—3

2a. Flowers yellow throughout; NM (20–50 cm; spring and early summer)—**Golden Corydalis**, *Corydalis aurea*

2b. Flowers pink, tipped with yellow (30–80 cm high; spring and summer)—**Pink Corydalis** or **Rock Harlequin**, *Corydalis sempervirens*

3a. A climbing vine with cauline leaves and panicles of white or pinkish flowers; woods (to 3 m high; summer)—**Climbing Fumitory**, *Adlumia fungosa*

3b. Low herbs with basal leaves and racemes of white (rarely pinkish) flowers; rich woods (scapes 10–30 cm high; early spring) (*Dicentra* spp.)—4

4a. Spurs of the corolla triangular, divergent—**Dutchman's-breeches**, *Dicentra cucullaria*

4b. Spurs of the corolla short, rounded, nearly parallel—**Squirrel-corn**, *Dicentra canadensis*

CAPPARACEAE, The Caper Family

Annual herbs with alternate, compound leaves. Flowers mostly irregular, perfect, in a terminal raceme; sepals 4, sometimes united below; petals 4, of equal lengths or not; stamens 6–20 or more, the anthers on elongate filaments; pistil 1, style 1, ovary superior, 1-celled. Fruit a long capsule opening by 2 valves.

1. Stamens more than six; leaflets three; native, sandy areas, mostly SLP (20–60 cm high; summer)—**Clammy-weed**, *Polanisia dodecandra*
1. Stamens six; leaflets five to seven; commonly cultivated, rarely escaping, SLP (to 1.5 m high; summer and early autumn)—**Spider Plant**, *Cleome hassleriana*

CRUCIFERAE (BRASSICACEAE), The Mustard Family

Herbs with alternate (rarely opposite), frequently lobed or divided leaves. Flowers regular, perfect, often in racemes; sepals 4; petals 4 (rarely 0), often yellow or white; stamens (rarely 2) 6, 4 long and 2 short; pistil 1, style 1, ovary superior, 2-celled. Fruit usually dehiscent, resembling a capsule: short, plump or flat fruits are known as *silicles*; elongate fruits as *siliques*.

> Many members of this large diverse family are best identified by examining the fruit, and this key makes considerable use of fruit characteristics. It is strongly recommended that specimens with both flowers and fruit present be used for identification.

1a. Petals yellow or yellowish—2
1b. Petals white, pink, or purple, or absent—30

2a. Leaves simple, entire or dentate, never lobed—3
2b. Leaves deeply lobed or compound (bracts at or near the inflorescence may be simple and unlobed)—12

3a. Leaves clasping the stem—4
3b. Leaves not clasping at the base—6

4a. Fruit obovoid, to about 1 cm long; sandy disturbed areas, LP (30–70 cm high; spring)—**False-flax**, *Camelina microcarpa*

4b. Fruit a long, slender, silique over 5 cm long—5

5a. Clasping base and apex of leaf obtuse or rounded; disturbed areas (30–80 cm high; summer)—**Hare's-ear Mustard**, *Conringia orientalis*

5b. Clasping base and apex of leaf acute; sandy fields and woodlands (30–150 cm high; spring)—**Tower Mustard**, *Arabis glabra*

6a. Flowers about 2 mm wide; leaves entire; sandy railroad banks and roadsides (5–25 cm high; spring)—**Pale Alyssum**, *Alyssum alyssoides*

6b. Flowers 5 mm wide or more; leaves entire or toothed—7

7a. Plants glabrous or pubescent with unbranched hairs—8

7b. Plants pubescent, the hairs either stellate (star-shaped) or forked—9

8a. Leaves lanceolate, gradually tapering to the base; two rows of seeds per locule; sandy or gravelly roadsides, railroads, etc., mostly LP (20–50 cm high; spring and summer)—**Sand Rocket**, *Diplotaxis muralis*

8b. Leaves ovate, acute at base; one row of seeds per locule; abundant in disturbed areas (20–80 cm high; spring and early summer)—**Charlock**, *Brassica kaber* [*Sinapis arvensis*—CQ]

9a. Fruit flattened, less than 10 mm long; gravels and rocks, UP (10–40 cm high; late spring)—*Draba arabisans*

9b. Fruit round or squarish, more than 10 mm long (*Erysimum* spp.)—10

10a. Siliques 25 mm long or less, on slender pedicels; petals less than 6 mm long; wet and disturbed areas (20–100 cm high; summer)—**Wormseed Mustard**, *Erysimum cheiranthoides*

10b. Siliques 20 mm long or more, on stout pedicels; petals longer than 6 mm—11

11a. Siliques to 50 mm long, pedicels ascending; leaves entire (30–80 cm high; spring and summer)—*Erysimum inconspicuum*

11b. Siliques 50 mm long or more, pedicels divergent; leaves toothed; roadsides, weedy areas, etc., LP (20–40 cm high)—**Treacle Mustard**, *Erysimum repandum*

12a. Leaves pinnately compound or bipinnately divided, the leaves dissected into very numerous divisions; roadsides, railroads, etc. (*Descurainia* spp.)—13

12b. Leaves pinnately divided—14

Figure 19: **Arabis glabra**: *A, silique; B, habit*

13a. Stems whitish with a close, fine non-glandular pubescence; pods about 20 mm long by 1 mm broad (30–80 cm high; spring and summer)— **Herb Sophia**, *Descurainia sophia*

13b. Stems green with sparse glandular pubescence; pods about 8 mm long by 2 mm broad (20–70 cm high)—**Tansy Mustard**, *Descurainia pinnata*

14a. Pedicels subtended by a pinnate bract; railroads, weedy areas, etc. (30–60 cm high; spring and summer)—**Dog Mustard**, *Erucastrum gallicum*

14b. Pedicels not subtended by a pinnate bract—15

15a. Fruit short, not more than three times as long as broad; often in wet areas (30–100 cm high)—**Yellow Cress**, *Rorippa palustris*

15b. Fruit elongated, more than four times as long as broad—16

16a. Upper cauline leaves clasping at the base—17
16b. Upper cauline leaves not clasping—18

17a. Open flowers overtop the terminal buds of the inflorescence; petals 9 mm long or less; fields and roadsides (to 80 cm high)—**Field Mustard**, *Brassica rapa*

17b. Open flowers do not overtop the terminal buds of the inflorescence; petals 9 mm long or longer; cultivated, sometimes escaping to disturbed areas—*Brassica napus*

> Several species with many varieties of **Brassica** are commonly cultivated, some of which may escape and occasionally be collected. Cultivars of **B. napus** include the rutabaga and rapeseed (the source of canola oil). Turnips, Chinese cabbage, bok-choy and rapini are all cultivars derived from **B. rapa. Brassica oleracea** is the source of cabbage, kale, broccoli, cauliflower, kohlrabi, Brussels sprouts, collards, and the ornamental kales and cabbages.

18a. Fruit terminating in a conspicuous beak; petals mostly 10 mm long or more—19

18b. Fruit not terminating in a conspicuous beak; petals mostly less than 10 mm long—23

19a. Fruit indehiscent, with spongy cross-partitions between the seeds, tipped with a conical beak; dry fields and roadsides (30–80 cm high)— **Wild Radish**, *Raphanus raphanistrum*

19b. Fruit dehiscent when ripe by two valves, tipped with a flat, angled, or round beak—20

20a. Beak of fruit flat; fruit densely pubescent; disturbed areas, mostly SLP (30–70 cm high)—**White Mustard**, *Brassica alba* [*Sinapis alba*—CQ]

20b. Beak of fruit angled or round; fruit glabrous; disturbed areas—21

21a. Leaves dentate or lobed (20–80 cm high; spring and early summer)—
Charlock, *Brassica kaber* [*Sinapis arvensis*—CQ]

21b. Leaves deeply pinnately divided—22

22a. Beak of fruit 3-4 mm long; fruiting pedicels erect; mostly SLP (up to 150 cm high; spring and summer)—**Black Mustard**, *Brassica nigra*

22b. Beak of fruit 5 mm long or more; fruiting pedicels spreading (30–100 cm high; summer)—**Indian Mustard**, *Brassica juncea*

23a. Terminal segment of the principal leaves equaling or smaller than the lateral ones—24

23b. Terminal segment of the principal leaves much larger than the lateral segments—27

24a. Petals less than 4 mm long; often in wet areas, sometimes in lawns (20–60 cm high; spring and summer)—**Yellow Cress**, *Rorippa sylvestris*

24b. Petals 5 mm or longer; dry areas—25

25a. Fruits 5 cm or longer; one row of seeds per locule; disturbed areas (50–100 cm high; summer)—**Tumble Mustard**, *Sisymbrium altissimum*

25b. Fruits 5 cm or less; two rows of seeds per locule; railroads, weedy places, etc. (Diplotaxis spp.; spring and summer)—26

26a. Leaves mostly basal, lobed or sometimes pinnately divided; mostly LP (20–50 cm high)—**Sand Rocket**, *Diplotaxis muralis*

26b. Leaves mostly cauline, deeply pinnately divided; LP (30–80 cm high)—*Diplotaxis tenuifolia*

27a. Flowers about 3 mm wide; pods erect and closely appressed to the stem; disturbed areas (30–80 cm high; spring and summer)—**Hedge Mustard**, *Sisymbrium officinale*

27b. Flowers about 7 mm wide; pods spreading or ascending, rarely appressed to the stem (*Barbarea* spp., **Winter Cress**)—28

28a. Cauline leaves include four to five (or up to eight) pairs of lateral leaf-segments; fields, LP (30–80 cm high; spring)—**Early Winter Cress**, *Barbarea verna*

28b. Cauline leaves include one to three pairs of lateral leaf segments—29

29a. Flowers bright yellow; petals 6 mm or longer; beak of fruit 1.5 mm or longer; abundant in disturbed areas (20–80 cm high; spring)—**Yellow Rocket**, *Barbarea vulgaris*

29b. Flowers pale yellow; petals 5 mm long or less; beak of fruit less than 1.5 mm; shores, Straits area (30–80 cm high; late spring and early summer)—**Northern Water Cress**, *Barbarea orthoceras*

30a. Principal cauline leaves compound or deeply lobed (the uppermost or bracteal leaves may be simple)—31
30b. Principal cauline leaves simple, toothed or entire (the basal leaves, at the surface of the ground, may be deeply lobed or compound)—44

31a. Leaves palmately divided or compound; rich woods (20–40 cm high; spring) (*Dentaria* spp., **Toothwort**)—32
31b. Leaves pinnately divided or compound—34

32a. Leaf segments lanceolate or narrowly oblong—**Cut-leaved Toothwort**, *Dentaria laciniata* [*Cardamine concatenata*—CQ]
32b. Leaf segments, ovate or ovate-oblong—33

33a. Cauline leaves two, opposite or virtually so—**Two-leaved Toothwort**, *Dentaria diphylla* [*Cardamine diphylla*—CQ]
33b. Cauline leaves three to four, alternate—*Dentaria maxima* [*Cardamine ×maxima*—CQ]

34a. Aquatic plant; aerial leaves merely serrate to pinnately divided; the submerged leaves pinnately dissected; cold waters (10–50 cm high; spring and summer)—**Lake Cress**, *Armoracia aquatica* [*A. lacustris*—CQ]
34b. Terrestrial or aquatic; if aquatic, submerged leaves not pinnately dissected—35

35a. Petals less than 6 mm long (or absent)—36
35b. Petals 6 mm long or more—41

36a. Plant often aquatic, roots form at stem nodes; aerial leaves distinctly compound; fruits curved, pedicels divergent; stream banks, damp woods (summer)—**Watercress**, *Nasturtium officinale* [*Rorippa nasturtium-aquaticum*—CQ]
36b. Plant terrestrial, those in wet areas not forming roots at stem nodes; leaves compound or not; fruits straight, pedicels erect (*Cardamine* spp. in part, **Bitter Cress**)—37

37a. Cauline leaves with ciliate basal auricles; petals either shorter than the sepals or absent; damp banks, SE—*Cardamine impatiens*
37b. Cauline leaves lack auricles; petals present, longer than the sepals—38

38a. Leaves chiefly basal; petioles of cauline leaves pubescent—39
38b. Leaves chiefly cauline; petioles of cauline leaves glabrous—40

39a. Stem glabrous, unbranched above; lawns, other disturbed areas, SLP
   (10–40 cm high; early spring)—***Cardamine hirsuta***
39b. Stem pubescent, branched above; moist areas, SE & UP—***Cardamine
   flexuosa***

40a. Plant of dry soil; cauline leaves compound; rocky areas, UP (10–30 cm
   high)—***Cardamine parviflora***
40b. Plant of moist or wet soil; cauline leaves deeply pinnately divided; wet
   seeps (20–60 cm high; spring)—***Cardamine pensylvanica***

41a. Leaf blade margins pubescent with stiff hairs; fruit indehiscent—42
41b. Leaf blades margins glabrous; fruit dehiscent—43

42a. Petals yellow at first, turning white with age; fruit with spongy cross-
   partitions between the seeds, tipped with a conical beak; dry fields and
   roadsides (30–80 cm high; summer)—**Wild Radish**, ***Raphanus
   raphanistrum***
42b. Petals pink or white; fruit lacks spongy cross-partitions; cultivated,
   sometimes escaping to fields and roadsides—**Radish**, ***Raphanus sativus***

43a. Fruit slender, more than twice as long as broad; bogs and marshes
   (20–50 cm high; spring)—**Cuckoo-flower**, ***Cardamine pratensis***
43b. Fruit globular, less than twice as long as broad; damp ditches, roadsides, etc. (50–100 cm high; spring and early summer)—**Horseradish**,
   Armoracia rusticana

44a. Fruit transversely divided into two sections; fleshy, much-branched
   plant of sandy shores of the Great Lakes (10–80 cm high; summer)—
   Sea-rocket, ***Cakile edentula***
44b. Fruit not transversely divided into two sections—45

45a. Fruit short, its length not more than three times its diameter—46
45b. Fruit long and slender, its length more than three times its diameter—61

46a. Fruits not conspicuously flattened, thick and plump, about circular in
   cross-section (summer) (***Armoracia*** spp.)—47
46b. Fruits distinctly flattened—48

47a. Plant terrestrial; basal leaves up to 30 cm long; damp ditches, roadsides, etc. (50–100 cm high; spring and early summer)—**Horseradish**,
   Armoracia rusticana

47b. Plant aquatic; lower leaves submerged, finely dissected; cold waters (10–50 cm high; spring and summer)—**Lake Cress**, *Armoracia aquatica* [*Armoracia lacustris*—CQ]

48a. Cauline leaves clasping the stem by an auricled base—49
48b. Cauline leaves sessile or petioled, not clasping, or none—52

49a. Apex of fruit not notched; rare along roadsides and in weedy areas, SLP (to 60 cm high; spring and early summer)—**Hoary Cress**, *Cardaria draba*
49b. Apex of fruit notched—50

50a. Stem and leaves glabrous; fruit very flat and circular, about 10 mm wide; disturbed areas (10–50 cm tall; spring)—**Penny Cress**, *Thlaspi arvense*
50b. Stem and leaves glabrous or pubescent; fruit not more than 5 mm wide (spring and early summer)—51

51a. Fruits oblong; stems densely pubescent, the hairs unbranched; disturbed areas (20–50 cm high)—**Field Cress**, *Lepidium campestre*
51b. Fruits triangular; stems pubescent, the hairs stellate; roadsides, gardens, weedy areas, etc. (10–60 cm high)—**Shepherd's Purse**, *Capsella bursa-pastoris*

52a. Petals unequal, of two sizes; sometimes escaping cultivation onto sandy shores and weedy areas—**Globe Candytuft**, *Iberis umbellata*
52b. Petals all of equal size—53

53a. Fruits nearly circular, or a very little longer than broad—54
53b. Fruits ovoid or oblong, broadest near the middle, and distinctly longer than broad—58

54a. Fruits 15–35 mm broad; escapes from cultivation to roadsides, etc., mostly LP (to 1 m high; spring and early summer)—**Money-plant**, *Lunaria annua*
54b. Fruits much less than 10 mm broad—55

55a. Stem and leaf pubescence consists of appressed, two-pronged hairs; commonly cultivated, sometimes escaping to disturbed areas (10–30 cm high, summer)—**Sweet Alyssum**, *Lobularia maritima*
55b. Stem and leaf pubescence lacks such hairs (spring)—56

56a. Leaves entire; stamens 6: sandy railroad banks and roadsides (5–25 cm high)—**Pale Alyssum**, *Alyssum alyssoides*

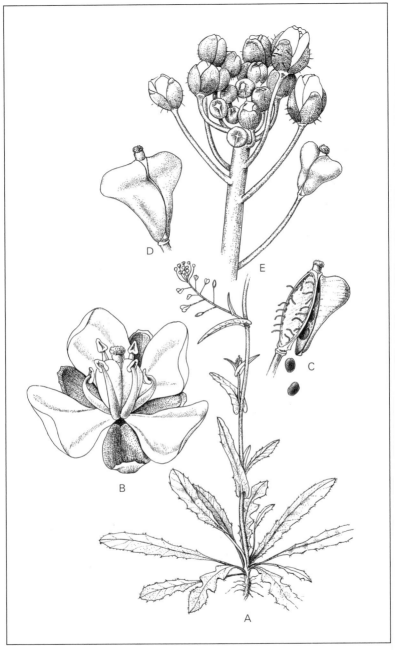

Figure 20: ***Capsella bursa-pastoris:*** *A, habit; B, flower;*
C, silicle (dehiscing); D, silicle (intact); E, raceme

56b. Leaves serrate to pinnately divided; stamens 2; dry disturbed areas and woods (10–50 cm high) (*Lepidium* spp. in part, **Pepper-grass**)—57

57a. Petals present, exceeding the sepals—**Pepper-grass**, *Lepidium virginicum*
57b. Petals absent or tiny (shorter than the sepals)—*Lepidium densiflorum*

58a. Petals deeply two-cleft—59
58b. Petals entire or barely notched at the tip (*Draba* spp.)—60

59a. Cauline leaves present; disturbed areas (to 70 cm high; summer and early autumn)—**Hoary Alyssum**, *Berteroa incana*
59b. All leaves in a basal rosette; sandy areas (5–20 cm high; early spring)—**Whitlow-grass**, *Erophila verna* [*Draba verna*—CQ]

60a. Leaves all or chiefly at or near the base; sandy areas, LP (5–20 cm high; early spring)—*Draba reptans*
60b. Cauline leaves few, but evident along the stem; gravels and rocks, UP (10–40 cm high; summer)—*Draba arabisans*

61a. Stem-leaves cordate or sagittate at the base and sessile, forming a more or less clasping leaf—62
61b. Stem-leaves sessile or somewhat petioled, but not clasping—71

62a. All leaves cauline; apex of leaf obtuse or rounded (30–80 cm high; spring and summer)—**Hare's-ear Mustard**, *Conringia orientalis*
62b. Both basal and cauline leaves present; apex of leaf acute (*Arabis* spp. in part, **Rock Cress**)—63

63a. Pedicels sharply reflexed, the fruits pendant; dunes ridges and rocks, NM (20–90 cm high; summer)—*Arabis holboellii*
63b. Pedicels not reflexed, the fruits erect to spreading, or sometimes arching—64

64a. Pedicels erect, the fruits erect and appressed to the stem—65
64b. Pedicels spreading or diverging, the fruits diverging away from the stem—67

65a. Fruits not flattened; sandy fields and woodlands (30–150 cm high; spring)—**Tower Mustard**, *Arabis glabra*
65b. Fruits flattened—66

66a. Basal leaves and the lower stem glabrous or nearly so; seeds in two rows in each locule; woods and dunes, NM & WM (30–90 cm high; spring and summer)—***Arabis drummondii***

66b. Basal leaves and the lower stem pubescent; seeds in one row in each locule (20–80 cm high; spring)—***Arabis hirsuta***

67a. Upper cauline leaves dentate, pubescent above and below—68

67b. Upper cauline leaves entire, glabrous—69

68a. Petals small, less than 5 mm long; floodplains, SLP (20–50 cm high; spring)—***Arabis perstellata*** [*A. shortii*—CQ]

68b. Petals larger, longer than 9 mm long; sometimes escaping from cultivation, NM—**Wall Rock Cress**, ***Arabis caucasica***

69a. Basal leaves densely pubescent; sandy areas, rock outcrops, mostly NM (to 1 m high; early summer)—***Arabis divaricarpa***

69b. Basal leaves sparsely pubescent or glabrous; SLP—70

70a. Petals about twice as long as the sepals; open sandy areas (20–50 cm high; spring and early summer)—***Arabis missouriensis***

70b. Petals equaling or slightly exceeding the sepals; forests and thickets (to 1 m high; spring)—***Arabis laevigata***

71a. Cauline leaves 6–10 cm long, or more—72

71b. Cauline leaves 2–6 cm long—73

72a. Leaves lanceolate or oblong; petals 3–5 mm, usually white; dry woods, SLP (30–100 cm high; spring and early summer)—**Sickle-pod**, ***Arabis canadensis***

72b. Leaves ovate or ovate-lanceolate; petals 15–20 mm long, often purple; damp woods, roadsides, etc. (50–100 cm high; late spring)—**Dame's Rocket**, ***Hesperis matronalis***

73a. Basal leaves ovate to orbicular or cordate, not more than twice as long as broad—74

73b. Basal leaves oblong, lanceolate, or oblanceolate, at least three times as long as broad—76

74a. Cauline leaves deltoid, coarsely toothed; moist, shaded disturbed areas (to 1 m high; late spring)—**Garlic Mustard**, ***Alliaria petiolata***

74b. Cauline leaves lanceolate to ovate, not coarsely toothed; wet woods, SLP (spring)—75

Many weedy species are sun-loving plants, but Garlic Mustard (**Alliaria petiolata**) can aggressively invade wooded areas. Nuzzo (1993) described the exponential spread of this species in northern Illinois. First collected in 1918, **A. petiolata** was known from sites in 44 counties by 1991, including about one-half of the natural areas and state parks in these counties. Most collections of this species in Michigan have been made in the SLP where some woodland floras are being threatened by this species as it becomes locally abundant.

75a. Sepals purple; petals purple or rose (rarely white) (10–40 cm high)—**Pink Spring Cress**, *Cardamine douglassii*
75b. Sepals green; petals white (20–60 cm high)—**Spring Cress**, *Cardamine bulbosa* [*C. rhomboidea*—CQ]

76a. Petals longer than 4 mm; basal leaves often lyre-shaped; sandy aeas esp. coastal dunes (10–40 cm high; spring and early summer)—**Sand Cress**, *Arabis lyrata*
76b. Petals shorter than 4 mm; basal leaves entire or toothed; sandy areas, LP (10–40 cm high; spring)—**Mouse-ear Cress**, *Arabidopsis thaliana*

RESEDACEAE, The Mignonette Family

Herbs with alternate, simple leaves. Flowers irregular, perfect, in a terminal raceme; sepals usually 6; petals usually 6; stamens 10–25; pistil 1, style 0, ovary superior, 1-celled and open at the top with 3 stigmas found along the upper rim. Fruit an open capsule. Summer.

One species in Michigan, with pinnately divided leaves and irregularly cleft yellowish petals; fields, roadsides, LP (to 80 cm high)—**Yellow Mignonette**, *Reseda lutea*

SARRACENIACEAE, The Pitcher-plant Family

Insectivorous herbs with hollow, basal, pitcher-shaped leaves. Flowers regular, perfect, solitary at the ends of naked scapes; sepals 5, reddish-purple (rarely yellow in forma *heterophylla*); petals 5, reddish-purple (rarely yellow); stamens many; pistil 1, styles 2, ovary superior, 5-celled. Fruit a capsule.

One species in Michigan, growing in bogs, swamps, and fens (scapes 30–50 cm high; late spring and summer)—**Pitcher-plant**, *Sarracenia purpurea*

DROSERACEAE, The Sundew Family

Insectivorous herbs with a rosette of basal leaves, each with gland-tipped bristles on the upper surface. Flowers regular, perfect, in terminal cymes; sepals 5; petals 4–8, often 5; stamens 4–8, often 5; pistil 1, styles 3, ovary superior, 1-celled. Fruit a capsule. Summer.

1a. Leaf blade round, about as long as wide; leaves usually spreading; petals white (scapes 7–35 cm high)—*Drosera rotundifolia*
1b. Leaf blade longer than wide; leaves usually erect; petals white or pink-ish—2

2a. Leaf blade narrowly linear, seven or more times as long as wide; petals pinkish (scapes 6–13 cm high)—*Drosera linearis*
2b. Leaf blade wider, two to six times as long as wide; petals white—3

3a. Petioles glabrous; base of floral stalk not erect—*Drosera intermedia*
3b. Petioles pubescent; base of floral stalk erect; mostly NM (scapes 6–25 cm high)—*Drosera ×anglica*

CRASSULACEAE, The Orpine Family

Perennial herbs with alternate, opposite, or whorled simple, succulent leaves. Flowers regular, perfect, in terminal cymes; sepals 4 or 5; petals 4 or 5; stamens as many or (usually) twice as many as the sepals; pistils and styles 4 or 5, each ovary superior, 1-celled. Fruit a follicle.

1a. Cross-section of leaf flattened; leaves alternate, opposite, or whorled (spring or summer)—2
1b. Cross-section of leaf elliptic or round; leaves alternate (summer)—5

2a. Leaves whorled in threes; petals white or yellow—3
2b. Leaves alternate or opposite; petals white or pink (summer)—4

3a. Petals four, white; SLP (10–20 cm high; spring)—*Sedum ternatum*
3b. Petals five, yellow; WM (to 10 cm high; summer)—*Sedum sarmentosum*

4a. Leaves alternate (20–60 cm high)—**Live-forever**, *Sedum telephium* [*S. purpureum*—CQ]
4b. Leaves opposite; SW & UP (10–20 cm high)—*Sedum spurium*

5a. Petals white; LP (10–20 cm high)—*Sedum album*
5b. Petals yellow (5–10 cm high)—6

6a. Leaves ovoid; widespread in sandy areas, cemeteries—**Mossy Stonecrop**, *Sedum acre*

6b. Leaves linear; Straits—***Sedum sexangulare***

Many species of **Sedum** are cultivated, and often escape to nearly disturbed areas. All species included here have been introduced to North America except for **Sedum ternatum**. **Sedum acre** is the most widely distributed in Michigan outside of cultivation.

PENTHORACEAE, The Ditch Stonecrop Family

Perennial herbs with alternate, simple, non-succulent leaves. Flowers regular, perfect, in a terminal cyme; sepals 5; petals 0; stamens 10; pistils 5, united below, styles 5, each ovary superior, 1-celled. Fruit a 5-angled cluster of follicles.

One species in Michigan. Flowers cream-colored, reddish in fruit; low wet areas (20–70 cm high; summer)—**Ditch Stonecrop**, *Penthorum sedoides*

SAXIFRAGACEAE, The Saxifrage Family

Perennial herbs with alternate, opposite, or basal, simple leaves. Flowers mostly regular, perfect; sepals (4)5; petals 5 or 0; stamens (as many) or twice as many as the sepals; pistils 1 or 2, styles 2 or 0 (then with 4 sessile stigmas), ovary superior or half-inferior, 1-celled. Fruit a capsule.

1a. Leaves opposite; flowers minute, in the axils of the leaves; sepals four, petals absent; wet areas (5–20 cm high; spring)—**Golden Saxifrage**, *Chrysosplenium americanum*

1b. Leaves alternate or all basal; flowers on terminal stalks; sepals five, petals five—2

2a. Flower solitary; petals white with green or yellow veins; wet, often calcareous areas (*Parnassia* spp., **Grass-of-Parnassus**)—3

2b. Flowers in terminal racemes, panicles, or clusters; petal veins not distinctly colored—5

3a. Flowers less than 2 cm wide; leaves narrowed to the base; NM (10–30 cm high; summer)—*Parnassia parviflora*

3b. Flowers 2–3.5 cm wide; leaves rounded to cordate at the base (20–40 cm high; late summer)—4

4a. A three-cleft scale-like staminode at the base of each petal—*Parnassia glauca*

4b. A nine- to many-cleft scale-like staminode at the base of each petal; UP—*Parnassia palustris*

5a. Leaves linear to oblanceolate, three times as long as broad or more, and pinnately veined (*Saxifraga* spp., **Saxifrage**)—6

5b. Leaves broadly ovate to nearly circular, frequently cordate at the base, and always palmately veined or lobed—7

6a. Petals white; sepals erect; leaves conspicuously toothed, 3–7.5 cm long; rocky areas, UP (10–40 cm high; spring)—**Early Saxifrage**, *Saxifraga virginiensis*

6b. Petals greenish; sepals reflexed; leaves minutely toothed or entire, 10–30 cm long or more; swamps and calcareous wet areas (30–100 cm high; spring)—**Swamp Saxifrage**, *Saxifraga pensylvanica*

7a. Stamens five (spring and early summer)(*Heuchera* spp., **Alum Root**)—8

7b. Stamens ten; rich or wet woods—10

8a. Flowers regular; petioles glabrous or sparsely hairy; woods, SLP (40–140 cm high)—*Heuchera americana*

8b. Flowers irregular, the calyx oblique, longer on the upper side than on the lower; petioles densely hirsute—9

9a. Perianth less than 6 mm long; stamens clearly projecting beyond the calyx; SLP—*Heuchera americana* var. *hirsuticalis* [*H.* ×*hirsuticaulis*—CQ]

9b. Perianth 6 mm or longer (rarely less); stamens not (or barely) projecting beyond the calyx; dry, open sites, mostly SLP (20–90 cm high)—*Heuchera richardsonii*

10a. Stem with a pair of opposite leaves (10–40 cm high; spring)—**Bishop's-cap**, *Mitella diphylla*

10b. Stem leaves alternate or none—11

11a. Petals deeply fringed (5–20 cm high; spring)—**Naked Miterwort**, *Mitella nuda*

11b. Petals entire (10–35 cm high; spring)—**Foamflower** or **False Miter-wort**, *Tiarella cordifolia*

GROSSULARIACEAE, The Gooseberry Family

Shrubs with alternate, simple, palmately lobed leaves. Flowers mostly regular, mostly perfect; sepals (4)5; petals (4)5; stamens (4)5; pistil 1, style 1 (often divided), ovary inferior, 1-celled. Fruit a fleshy berry. Spring.

> **Ribes**, both wild and cultivated species, are the alternate host for **Cronartium ribicola**, the fungus which causes white pine blister rust. Eradication campaigns were conducted against these plants early in the 20th century in an effort to save North America's white pine plantations (Large, 1962). Even now, there are restrictions on planting them in some states.

1a. Stems thorny, especially at nodes—2
1b. Stems not thorny—5

2a. Flowers and fruits in racemes; flowers saucer-shaped; damp woods, NM—**Swamp Black Currant**, *Ribes lacustre*
2b. Flowers and fruit solitary or up to four in small clusters; flowers cylindrical to bell-shaped—3

3a. Ovary and fruit prickly and bristly; deciduous woods—**Wild Gooseberry**, *Ribes cynosbati*
3b. Ovary and fruit smooth, or sometimes a little glandular—4

4a. Stamens shorter than the sepals; rocky areas, UP—**Northern Gooseberry**, *Ribes oxyacanthoides*
4b. Stamens about equal to the sepals in length; swamps and wet woods—**Swamp Gooseberry**, *Ribes hirtellum*

5a. Flowers and fruit solitary or up to four in small clusters; swamps and wet woods—**Swamp Gooseberry**, *Ribes hirtellum*
5b. Flowers and fruits in racemes—6

6a. Tissue around the ovary prolonged into a tube which is longer than the sepals; sepals bright yellow; flowers intensely clove-scented; sometimes escaping from cultivation to roadsides and fields—**Golden Currant**, *Ribes odoratum*
6b. Tissue around the ovary shorter than the sepals; sepals yellowish to green or reddish; flowers not fragrant—7

7a. Leaves dotted beneath with resinous glands; fruit black—8
7b. Leaves glabrous or pubescent beneath, without resinous glands; fruit red—10

8a. Racemes erect to ascending; swamps, NM—**Northern Black Currant**, *Ribes hudsonianum*
8b. Racemes pendent or seldom spreading—9

9a. Calyx pubescent; pedicels longer than inflorescence bracts; occasional escape from cultivation to wet areas—**Black Currant**, *Ribes nigrum*
9b. Calyx glabrous or with just a few hairs; pedicels shorter than inflorescence bracts; damp woods—**Wild Black Currant**, *Ribes americanum*

10a. Ovary and berry bristly with glandular hairs; woods and swamps, NM—**Skunk Currant**, *Ribes glandulosum*
10b. Ovary and fruit smooth, or with sessile glands—11

11a. Flowers greenish; pedicels smooth—**Red Currant**, *Ribes rubrum* [*R. sativum*—CQ]
11b. Flowers pink or red; pedicels with glands, sometimes also pubescent—**Swamp Red Currant**, *Ribes triste*

HAMAMELIDACEAE, The Witch Hazel Family

Shrubs with alternate, simple leaves. Flowers regular, perfect, in axillary clusters; sepals 4; petals 4, yellow, linear; stamens 4; pistil 1, styles 2, ovary half-inferior, 1-celled. Fruit a capsule.

One species in Michigan; leaf blades with rounded teeth and an asymmetrical base; rich or sandy woods (to 5 m high; late autumn)—**Witch Hazel**, *Hamamelis virginiana*

PLATANACEAE, The Plane-tree Family

Monoecious trees with alternate, simple, palmately veined and lobed leaves. Flowers regular, unisexual, minute, in dense spherical heads; sepals often 3 or 4, sometimes united below; petals often 3 or 4 or 0; stamens 3 or 4; pistils and styles usually 5–8, each ovary superior, 1-celled. Fruit a spherical head of achenes.

One species in Michigan, with distinctive mottled bark falling to the ground as "jigsaw puzzle pieces", often in woods along watercourses; SLP (to 50 m high)—**Sycamore**, *Platanus occidentalis*

Single heads of flowers or fruit on long peduncles are characteristic of ***Platanus occidentalis***. Specimens with two or more heads in each inflorescence are the cultivated ***Platanus ×hybrida***, the London Plane Tree.

ROSACEAE, The Rose Family

Trees, shrubs, or herbs with alternate (seldom basal), simple or frequently compound leaves. Flowers regular, perfect (rarely unisexual, the plants then monoecious or dioecious); sepals (4)5; petals 5 (rarely 10 or 0); stamens (4)many; pistils and styles 1–many, ovaries 1–many and superior or 1 and inferior, each 1-celled; receptacle often expanded into a saucer-shaped or cup-shaped organ (*hypanthium*), bearing the sepals, petals, and stamens at its margin, the pistils at its center, and resembling a calyx-tube or flattened calyx. Fruits are follicles, achenes, drupes, clusters of drupelets, or pomes.

1a. Trees, shrubs, or woody vines—2
1b. Herbaceous plants (rarely with a woody base)—56

2a. Leaves compound—3
2b. Leaves simple—27

3a. Flowers in large panicles or corymbs, each flower 5–10 mm wide; leaflets 12 or more—4
3b. Flowers solitary or in small clusters, each flower over 20 mm wide; leaflets frequently three to five or up to seven—7

4a. Shrub; flowers in a pyramidal or oblong panicle; ovaries five, superior, each maturing as a follicle; escape from cultivation to roadsides and fields (1–2 m high; summer)—**False Spiraea**, *Sorbaria sorbifolia*
4b. Tree; flowers in a broad corymb; ovary one, inferior, maturing as a orange to red drupe (to 10 m high; late spring) (*Sorbus* spp., **Mountain-ash**)—5

5a. Leaflets pubescent beneath; escape from cultivation, SLP—**European Mountain-ash** or **Rowan**, *Sorbus aucuparia*
5b. Leaflets glabrous beneath (except sometimes along the midvein) when mature—6

6a. Leaflets about four times as long as broad, the apex acuminate; swamps and woods, NM—*Sorbus americana*
6b. Leaflet about three times as long as broad, the apex obtuse or acute; woods and shores—*Sorbus decora*

7a. Flowers yellow; stem smooth; fruit a pubescent achene; wet thickets, bogs, fens (to 1 m high; summer)—**Shrubby Cinquefoil**, *Potentilla fruticosa*
7b. Flowers pink, red, white, or rarely yellow; stem usually prickly or thorny; fruit either achenes enclosed in a fleshy receptacle or a cluster of fleshy drupelets—8

8a. Flowers pink or red, rarely white or yellow, 4–10 cm across; fruit of achenes enclosed in a fleshy receptacle (early summer) (***Rosa*** spp., **Rose**)—9

8b. Flowers white, 1–3 cm across; fruit a cluster of fleshy drupelets (late spring) (***Rubus*** spp. in part, **Bramble**)—19

9a. Styles glabrous, cohering in a column which protrudes from among the stamens (plants vinelike, climbing to 4 m)—10

9b. Styles pubescent, not cohering in a protruding column (shrubs 1–3 m high)—11

> Roses are commonly cultivated in Michigan. Many cultivars have "doubled" flowers, with some stamens transformed into petals; none of the species of **Rosa** in this key have that feature. Interspecific hybridization is common in this genus, often complicating identification. See Voss (1985) for additional comments on **Rosa** in Michigan, Beales (1992) for information on the origins of cultivated roses.

10a. Leaflets mostly three or as many as five, each 3 cm or longer; thickets, SLP—**Prairie Rose**, *Rosa setigera*

10b. Leaflets five to eleven, each less than 3 cm long; roadsides and fields—**Multiflora** or **Japanese Rose**, *Rosa multiflora*

11a. Flowers solitary at tips of branches; pedicel not subtended by a bract; escape from cultivation to fields, roadsides, etc.—12

11b. Flowers solitary or several in a corymb; if solitary, bracts subtend the pedicel—13

12a. Flowers pink; leaflets three to seven, over 2 cm long; LP (to 1 m high)—**French Rose**, *Rosa gallica*

12b. Flowers white, yellow, or pink; leaflets seven to eleven, less than 2 cm long (to 1 m high)—**Scotch Rose**, *Rosa spinosissima* [*R. pimpinellifolia*—CQ]

13a. Sepals erect, persistent on the fruit after flowering—14

13b. Sepals spreading, soon deciduous from the young fruit after flowering—18

14a. Hypanthium glandular-pubescent; escape from cultivation to fields, roadsides, shores, etc. (1–3 m high)—**Sweetbrier**, *Rosa eglanteria*

14b. Hypanthium glabrous—15

15a. Stems with few thorns on lower internodes or none at all (to 2 m high)—**Wild Rose**, *Rosa blanda*

15b. Stems prickly—16

16a. The pair of spines at the base of each leaf distinctly recurved or hooked; escape from cultivation to fields, roadsides, etc. (1–2 m high)—**Cinnamon Rose, *Rosa cinnamomea* [*R. majalis*—CQ]**

16b. The pair of spines at the base of each leaf straight or nearly so—17

17a. Leaflets usually nine to eleven, the margins serrate; fields, roadsides, shores, etc. (to 1 m high)—**Prairie Rose, *Rosa arkansana***

17b. Leaflets usually five to seven, the margins doubly-serrate; NM & WM (to 1 m high, or rarely up to 2 m)—**Wild Rose, *Rosa acicularis***

18a. The pair of spines at the base of each leaf straight or nearly so; SLP (to 1 m high)—**Pasture Rose, *Rosa carolina***

18b. The pair of spines at the base of each leaf distinctly recurved or hooked; wet areas (to 2 m high)—**Swamp Rose, *Rosa palustris***

> The variation within **Rubus** makes the taxonomy of the genus very difficult. While the raspberries can easily be identified, identification of the blackberries and dewberries to species is extremely complicated as a result of extensive hybridization. We follow Voss (1985) in adopting *species complexes*, each of which encompasses several described species.

19a. Stems almost herbaceous, mostly without thorns (15–50 cm high)—**Dwarf Raspberry, *Rubus pubescens***

19b. Stems distinctly shrubby and thorny—20

20a. Stems trailing or creeping; mature fruit black (***Rubus*** subg. ***Rubus*** in part, **Dewberries**)—21

20b. Stems erect, ascending, or arched; mature fruit red or black—22

21a. Leaflets thin, dull above (when fresh), the apex acute to acuminate; prickles broad-based; dry, often sandy areas—**Northern Dewberry, *Rubus flagellaris* complex**

21b. Leaflets firm or thick, shining above (when fresh), the apex rounded; prickles slender, the base not expanded; damp woods and swamps—**Swamp Dewberry, *Rubus hispidus* complex**

22a. Ripe fruit red or black, dropping away from the white receptacle (core); leaves of flowering stems whitened beneath (***Rubus*** subg. ***Idaeobatus***, **Raspberries**)—23

22b. Ripe fruit black, the fruit and receptacle (core) dropping together; leaves of flowering stems not whitened beneath (***Rubus*** subg. ***Rubus*** in part, **Blackberries**)—25

23a. Pedicels with slightly curved, non-glandular prickles; fruit black; mostly LP—**Black Raspberry, *Rubus occidentalis***

23b. Pedicels with straight, glandular bristles; fruit red (to 2 m high)—**Red Raspberry**, *Rubus strigosus* [*R. idaeus* var. *strigosus*—CQ]

24a. Stems with either bristles or slender prickles; sandy or moist areas (to 1 m high)—*Rubus setosus* complex
24b. Stems with broad-based prickles; mostly in dry fields and woods (1–3 m high)—25

25a. Pedicels with gland-tipped hairs—*Rubus allegheniensis* complex
25b. Pedicels without gland-tipped hairs—26

26a. Leaves pubescent beneath; stems with many prickles—*Rubus pensilvanicus* complex
26b. Leaves glabrous beneath; stems with few prickles—*Rubus canadensis* complex

27a. Ovary one to many, superior (attached to the surface of the receptacle, not concealed within it or united to it)—28
27b. Ovary one, inferior (permanently enclosed within the receptacle, with only the styles protruding)—46

28a. Ovaries more than one—29
28b. Ovary one (*Prunus* spp., **Cherry** and **Plum**)—34

29a. Petals less than 5 mm long; ovaries three to five; fruits dry and dehiscent—30
29b. Petals more than 15 mm long; ovaries many; fruit a cluster of fleshy drupelets (*Rubus* subg. *Anoplobatus*)—33

30a. Leaves lobed and toothed; stipules (or scars) present; stream banks and swamps (1–3 m high; early summer)—**Ninebark**, *Physocarpus opulifolius*
30b. Leaves toothed, but not lobed; stipules absent (*Spiraea* spp.)—31

31a. Inflorescence a dense, unbranched raceme; commonly cultivated (to 2 m high; spring)—**Bridal-wreath**, *Spiraea ×vanhouttei*
31b. Inflorescence a branched panicle; wet areas (summer)—32

32a. Leaves glabrous or very nearly so; flowers white or pinkish (0.8–2 m high)—**Meadow-sweet**, *Spiraea alba*
32b. Leaves closely pubescent beneath; petals pink (0.5–1.2 m high)—**Hardhack**, *Spiraea tomentosa*

33a. Petals purple; sepals with purple glandular hairs; forest edges, eastern
    shore of LP (1–2 m high; summer)—**Flowering Raspberry**, *Rubus*
    odoratus
33b. Petals white; sepals with yellow glandular hairs; woods, NM (1–2 m
    high; spring and early summer)—**Thimbleberry**, *Rubus parviflorus*

34a. Inflorescence a raceme of at least twelve flowers (late spring)—35
34b. Inflorescence a small umbel or corymb; flowers one to eleven
    (spring)—36

35a. Leaf blades broadest at or below the middle; marginal teeth incurved;
    calyx teeth persisting in fruit; fields and woods (to 25 m high)—**Wild**
    Black Cherry, *Prunus serotina*
35b. Leaf blades broadest at or above the middle; marginal teeth spreading;
    calyx teeth deciduous before the fruit matures (to 10 m high)—**Choke**
    Cherry, *Prunus virginiana*

36a. Flowers and fruits sessile, usually solitary; ovary (and fruit) pubes-
    cent—37
36b. Flowers and fruits borne on pedicels, usually more numerous; ovary
    (and fruit) glabrous—38

37a. Shrub; petals white; fruit less than 1.5 cm in diameter; escape from cul-
    tivation, SLP—**Nanking Cherry**, *Prunus tomentosa*
37b. Tree; petals pink; fruit often over 6 cm in diameter; often cultivated,
    LP (to 10 m high)—**Peach**, *Prunus persica*

38a. Fruit globose, lacking a longitudinal furrow and glaucous surface; calyx
    lobes glabrous or at most with glandular margins (**Cherries**)—39
38b. Fruit with both a longitudinal furrow and glaucous surface; upper sur-
    face of calyx lobes usually pubescent (**Plums**)—43

39a. Low shrubs; leaves spathulate or oblong, margin toothed above the
    middle; sandy or rocky areas (to 1 m, or rarely 3 m high)—**Sand**
    Cherry, *Prunus pumila*
39b. Erect tall shrubs or small trees; leaves ovate, entire margin toothed—40

40a. Petals less than 8 mm long; fruit small, the diameter less than 1 cm—
    41
40b. Petals 9 mm or longer; diameter of fruit about 2 cm—42

41a. Leaves very broadly ovate, almost as wide as long; fruit almost black;
    escape from cultivation, LP (to 10 m high)—**Perfumed Cherry**,
    Prunus mahaleb

41b. Leaves oblong-lanceolate, two or more times as long as broad; sandy woods and fields (shrub or small tree, to 15 m high)—**Pin Cherry,** *Prunus pensylvanica*

42a. Leaves pubescent below along the midrib; the blades 7 cm or longer; fruit dark red (to 20 m high)—**Sweet Cherry,** *Prunus avium*

42b. Leaves glabrous beneath, the blades up to 8 cm long; fruit bright red; often cultivated (to 10 m high)—**Pie** or **Sour Cherry,** *Prunus cerasus*

43a. Sepal margins glandular; leaf teeth gland-tipped (or rounded with a scar)—44

43b. Sepal margin not glandular; leaf teeth sharply pointed, not gland-tipped—45

44a. Leaf blade apex acuminate; petals pink; fruit red to yellow; forest edges and stream banks (to 10 m high)—**Canada Plum,** *Prunus nigra*

44b. Leaf blade apex rounded or obtuse; petals white; fruit dark blue to black; escape from cultivation to fields, roadsides, etc.—**Common Plum,** *Prunus domestica*

45a. Leaf blade apex abruptly acuminate; fruit red to purple; dry woods or stream banks (to 8 m high)—**Wild Plum,** *Prunus americana*

45b. Leaf blade apex acute to slightly acuminate; fruit purple to black; often in dry woods and fields, LP (to 4 m high)—**Alleghany Plum,** *Prunus alleghaniensis*

46a. Mid-vein of leaf blade glandular above; shrubs without thorns; swamps and bogs (often 1–2 m high; spring and early summer)—**Chokeberry,** *Aronia prunifolia* [*A. melanocarpa*—CQ]

46b. Mid-vein of leaf blade not glandular (or flowers present before leaves); shrubs or trees with or without thorns—47

47a. Flowers white or pink, solitary or in cymes, corymbs, umbels, or fascicles; opening with or after the leaves; petals less than twice as long as broad—48

47b. Flowers white, in racemes; opening with or before the leaves; petals at least twice as long as broad; often in dry, sandy woods and on dunes (spring) (most *Amelanchier* spp., **Serviceberry** or **Juneberry**)—52

48a. Trees, in cultivation or escaped from cultivation near roads or dwellings; with flowers showy, 2.5–5 cm across; fruits more than 4 cm in diameter; stem lacks thorns (to 15 m high; spring)—49

48b. Trees or shrubs, primarily native and growing in woods, fields, or thickets; flowers generally 1.5–2.5 cm across; fruits 3 cm or less in diameter; stem with thorns or not—50

49a. Styles separate to the base; petals white; LP—**Pear**, *Pyrus communis*
49b. Styles connate at the base; petals pinkish—**Apple**, *Malus pumila*
   [*Pyrus malus*—CQ]

50a. Shrub; stem without thorns; flowers in leafy fascicles; also in swamps
   and bogs, mostly UP (to 2 m high; spring-summer)—**Mountain
   Juneberry**, *Amelanchier bartramiana*
50b. Bushy trees or shrubs, with thorns or stiff thorn-like branches; flowers
   in cymes, corymbs, or umbels—51

51a. Petals pink; styles connate at the base; flowers often fragrant; mostly
   SLP (to 10 m high; spring)—**Wild Crab**, *Malus coronaria* [*Pyrus
   coronaria*—CQ]

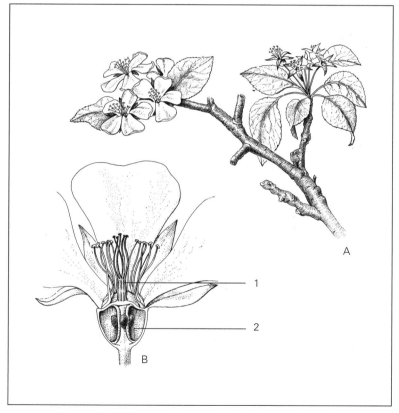

Figure 21: ***Malus pumila****: A, habit; B, flower, dissected;*
1, styles (connate), 2, ovary (inferior)

51b. Petals white; styles separate to the base—(*Crataegus* spp., **Hawthorn**)

> Hawthorns (***Crataegus*** spp.) are often found in pastures and thickets in Michigan. Identification of species is difficult because of extensive hybridization and apomixis (seed set without pollination). Morphology of individual plants varies, which has produced a confused taxonomy; approximately 1000 species have been described from North America, many with reference to a single specimen only. Voss (1985) listed 45 species, noting that most collections could be assigned to only 19 species. We have not attempted to provide a key to ***Crataegus*** species; refer to Voss (1985).

52a. Apex of the ovary glabrous or at most with a few hairs—53
52b. Apex of the ovary tomentose, even in fruit—54

> The genus ***Amelanchier*** has many common names, including Juneberry (because the edible berries are produced in June) and Serviceberry. Another name is "Shadblow", so named because the brief flush of bloom on these small trees coincides with the shad (fish) run in spring. Like ***Crataegus*** and ***Rubus***, this genus is taxonomically complex as a result of extensive hybridization and asexual propagation. The *species complexes* used here follow Voss (1985).

53a. Leaf blades reddish, glabrous below, and about half-grown when flowers open (to 15 m high)—*Amelanchier laevis*
53b. Leaf blades green, white-tomentose below, and much less than half-grown when flowers open (2–15 m high)—*Amelanchier arborea*

54a. Leaf blades open at flowering time; glabrous below; marginal teeth more than twice as many as the lateral veins—*Amelanchier interior* complex
54b. Leaf blades folded at flowering time; white-tomentose below; marginal teeth twice as many as the lateral veins or less—55

55a. Petals less than 10 mm long; lateral veins entering marginal teeth of the leaf blade weak (30 cm–1 m, or rarely 1.5 m high)—*Amelanchier spicata* complex
55b. Petals more than 10 mm long; lateral veins entering marginal teeth of the leaf blade distinct (to 3 m, or rarely 6 m, high)—*Amelanchier sanguinea* complex

56a. Leaves simple, not lobed; coniferous woods (10–20 cm high; summer)—**Dewdrop**, *Dalibarda repens*
56b. Leaves compound or pinnately lobed—57

57a. Petals yellow—58
57b. Petals white, pink, purple, or rose, never yellow—76

58a. Plant with basal trifoliolate leaves—59
58b. Plants with leafy stems—60

59a. Calyx with three- or up to five-lobed bractlets between adjacent sepals; leafy stolons often present; disturbed areas, SLP (spring and summer)— **Indian-strawberry**, *Duchesnea indica*

59b. Calyx without bractlets (or bractlets minute and not lobed); leafy stolons not present; woods (10–30 cm high; spring)—**Barren-strawberry**, *Waldsteinia fragarioides*

60a. Calyx without bractlets between adjacent sepals; flowers in terminal corymbs or narrow spike-like racemes—61

60b. Calyx with bractlets between adjacent sepals; flowers in irregular or spreading clusters (most *Potentilla* spp., **Cinquefoil** and *Geum* spp., **Avens**)—66

61a. Flowers in corymbs; pistils many; style jointed near the middle, the lower half persisting in fruit; moist woods, SE (30–60 cm high; spring)—**Spring Avens**, *Geum vernum*

61b. Flowers in narrow spike-like racemes; pistils 2; style not jointed near the middle (summer) (*Agrimonia* spp., **Agrimony**)—62

62a. Principal leaflets mostly thirteen or more, more than three times as long as wide; SLP (to 1.2 m high)—*Agrimonia parviflora*

62b. Principal leaflets mostly nine or fewer, less than twice as long as wide—63

63a. Axis of the inflorescence covered with glands—64
63b. Axis of the inflorescence pubescent, a few glands may be present—65

64a. Long spreading hairs also present on inflorescence axis (0.5–1.5 m high)—*Agrimonia gryposepala*

64b. Inflorescence axis just glandular; SLP (to 1 m high)—*Agrimonia rostellata*

65a. Leaves distinctly glandular beneath, the veins sparsely hairy; woods and thickets, UP & Straits (to 1 m high)—*Agrimonia striata*

65b. Leaves with spreading hairs, but not glandular beneath; SLP (to 1 m high)—*Agrimonia pubescens*

66a. Flowers solitary in the axils of foliage leaves, on long peduncles—67
66b. Flowers in cymes or terminal corymbs—68

67a. Leaves palmately compound, both basal and along the stem; leaflets five; roadsides and sandy woods (20–30 cm high; spring)—**Five-finger** or **Common Cinquefoil**, *Potentilla simplex*

67b. Leaves pinnately compound, all in basal tufts; leaflets thirteen or more; damp, often sandy ground (spring and summer)—**Silverweed**, *Potentilla anserina*

68a. Principal leaves palmately compound with five to seven leaflets; fields and roadsides (summer)—69
68b. Principal stem-leaves with three leaflets, or pinnately compound with several to many leaflets—71

69a. Leaves silvery-white beneath, laciniately toothed (10–50 cm high)— **Silvery Cinquefoil**, *Potentilla argentea*
69b. Leaves not silvery-white beneath—70

70a. Leaves with long hairs beneath; petals longer than (6.5) 8 mm (40–80 cm tall)—**Rough-fruited Cinquefoil**, *Potentilla recta*
70b. Leaves gray-tomentose beneath; petals less than 6 mm long—*Potentilla inclinata*

71a. Principal leaves with lobed leaflets, of which the terminal is the largest; leaf axis bearing also some small leaflets between those of usual size (late spring and summer) (*Geum* spp. in part, **Avens**)—72
71b. Principal leaves with toothed or pinnately cleft leaflets, the lateral ones about equaling the terminal one in size, and without any small scattered leaflets (summer) (*Potentilla* spp. in part, **Cinquefoil**)—74

72a. Terminal leaflet of the basal leaves cordate at base; woods, UP (to 1 m high)—*Geum macrophyllum*
72b. Terminal leaflet of the basal leaves wedge-shaped or acute at base—73

73a. Petals shorter than the sepals; disturbed areas, SE—*Geum urbanum*
73b. Petals equalling or exceeding the sepals (to 1 m high); swamps and wet woods—*Geum aleppicum*

74a. Leaflets three; fields and roadsides—**Rough Cinquefoil**, *Potentilla norvegica*
74b. Leaflets five to eleven—75

75a. Leaflets white or silvery tomentose beneath; rocky areas, UP (20–80 cm high)—*Potentilla pensylvanica*
75b. Leaflets green beneath, hairy but not tomentose; sandy or rocky areas (30–100 cm high)—**Prairie Cinquefoil**, *Potentilla arguta*

76a. Leaves all basal, trifoliolate; the flowers on leafless stalks (10–20 cm high; spring) (*Fragaria* spp., **Strawberry**)—77
76b. Cauline leaves present—78

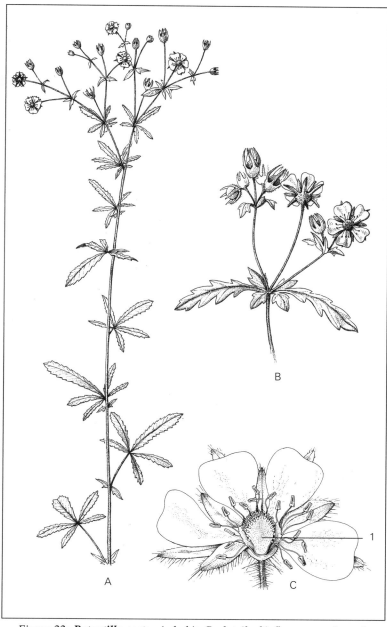

Figure 22: **Potentilla recta**: *A, habit; B, detail of inflorescence (cyme);*
C, flower; 1, receptacle

77a. Leaflets thick and firm, each usually with a short stalk; petioles and pedicels pubescent with spreading or ascending hairs; fruit subglobose, the achenes embedded in pits on its surface; woods, swamps, fields, roadsides—**Wild Strawberry**, *Fragaria virginiana*

77b. Leaflets thin, each sessile; petioles and pedicels nearly glabrous or with appressed hairs; fruit conic, the achenes on its surface; woods and swamps—**Woodland Strawberry**, *Fragaria vesca*

78a. Leaves pinnately compound—79

78b. Leaves trifoliolate or palmately compound with five leaflets—86

79a. Leaves twice or thrice compound (i.e., some leaflets are also compound); flowers unisexual; escape from cultivation, UP (1–2 m high; spring)—**Goatsbeard**, *Aruncus dioicus*

79b. Leaves only once compound; flowers perfect—80

80a. Leaflets laciniate or deeply lobed; flowers pink or purple—81

80b. Leaflets merely toothed; flowers white or red-purple—82

81a. Calyx with bractlets between adjacent sepals; calyx lobes erect; mostly SLP (20–40 cm tall; spring)—**Purple Avens** or **Prairie Smoke**, *Geum triflorum*

81b. Calyx lacks additional bractlets; calyx lobes reflexed; prairies, LP (esp. SW) (50–200 cm tall; summer)—**Queen-of-the-Prairie**, *Filipendula rubra*

82a. Individual flowers small, not exceeding 6 mm across, in large spikes or panicles—83

82b. Individual flowers more than 10 mm wide, in few-flowered clusters—85

83a. Flowers in panicles; stamens many; petals white; escape from cultivation to roadsides and fields (1–2 m high; summer)—**False Spiraea**, *Sorbaria sorbifolia*

83b. Flowers in dense spikes; stamens four or many; petals absent (*Sanguisorba* spp., **Burnet**)—84

84a. Leaflets 3 cm long or longer; stamens 4; pistil 1; moist, calcareous areas, SLP (50–150 cm high; late summer)—**American Burnet**, *Sanguisorba canadensis*

84b. Leaflets 2 cm long or less; stamens many; pistils 2; dry, sandy areas, mostly LP (20–70 cm high; spring)—**Garden Burnet**, *Sanguisorba minor*

85a. Flowers red or purple; swamps and bogs (20–60 cm high; summer)— **Marsh Cinquefoil**, *Potentilla palustris*

85b. Flowers white; sandy areas (30–100 cm high; early summer)—**Prairie Cinquefoil**, *Potentilla arguta*

86a. Pistils five; flowers pink or white; fruits are follicles; oak woods, SLP (50–100 cm high; early summer)—**Bowman's Root**, *Porteranthus trifoliatus*

86b. Pistils numerous, in a head or close group; flowers pink to purple or white; fruits are achenes or a cluster of fleshy drupelets—87

87a. Flowers pink, red, or purple; swamps, bogs, and fens (30–60 cm high; early summer)—**Purple Avens**, *Geum rivale*

87b. Flowers white—88

88a. Leaflets entire below, three-toothed at the apex; sandy areas, NM (10–30 cm high; summer)—**Three-toothed Cinquefoil**, *Potentilla tridentata*

88b. Leaflets toothed all around the margin—89

89a. Leaves all trifoliolate; woods, swamps, rocky areas (15–50 cm high; late spring)—**Dwarf Raspberry**, *Rubus pubescens*

89b. Some of the upper leaves merely lobed or dentate (40–100 cm high) (*Geum* spp. in part, **Avens**)—90

90a. Stem glabrous or with scattered hairs; petals equalling or longer than the sepals; rich woods and stream banks (spring)—*Geum canadense*

90b. Stem bristly-hairy; petals shorter than the sepals—91

91a. Pedicels hirsute, the hairs either reflexed or spreading; receptacle glabrous; damp woods and fields, SLP (spring)—**Rough Avens**, *Geum laciniatum*

91b. Pedicels pubescent, the hairs long, scattered; receptacle pilose; woods, SLP (early summer)—*Geum virginianum*

LEGUMINOSAE (FABACEAE), The Legume Family

Trees, shrubs, or herbs, with alternate (seldom basal), simple, or most often trifoliolate (three leaflets) or pinnately compound, stipulate, leaves. Flowers usually irregular, perfect (rarely unisexual, the plants then monoecious or dioecious); sepals (3–) 5, often connate and forming a two-lipped tube; petals (1 in *Amorpha*) 5, small and sepal-like or usually with a large upper petal (the *banner* or *standard*) and four smaller ones (two laterals called *wings* and two lower petals apically united to form a *keel* enclosing the stamens and pistil); stamens (5–)10, most or all of the filaments often united; pistil 1, style 1, ovary superior, 1-celled. Fruit a *legume*, often dehiscent, sometimes jointed.

The Legume family, which includes nearly 18,000 species (Zomlefer 1995), is treated either as one family including three subfamilies (as here) or sometimes as three separate families. Two subfamilies, the Caesalpinioideae and the Faboideae, are represented in Michigan; all but species of *Cassia*, *Cercis*, *Gleditsia*, and *Gymnoclados* are members of the Faboideae. The differences between these two subfamilies are most easily seen in the flower. Members of the Caesalpinioideae usually have distinct petals with the uppermost petal smaller than the (lower) lateral petals and stamens which are neither fused nor hidden by the petals. The Faboideae are characterized by a large uppermost petal and the stamens hidden by the petals and either all distinct or, more often, with filaments fused.

1a. Trees or shrubs—2
1b. Herbs—10

2a. Leaves simple, broadly cordate; flowers pink (white in some cultivars), appearing before the leaves; moist woods, stream banks, SLP (to 12 m; early spring)—**Redbud**, *Cercis canadensis*
2b. Leaves pinnately compound; flowers white to pink or blue, appearing with or after the leaves—3

3a. Twigs or branches thorny—4
3b. Thorns absent—7

4a. Thorns branched, scattered on the trunk and along the branches; flowers regular, greenish-yellow; fruits 18 cm or longer; moist woods (SLP), also escape from cultivation (LP) (to 20(–40) m high; spring)—**Honey Locust**, *Gleditsia triacanthos*
4b. Thorns unbranched, a pair of them at the base of each leaf; flowers irregular, white to rose; fruits 10 cm or shorter; escape from cultivation to roadsides, fields, etc. (*Robinia* spp., **Locust**)—5

5a. Flowers white; branches glabrous or nearly so; tree (to 25 m; spring)—**Black Locust**, *Robinia pseudoacacia*
5b. Flowers pale pink to rose; branches glandular-pubescent or bristly; tree or shrub; LP—6

6a. Leaflets thirteen or more; branches glandular-pubescent; tree or shrub (to 5 or rarely 12 m high; late spring)—**Clammy Locust**, *Robinia viscosa*

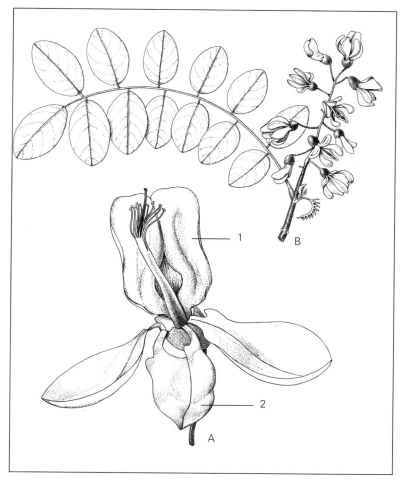

Figure 23: **Robinia pseudoacacia**: *A, flower (partially dissected so that stamen tube is not hidden by keel); 1, banner; 2, keel; B, raceme*

6b. Leaflets thirteen or fewer; branches bristly with stiff hairs; shrub (1–3 m high; early summer)—**Bristly Locust**, *Robinia hispida*

7a. Trees; leaves once- to thrice-pinnate; flowers regular with five petals, greenish-white (spring)—8
7b. Low shrubs; leaves once-pinnate; flowers irregular with one petal, blue to purple (spring and summer) (*Amorpha* spp.)—9

8a. Leaflets 2 cm or more broad; fruits 3 cm or more broad, less than 15 cm long; river-bank woods, SLP (to 30 m high)—**Kentucky Coffee-tree**, *Gymnocladus dioica*

8b. Leaflets about 1 cm broad; fruits up to 3 cm broad, more than 18 cm long; a thornless variety escaping from cultivation (LP)—**Honey Locust**, *Gleditsia triacanthos*

9a. Leaflets pubescent, not (or scarcely) glandular below; fruit pubescent; sandy areas, mostly SW (0.5–1 m high; summer)—**Lead-plant**, *Amorpha canescens*

9b. Leaflets glabrous and glandular below; fruit glandular; escape from cultivation to roadsides, fields, etc., SLP & UP (to 4 m high; spring)—**False Indigo**, *Amorpha fruticosa*

10a. Stem vine-like, twining or trailing; terminal leaflet not modified into a tendril—11

10b. Stem twining, climbing, or prostrate to erect; if twining or climbing, the terminal leaflet is modified into a tendril—15

11a. Leaflets five to seven; flowers brownish-purple; wet woods and fields (summer)—**Wild Bean**, *Apios americana*

11b. Leaflets three; flowers greenish, purple, or white—12

12a. Plants prostrate; leaflets nearly circular, the apex rounded; fruit segmented into single-seeded units; dry oak woods, SLP (summer)— *Desmodium rotundifolium*

12b. Plants twining; leaflets longer than broad, the apex acute; fruit not segmented—13

13a. Flowers purple to white, not immediately subtended by a pair of tiny bracts; apex of keel petals straight; woods and shores (to 1.5 m high; late summer)—**Hog Peanut**, *Amphicarpaea bracteata*

13b. Flowers purple, greenish, or white, immediately subtended by a pair of tiny bracts; apex of keel petals arched upward or coiled—14

14a. Flowers in small heads; apex of keel petals arched upward; thickets, SLP (to 1 m high; summer)—**Wild Bean**, *Strophostyles helvula*

14b. Flowers in racemes; apex of keel petals coiled; commonly cultivated, rarely escapes; LP (summer)—**Common Bean**, *Phaseolus vulgaris*

15a. Terminal leaflet is a tendril—16

15b. Terminal leaflet is not a tendril—31

16a. Style round in cross-section, with a tuft of hairs at the apex or only along the outer (lower) side; wings of the corolla adherent to the keel for up to one-half its length; leaflets four or more (plus the tendril); flowers purple to white or rarely pink (spring and summer) (*Vicia* spp., **Vetch**)—17

16b. Style flat in cross-section, hairy along the inner (upper) side; wings of the corolla free from or adherent only at the base of the keel; leaflets two or four to more (plus the tendril); flowers red-purple, rose, yellow, or white (late spring and summer) (*Lathyrus* spp.)—23

17a. Flowers paired in the leaf axils, sessile or nearly so; cultivated and escaping to roadsides, etc. (to 1 m high; summer)—**Spring Vetch**, *Vicia sativa* [incl. *V. angustifolia*—CQ]

17b. Flowers solitary on a long peduncle or (more often) in peduncled racemes—18

18a. Flowers less than 8 mm long, white; one to seven flowers per raceme; roadsides, fields, etc. (spring and summer)—19

18b. Flowers 8 mm or more long, purple to white or rarely pink; more than eight flowers per raceme—20

19a. Calyx lobes of equal lengths; fruit hairy, two-seeded; LP (30–60 cm high)—*Vicia hirsuta*

19b. Calyx lobes of different lengths; fruit glabrous, four-seeded; LP & eastern UP (30–60 cm)—**Sparrow Vetch**, *Vicia tetrasperma*

20a. Base of calyx swollen; pedicel attachment appears to be lateral; roadsides, fields, etc. (to 1 m high; spring and summer)—**Hairy Vetch**, *Vicia villosa* [incl. *V. dasycarpa*—CQ]

20b. Base of calyx not swollen, pedicel attachment is terminal—21

21a. Margin of stipules serrate; flowers 15 mm or longer (to 1 m high; spring and early summer)—**American Vetch**, *Vicia americana*

21b. Margin of stipules entire; flowers less than 15 mm long—22

22a. Racemes dense, one-sided; flowers blue (rarely white); calyx lobes of equal lengths; roadsides, fields, etc. (to 1 m high; summer)—**Bird Vetch**, *Vicia cracca*

22b. Racemes loosely flowered, not one-sided; flowers whitish, the keel petals tipped with blue; calyx lobes of unequal lengths; dry woods, SLP (to 1 m high; spring)—**Wood Vetch**, *Vicia caroliniana*

23a. Leaflets four or more; native species—24

23b. Leaflets two; escapes from cultivation to roadsides, fields, etc.—28

24a. Flowers yellowish-white; woods (to 80 cm high; spring and early summer)—**Pale Vetchling, *Lathyrus ochroleucus***

24b. Flowers purple to red-purple or rarely white (to 1 m high; summer)—25

25a. Stipules with two basal lobes, almost as large as the leaflets; Great Lakes beaches and dunes—**Beach Pea, *Lathryrus japonicus* [*L. maritimus*—CQ]**

25b. Stipules with one basal lobe, apparently attached laterally near the middle—26

26a. Leaflets in four to six (rarely seven) pairs, ovate; racemes usually with ten or more flowers; sandy areas—***Lathyrus venosus***

26b. Leaflets in two to four (rarely five) pairs, linear to oblong or elliptical; racemes with two to nine flowers; marshes, wet areas—27

27a. Stems with a membranous wing on the margins—**Marsh Pea, *Lathyrus palustris***

27b. Stems angled, but not winged—**Marsh Pea, *Lathyrus palustris* var. *myrtifolius***

28a. Stems with flat green wings; flowers red-purple, rose, or white—29

28b. Stems may be angled, but wings absent; flowers red-purple or yellow—30

29a. Primary stipule lobe half as wide as the winged stem or more; leaflets and stipules narrow, lance-linear to lanceolate—**Everlasting Pea, *Lathyrus sylvestris***

29b. Primary stipule lobe wider than half the width of the winged stem; leaflets and stipules broader, elliptic to narrowly ovate (to 2 m high; summer)—**Everlasting Pea, *Lathyrus latifolius***

30a. Flowers yellow; stipules with two basal lobes (30–80 cm high)—**Yellow Vetchling, *Lathyrus pratensis***

30b. Flowers red-purple; stipules with one basal lobe (to 80 cm high)—**Tuberous Vetchling, *Lathyrus tuberosus***

31a. Leaves simple; flowers yellow; dry areas, SLP (10–40 cm high; summer)—**Rattlebox, *Crotalaria sagittalis***

31b. Leaves palmately or pinnately compound; flowers of various colors—32

32a. Leaves palmately compound, leaflets seven or more; flowers blue (or less often pink, yellow, or white), in showy terminal racemes—33

32b. Leaves with three leaflets or pinnately compound; flowers of various colors—34

33a. Leaflets seven to eleven, less than 5 cm long, the apex rounded; sandy areas, LP (20–60 cm high; spring)—**Wild Lupine**, *Lupinus perennis*
33b. Leaflets twelve to seventeen, more than 6 cm long, the apex acute; roadsides, western UP (to 1 m high; summer)—*Lupinus polyphyllus*

34a. Leaves pinnately compound; leaflets five to many—35
34b. Leaves compound; leaflets three (rarely four)—43

35a. Leaflets often five, linear, less than 3 mm wide; flowers rose-purple in a cone-like spike; sandy areas, SLP (20–100 cm high; summer)—**Purple Prairie-clover**, *Dalea purpurea*
35b. Leaflets five or more, more than 3 mm wide; flowers yellow, rose, or white—36

36a. Leaves with an even number of leaflets, a terminal leaflet absent; flowers bright yellow; stamens separate; SLP (*Cassia* spp., **Senna**)—37
36b. Leaves with an odd number of leaflets, a terminal leaflet present; flowers white, yellow, or pink; stamens united, a sheath formed by 9 connate filaments—38

37a. Leaflets lanceolate-oblong, 2–5 cm long; stamens ten, seven with normal anthers and three with abortive anthers; moist banks and woods (0.5–2 m high; summer)—**Wild Senna**, *Cassia hebecarpa* [*Senna hebecarpa*—CQ]
37b. Leaflets linear-oblong, 2 cm long or less; stamens ten, with ten normal anthers; sandy roadsides and railroads (10–60 cm high; summer)—**Partridge-pea**, *Cassia chamaecrista* [*Chamaecrista fasciculata*—CQ]

38a. Flowers in a spike or raceme—39
38b. Flowers in an umbel—42

39a. Stem silky-hairy with whitish spreading hairs; flowers with a yellowish banner, pink wings in a terminal raceme; sandy fields and woods, SLP (20–70 cm high; summer)—**Goat's Rue**, *Tephrosia virginiana*
39b. Stem glabrous or nearly so; flowers white or yellowish in axillary racemes (*Astragalus* spp., **Milk-vetch**) (summer)—40

40a. Ovary and fruit pubescent; escape from cultivation, East Lansing (25–70 cm high)—**Chick-pea Milk-vetch**, *Astragalus cicer*
40b. Ovary and fruit glabrous—41

41a. Stipules connate around the stem; leaves pubescent beneath with T-shaped hairs (to 1.5 m high)—**Canadian Milk-vetch**, *Astragalus canadensis*

41b. Stipules not connate; leaves pubescent beneath with simple hairs (30–90 cm high)—**Cooper's Milk-vetch**, *Astragalus neglectus*

42a. Flowers yellow or orange; leaflets five; a forage plant, escaping to roadsides, etc. (to 60 cm high; summer)—**Birdfoot Trefoil**, *Lotus corniculata*

42b. Flowers pink; leaflets ten to many; often planted for erosion control, mostly LP (30–100 cm high; spring and summer)—**Crown-vetch**, *Coronilla varia*

43a. Leaflet margins finely toothed—44

43b. Leaflet margins entire—57

44a. Inflorescence an elongate raceme, mostly four or more times longer than wide; stipules one-veined; roadsides, etc. (summer) (*Melilotus* spp., **Sweet Clover**)—45

44b. Inflorescence a head, umbel, spike, or short raceme, less than three times longer than wide; stipules two- to three-veined—46

45a. Flowers yellow (0.5–2 m high)—**Yellow Sweet-clover**, *Melilotus officinalis*

45b. Flowers white (1–3 m high)—**White Sweet-clover**, *Melilotus alba*

46a. Stalk of the terminal leaflet distinctly longer than that of the two lateral leaflets—47

46b. Stalk of the terminal leaflet the same length as that of the two lateral leaflets, or all three leaflets sessile; roadsides, lawns and fields, etc. (*Trifolium* spp. in part, **Clover**)—51

47a. Fruit curved or coiled, the corolla not persistent; calyx lobes of similar lengths (*Medicago* spp.)—48

47b. Fruit straight, the corolla persistent and sometimes enclosing the fruit; calyx lobes of two different lengths; roadsides, lawns and fields, etc. (10–40 cm high) (*Trifolium* spp. in part, **Clover**)—50

48a. Plant prostrate to ascending; flower yellow, less than 4 mm long; pod curved, black; roadsides, fields, etc. (spring and summer)—**Black Medick**, *Medicago lupulina*

48b. Plant erect; flowers violet, blue, yellow, or white, more than 7 mm long; pod coiled, green; widely cultivated for forage, escaping to roadsides, etc. (30–100 cm high; summer)—49

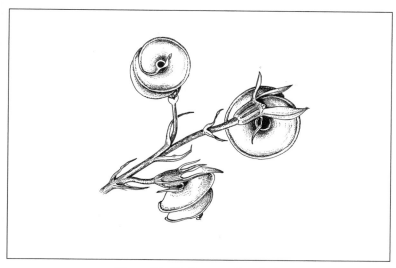

*Figure 24: **Medicago sativa**, fruit*

49a. Flowers violet or blue (rarely white); fruits coiled tightly into more
   than 1fi coils, the opening in the center tiny—**Alfalfa**, *Medicago sativa*
   subsp. *sativa*

49b. Flowers blue, yellow, white, or variegated; fruits coiled loosely into 1
   to 1fi (or rarely 2) coils, the opening in the center large—**Alfalfa**, *Medicago sativa* subsp. ×*varia*

Alfalfa is an important forage crop and may often be collected as an escape. Much of the
acreage devoted to this crop is planted with hybrid strains derived from **Medicago sativa** (blue
flowers and tightly coiled fruit) and **Medicago falcata** (yellow flowers and curved or nearly
straight fruit). The variability exhibited in **M. sativa** subsp. ×**varia** reflects this heritage. See Rabeler and Gereau (1984) and Voss (1985).

50a. Flowers twenty or more, each 4.5 mm or longer; banner striate when
   dry (spring and summer)—**Low Hop Clover**, *Trifolium campestre*

50b. Flowers usually fifteen or fewer, each less than 4 mm long; banner
   scarcely striate or not at all (summer)—**Little Hop Clover**, *Trifolium
   dubium*

51a. Flowers bright yellow—52

51b. Flowers white, cream, purple or red; never yellow—54

52a. Stipules lanceolate, as long as the petiole (20–50 cm high; spring and summer)—**Hop Clover**, *Trifolium aureum*

52b. Stipules ovate, half the length of the petiole (10–40 cm high)—53

53a. Flowers twenty or more, each 4.5 mm or longer; banner striate when dry (spring and summer)—**Low Hop Clover**, *Trifolium campestre*

53b. Flowers usually fifteen or fewer, each less than 4 mm long; banner scarcely striate or not at all (summer)—**Little Hop Clover**, *Trifolium dubium*

54a. Individual flowers sessile, or on very short pedicels (spring and summer)—55

54b. Individual flowers distinctly pedicelled (summer)—56

55a. Heads oblong, on distinct peduncles; calyx densely hairy, longer than the white or pink corolla; sandy areas (10–40 cm tall)—**Rabbitfoot Clover**, *Trifolium arvense*

55b. Heads nearly globose, almost sessile, closely subtended by a pair of leaves; corolla usually red-purple, longer than the sparsely-hairy calyx (20–80 cm high)—**Red Clover**, *Trifolium pratense*

56a. Stems prostrate or creeping; heads long-peduncled, arising from the creeping branches; corolla white (flower peduncles 10–20 cm high)—**White Clover**, *Trifolium repens*

56b. Some or all of the stems erect; heads arising from the leafy stems; corolla white and pink (30–80 cm high)—**Alsike Clover**, *Trifolium hybridum*

57a. Stalk of the terminal leaflet the same length as that of the two lateral leaflets, or all three leaflets sessile—58

57b. Stalk of the terminal leaflet distinctly longer than that of the two lateral leaflets—61

58a. Inflorescence an umbel of yellow or orange flowers; leaflets five, the lowest pair basal and stipule-like; a forage plant, escaping to roadsides, etc. (to 60 cm high; summer)—**Birdfoot Trefoil**, *Lotus corniculata*

58b. Inflorescence a terminal raceme or axillary clusters of one to three flowers; flowers yellow or white; leaflets three—59

59a. Inflorescence axillary, flowers either solitary or in two- or three-flowered clusters; flowers whitish; escape from cultivation to roadsides, fields, etc., LP (to 15 cm high; autumn)—**Sericea**, *Lespedeza cuneata*

59b. Inflorescence a terminal raceme; flowers yellow or white; sandy woods and fields, SLP (summer) (*Baptisia* spp., **False Indigo**)—60

60a. Flowers yellow (50 cm–1 cm high)—***Baptisia tinctoria***

60b. Flowers white, banner may be purple-tinged (1–2 m high)—***Baptisia lactea***

61a. Calyx lobes less than one-half the length of the calyx tube; flowers pink or purple, rarely white—62

61b. Calyx lobes as long or longer than the calyx tube; flowers various—63

62a. Raceme on a leafless stem arising near the base of the plant; pedicels 1 cm or longer; woods, mostly SLP (40–100 cm high; summer)— ***Desmodium nudiflorum***

62b. Raceme tops a leafy stem, most leaves near the apex; pedicels shorter than 1 cm; woods, mostly LP (30–80 cm high; summer)—***Desmodium glutinosum***

63a. Individual leaflets bear stipule-like structure (stipels); calyx two-lipped (upper and lower lobes of different lengths; the upper two connate); fruit two- to several-seeded, elongate, often transversely jointed—64

63b. Individual leaflets lack stipels; calyx not two-lipped, all five lobes separate; fruit one-seeded, ovate or ovoid, not transversely jointed; dry sandy woods and fields, mostly SLP (late summer) (***Lespedeza*** spp. in part, **Bush-clover**)—74

64a. Fruit transversely segmented into two or more joints; stems glabrous or with short, hooked hairs; flowers pinkish, often becoming green; noncrop plants (summer) (***Desmodium*** spp. in part, **Tick-trefoil** or **Beggarticks**)—65

Desmodium is a large genus of legumes in which the fruit is modified for dissimination via animals. The legume pod breaks up into one-seeded segments at maturity. Hooked hairs facilitate attachment to fur or clothing. Morphology of the mature fruit is important in identification of species.

64b. Fruit not transversely segmented; stems glabrous or with long simple hairs; flowers white, yellowish, or purple; commonly cultivated, rarely escapes (summer)—73

65a. Leaflets ovate-oblong to suborbicular, the apices rounded; terminal leaflet less than 3 mm long; oak woods and clearings, SLP—66

65b. Leaflets ovate to lanceolate or nearly linear, the apices acute; terminal leaflet more than 3.5 mm long—67

66a. Stem and leaves glabrous or very nearly so (60–120 cm high)— ***Desmodium marilandicum***

66b. Stem and leaves conspicuously pubescent (40–100 cm high)—***Desmodium ciliare***

67a. Leaves sessile or nearly so; leaflets more or less linear; sandy areas, SLP (60–150 cm high)—***Desmodium sessilifolium***
67b. Leaves obviously petioled; leaflets ovate to lanceolate—68

68a. Leaflets pubescent beneath, some of the hairs hooked; SLP—69
68b. Leaflets either glabrous or pubescent beneath; if pubescent, none of the hairs hooked—70

69a. Stem often branched, with more than one raceme; stem hairs long, straight and non-glandular; fruit segments 7 mm or longer; open sandy areas (60–150 cm high)—***Desmodium canescens***
69b. Stem often simple, with a single raceme; stem hairs shorter, glandular or hooked; fruit segments less than 7 mm long; prairies and sandy woods (1–2 m high)—***Desmodium illinoense***

70a. Lower margins of fruit segments rounded—71
70b. Lower margins of fruit segments angled, the segments distinctly triangular; SLP—72

71a. Flowers longer than 8 mm, the calyx longer than 3 mm; fields and shores (to 2 m high)—***Desmodium canadense***

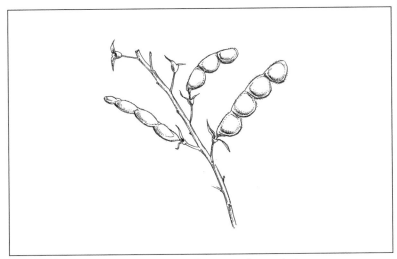

Figure 25: ***Desmodium canadense***, *fruit*

71b. Flowers smaller, less than 6 mm long, the calyx 3 mm or less; dry sandy areas, SLP (to 1.2 m high) ***Desmodium obtusum* [*D. rigidum*— CQ]**

72a. Stipules lanceolate to ovate, 1 cm long or more; terminal leaflet ovate, the apex acuminate; oak woods and openings (to 2 m high)—***Desmodium cuspidatum***

72b. Stipules narrowly lanceolate to subulate, less than 1 cm long; terminal leaflet elliptic to lanceolate, not acuminate; oak woods and sandy fields (50–120 cm high)—***Desmodium paniculatum* [incl. *D. glabellum*— CQ]**

73a. Stems densely covered with long, sharp hairs; SLP—**Soybean**, *Glycine max*

73b. Stems either glabrous or with scattered hairs; LP—**Common Bean**, *Phaseolus vulgaris*

74a. Flowers yellowish-white—75
74b. Flowers purple—76

75a. Leaflets less than twice as long as broad; flowers in short spikes, each spike on a peduncle about as long as the inflorescence (to 1.5 m high)—***Lespedeza hirta***

75b. Leaflets more than twice as long as broad; flowers in globose heads, each head sessile or nearly so (60 cm–1.5 m high)—***Lespedeza capitata***

76a. Flower clusters mostly on peduncles longer than the leaves; keel petals longer than the wings (40–80 cm high)—***Lespedeza violacea***

76b. Flower clusters sessile, or on peduncles shorter than the subtending leaves; keel petals shorter than the wings—77

77a. Leaflets linear-oblong, usually over four times as long as broad (30–100 (rarely to 150) cm high)—***Lespedeza virginica***

77b. Leaflets ovate or oval, at most three times as long as broad (to 1 m high)—***Lespedeza intermedia***

Specimens of **Lespedeza** spp. which are probable hybrids between different pairs of our five native species have been collected in Michigan. Any such hybrid will exhibit intermediate combinations of characteristics.

LINACEAE, The Flax Family

Herbs with alternate or opposite, simple, sessile leaves. Flowers regular, perfect, in cymes; sepals 5; petals 5, often falling soon after flower opens; stamens 5, the filaments united below, 5 staminodes sometimes present; pistil 1, styles 5, ovary superior, 5-celled. Fruit a capsule which splits into 10 *mericarps* (segments). Summer.

1a. Petals blue or white with a yellow base (summer)—2
1b. Petals yellow—4

2a. Petals white with a yellow base; cauline leaves opposite; calcareous areas, NM (10–30 cm high)—**Fairy Flax**, *Linum catharticum*
2b. Petals blue (rarely all white); cauline leaves alternate—3

3a. Leaves three-nerved; inner sepals with ciliate margins; disturbed areas (30–100 cm high)—**Common Flax**, *Linum usitatissimum*
3b. Leaves one-nerved (sometimes faintly three-nerved at the base); inner sepal margins glabrous; escapes from cultivation to disturbed areas, LP (30–70 cm high)—**Perennial Flax**, *Linum perenne*

4a. Cauline leaves below the branches opposite; sandy shores, WM (30–90 cm high)—*Linum striatum*
4b. Cauline leaves below the branches alternate; open, sandy areas—5

5a. All sepals glandular-ciliate; one pair of dark glands at the base of each leaf; mostly LP (20–80 cm high)—*Linum sulcatum*
5b. Outer sepals glandless, inner sepals glandular-ciliate; glands at the base of each leaf absent; SLP (20–70 cm high)—*Linum medium*

OXALIDACEAE, The Wood Sorrel Family

Perennial herbs with alternate or basal, compound leaves with 3 obcordate leaflets. Flowers regular, perfect; sepals 5; petals 5; stamens 10, of two lengths, all filaments united below; pistil 1, styles 5, ovary superior, 5-celled. Fruit a capsule.

1a. Leaves all basal; petals white to pink; woods (esp. coniferous), NM—*Oxalis acetosella*
1b. Cauline leaves present; petals yellow; disturbed areas or woods (10–50 cm high; spring and summer)—2

2a. Pedicels with spreading, septate (segmented) hairs; stipules absent; disturbed areas and woods—*Oxalis fontana* [*O. stricta*—CQ]

2b. Pedicels with appressed, non-septate hairs; stipules present—3

3a. Stem erect or at most decumbent; hairs dense and downward-pointing; leaves whorled or fascicled; disturbed areas and woods—***Oxalis stricta*** [***O. dillenii***—CQ]
3a. Stem prostrate and creeping; hairs sparse and spreading; leaves alternate; lawns, SLP—***Oxalis corniculata***

GERANIACEAE, The Geranium Family

Herbs with alternate, opposite, or basal, deeply lobed, divided, or compound leaves. Flowers regular, perfect; sepals 5; petals 5, sometimes of two sizes; stamens 5 or 10, up to 5 staminodes sometimes replacing the outer fertile anthers; pistil 1, style 1, ovary superior, 5-celled. Fruit a schizocarp, splitting into 5 mericarps.

1a. Leaves pinnately compound, leaf blades pinnately dissected, mostly basal; sandy or rocky areas (to 40 cm high; spring and summer)— **Stork's-bill**, ***Erodium cicutarium***
1b. Leaves palmately divided (lobed) or compound, mostly cauline and opposite (**Crane's-bill**, ***Geranium*** spp.)—2

2a. Leaves palmately compound, the leaflets pinnately divided; rich woods (20–60 cm high; spring and summer)—**Herb Robert**, ***Geranium robertianum***
2b. Leaf blades palmately lobed, not divided to the base—3

3a. Petals 10 mm long or more, exceeding the sepals—4
3b. Petals less than 10 mm long, about equalling the sepals—5

4a. Petals pink, the apex entire or nearly so; basal leaves conspicuous, the blades larger than those of the 1 or 2 pairs of cauline leaves; rich woods (30–70 cm high; spring)—**Wild Geranium**, ***Geranium maculatum***
4b. Petals reddish purple or white, the apex notched; basal leaves few (or absent), the blades of similar size to those of the many cauline leaves; escapes from cultivation to disturbed areas (summer)—***Geranium sanguineum***

5a. Sepal apex an awl-shaped awn up to 2 mm long; sandy or rocky areas (spring and summer)—6
5b. Sepal apex not an awn, at most a tiny callous spot; disturbed areas—7

6a. Pedicels shorter than or equalling the calyx, less than twice the length of the calyx in fruit; hairs on the pedicels not glandular; SLP and UP (to 60 cm high)—***Geranium carolinianum***

6b. Pedicels longer than the calyx, more than twice the length of the calyx in fruit; most hairs on the pedicels glandular (to 50 cm high)—*Geranium bicknellii*

7a. Mature mericarps glabrous, often wrinkled; WM (20–50 cm high; spring and summer)—*Geranium molle*
7b. Mature mericarps pubescent, not wrinkled (to 50 cm high; summer)—*Geranium pusillum*

ZYGOPHYLLACEAE, The Caltrop Family

Annual herbs with opposite, pinnately compound leaves. Flowers regular, perfect, axillary; sepals 5, falling soon after opening; petals 5; stamens 10; pistil 1, style 1, ovary superior, 5-celled. Fruit a schizocarp, splitting into 4 or 5 segments, each with two large spines.

One species in Michigan, a prostrate mat-forming herb with yellow flowers; disturbed areas, LP—**Caltrop** or **Puncture Vine**, *Tribulus terrestris*

RUTACEAE, The Rue Family

Herbs, shrubs, or small trees with alternate, compound leaves frequently dotted with translucent glands. Flowers regular, perfect or unisexual (the plants then dioecious or with both unisexual and perfect flowers); sepals 4 or 5; petals 4 or 5; stamens 4, 5, 8, or 10, the filaments sometimes united below; pistil and style 1 or 2–5; each ovary superior, 1- or 2-celled. Fruit a follicle, capsule, or samara.

1a. Herb; flowers showy, white or pink; stamens ten; escape from cultivation, Straits Area (to 1 m high; summer)—**Gas-plant**, *Dictamnus albus*
1b. Shrub or small tree; flowers small, greenish-white; stamens four or five—2

2a. Leaflets five to eleven; stems thorny; ovaries two to five, fruit two to five short follicles; woods and thickets (to 8 m high; spring)—**Prickly-ash**, *Zanthoxylum americanum*
2b. Leaflets three; stems not thorny; ovary 1, fruit a flat samara; sandy fields and dunes, SLP (spring)—**Hop-tree** or **Wafer-ash**, *Ptelea trifoliata*

SIMAROUBACEAE, The Quassia Family

Trees with alternate, pinnately compound leaves. Flowers regular, mostly unisexual (the plants then dioecious with occasional perfect flowers), small, greenish-yellow in large pyramidal panicles; sepals 5; petals 5; stamens 10 or fewer; pistils and styles 2–5, each ovary superior, 1-celled. Fruit a samara.

One species in Michigan, escaped from cultivation; often in urban areas, LP (early summer)—**Tree-of-heaven**, *Ailanthus altissima*

POLYGALACEAE, The Milkwort Family

Small herbs with alternate or whorled simple leaves. Flowers irregular, perfect; sepals 5, the 2 lateral sepals (the *wings*) enlarged and petaloid; petals 3, united below, the lower one keeled and enclosing the stamens and style; stamens (6) 7 or 8, the filaments united into a sheath with the petals; pistil 1, style 1, ovary superior, 2-celled. Fruit a capsule.

1a. Flowers few, loosely clustered, mostly 15–20 mm long, magenta to purple (or white); leaves clustered near the stem apex; woods (10–15 cm high; spring)—**Fringed Polygala**, **Flowering-wintergreen**, or **Gaywings**, *Polygala paucifolia*
1b. Flowers many, in a spike or raceme, less than 10 mm long, magenta to purple or white to greenish; leaves not clustered near the stem apex—2

2a. All of the leaves alternate—3
2b. Some or all of the leaves whorled or opposite—5

3a. Inflorescence a loose raceme; flowers magenta, the lower ones on 1–2 mm pedicels; sandy areas (10–25 cm high)—*Polygala polygama*
3b. Inflorescence a dense raceme resembling a spike; flowers magenta, greenish, or white, the lower ones nearly sessile—4

4a. Stem glabrous, solitary from a tap root; inflorescence about twice as long as broad; flowers magenta, greenish, or white; fields and ditches, LP (10–40 cm high; summer)—*Polygala sanguinea*
4b. Stems glandular-pubescent, several from woody base; inflorescence three to five times as long as broad; flowers white (10–50 cm high; spring)—**Seneca Snakeroot**, *Polygala senega*

5a. Raceme 10 mm or more broad, the apex blunt; flowers purplish; sandy shores, SLP (10–30 cm high; summer)—***Polygala cruciata***

5b. Raceme 5 mm or less broad, tapering to an acute apex; flowers greenish-white to purple; SLP (10–40 cm high; summer and autumn)—***Polygala verticillata***

EUPHORBIACEAE, The Spurge Family

Monoecious herbs with alternate, opposite, or whorled, simple leaves and usually with milky juice. Flowers small, regular, unisexual, small, often in a specialized inflorescence, the *cyathium*; sepals 3–7 or often 0; petals 5–7 or usually 0; stamens 1 or 4–9; pistil 1, styles 2 or 3, ovary superior, 1–3-celled. Fruit a capsule. Summer.

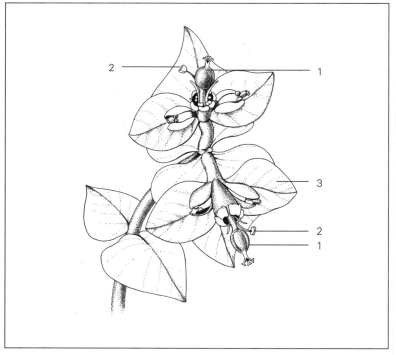

*Figure 26: Inflorescence of **Euphorbia** spp.(cyathium): 1, pistillate flower; 2, staminate flower; 3, bract*

Most of our species of the Euphorbiaceae are members of the large genus **Euphorbia**. They possess a unique, cup-shaped inflorescence known as a *cyathium*. Several staminate flowers, each consisting of only a single stamen, and one pistillate flower, consisting of a single, pedicelled, three-lobed ovary, are enclosed in a four- to five-lobed involucre. The involucre resembles a calyx, and sometimes bears one, four, or five glands which may be colored (resembling a corolla). Thus, a single cyathium may resemble a flower, or clusters of cyathia surrounded by showy bracts may also be flower-like in appearance.

1a. Inflorescences axillary or terminal cyathia; cauline leaves opposite or alternate (***Euphorbia*** spp., **Spurge**)—2
1b. Inflorescences axillary spikes, terminal spike-like clusters, or flowers solitary, never in cyathia; cauline leaves alternate—15

2a. All cauline leaves opposite, the leaf blade bases usually asymmetrical; inflorescence of axillary or terminal clusters, not umbellate—3
2b. Most cauline leaves alternate (the uppermost directly beneath the inflorescence may be opposite or whorled), the leaf blade bases symmetrical; inflorescence a terminal umbel subtended by a whorl of leaf-like bracts—10

3a. Stem and foliage glabrous; stems prostrate (subg. ***Chamaesyce*** in part)—4
3b. Stem and foliage more or less pubescent; stems prostrate or erect; disturbed areas—7

4a. Leaves entire; seeds rounded, smooth—5
4b. Leaves serrulate, at least apically; seeds angled, with roughened or pitted surfaces; open sandy areas—6

5a. Cyathium glands lack petaloid appendages; seeds over 2 mm long; sandy shores of the Great Lakes (not Lake Superior)—**Seaside Spurge**, ***Euphorbia polygonifolia***
5b. Cyathium glands have petaloid appendages; seeds 1.5 mm or shorter; disturbed sandy areas, NM—***Euphorbia geyeri***

6a. Leaves broadly oblong or obovate; seeds faintly wrinkled; NM—***Euphorbia serpyllifolia***
6b. Leaves narrowly oblong; seeds with three to four prominent transverse ridges—***Euphorbia glyptosperma***

7a. Leaf bases symmetrical; involucral gland one, lacking any petaloid appendages; mostly LP (20–60 cm high)—***Euphorbia dentata***
7b. Leaf bases asymmetrical; involucral glands four, each with a petaloid appendage (subg. **Chamaesyce** in part)—8

8a. Stems erect with ascending branches; largest leaf blade over 16 mm long
   (20–40 cm high)—***Euphorbia nutans***

8b. Stems prostrate or virtually so; largest leaf blade less than 16 mm
   long—9

9a. Ovary and capsule pubescent; lawn weed—***Euphorbia maculata***

9b. Ovary and capsule glabrous—***Euphorbia vermiculata***

One of the essential tasks in compiling a state flora (list of plants found in one state) is check-
ing to determine that specimens of each species reported have actually been collected in the
state. This involves careful review of herbarium specimens to verify the identification and the
location where collected. Gleason, in his original edition of *Plants of Michigan*, listed **Euphorbia humistrata** as present in Michigan. Voss (1985) noted that early reports of this
species in Michigan were based on misidentified collections which were actually specimens
of **E. vermiculata** or **E. maculata**. Species included here are those whose occurrence in Michigan is supported by at least one herbarium specimen. Most of these were verified by Voss in
preparation of *Michigan Flora* (1972, 1985, 1996).

10a. Cyathium glands five, each with a conspicuous, white, petaloid ap-
    pendage (subg. ***Agaloma***)—11

10b. Cyathium glands four, lacking petaloid appendages (subg. ***Esula***)—12

11a. Upper leaves with conspicuous white margins; capsules pubescent; es-
    capes from cultivation to disturbed areas (30–80 cm high)—**Snow-on-
    the-mountain**, ***Euphorbia marginata***

11b. Leaves lack a white margin; capsules glabrous; prairies, sandy fields,
    and open dry woodlands, mostly LP (30–100 cm high)—**Flowering
    Spurge**, ***Euphorbia corollata***

12a. Cauline leaves below the inflorescence serrulate; disturbed areas and
    shores (20–50 cm high)—***Euphorbia helioscopia***

12b. Cauline leaves below the inflorescence entire—13

13a. Cauline leaves obovate to nearly circular, not more than three times as
    long as wide; bracts immediately subtending the flowers green; seeds
    pitted; taprooted annual, yards (10–30 cm high)—***Euphorbia peplus***

13b. Cauline leaves linear to spathulate, over six times as long as wide;
    bracts immediately subtending the flowers yellowish; seeds smooth;
    aggressive European perennial—14

14a. Cauline leaves linear, less than 3 mm wide; floral bracts less than 7 mm
    wide; disturbed areas (15–40 cm high)—**Cypress Spurge**, ***Euphorbia
    cyparissias***

*Figure 27: **Euphorbia cyparissias**: note complex inflorescence
(umbel of cyathia)*

14b. Cauline leaves narrowly oblong-spathulate, more than 3 mm wide; floral bracts more than 8 mm wide; a serious local pest of disturbed areas
and fields, esp. northern LP (30–70 cm high)—**Leafy Spurge**, *Euphorbia esula*

15a. Inflorescences axillary, pistillate flowers subtended by a lobed bract;
stem with incurved, simple hairs; disturbed areas and shores, mostly
SLP (20–60 cm high)—**Three-seeded Mercury**, *Acalypha rhomboidea*

15b. Inflorescences terminal, pistillate flowers lack a subtending bract; stem
and leaves with star-shaped hairs; disturbed areas, SLP (20–60 cm
high)—*Croton glandulosus*

CALLITRICHACEAE, The Water-starwort Family

Small monoecious aquatic herbs with opposite entire leaves. Flowers regular, unisexual, small, 1 or 2 per leaf axil; sepals 0; petals 0; stamens 1(–3); pistil 1, styles 2, ovary superior, 4-celled. Fruit splitting into 4 nutlets. Summer.

1a. Plant entirely submerged; all leaves linear, dark green; UP—***Callitriche hermaphroditica***
1b. Submerged leaves linear, terrestrial and/or floating leaves obovate, both light green; wet, often muddy areas—***Callitriche verna*** [***C. palustris***—CQ]

EMPETRACEAE, The Crowberry Family

Low evergreen shrubs with alternate, simple, linear leaves completely rolled into a tube. Flowers regular, perfect, axillary, inconspicuous; sepals 4; petals 4; stamens 4; pistil 1, style 1, ovary superior, 6–9-celled. Fruit a berry-like drupe. Summer.

One species; leaves less than 1 cm long; fruit black; rock outcrops, shaded dunes, bogs, UP—**Black Crowberry**, ***Empetrum nigrum***

LIMNANTHACEAE, The False Mermaid Family

Low annual herbs with alternate, deeply pinnately lobed or compound leaves. Flowers regular, perfect, small, axillary; sepals 3; petals 3, white; stamens (3–)6; pistils 2 or 3, style 1, arising amidst the ovaries, each ovary superior, 1-celled. Fruit 2 or 3 indehiscent mericarps. Spring.

One species in Michigan; rich, moist woods (5–30 cm high)—**False Mermaid**, ***Floerkea proserpinacoides***

ANACARDIACEAE, The Cashew Family

Shrubs, vines, or small trees with alternate, pinnately compound or trifoliolate leaves and milky or resinous juice. Flowers regular, mostly unisexual (the plants then dioecious with occasional perfect flowers), small; sepals 5, united below; petals 5; stamens 5; pistil 1, style 1, ovary superior, 1-celled. Fruit a drupe.

1a. Leaves pinnately compound, leaflets (five) seven to many—2
1b. Leaflets three—5

2a. Inflorescence axillary, loose; fruits whitish, glabrous; leaflet margins entire; bogs and swamps, LP (to 5 m high)—**Poison Sumac**, *Toxicodendron vernix*
2b. Inflorescence terminal, crowded; fruits red, hairy; leaflet margins entire or serrate; dry areas—3

3a. Leaflets entire; axis of the leaves wing-margined between the leaflets; sandy fields and woods, LP (to 6 m high; summer)—**Shining Sumac**, *Rhus copallina*
3b. Leaflets serrate; axis of leaves not wing-margined; fields and slopes—4

4a. Younger stems, petioles, and the underside of leaflets glabrous; hairs on the fruits longer than 1 mm, sharply pointed (to 6 m high; summer)—**Smooth Sumac**, *Rhus glabra*
4b. Younger stems and petioles densely velvety-hairy; underside of leaflets pubescent, especially on the veins; hairs on the fruits shorter than 1 mm, not sharply pointed (to 10 m high; summer)—**Staghorn Sumac**, *Rhus typhina*

Many hybrids between **Rhus typhina** and **R. glabra** occur in Michigan. Specimens in which pubescence of either the stem or fruit appears intermediate are probably hybrids, referred to as **Rhus ×pulvinata**.

5a. Terminal leaflet narrowed to a sessile base; inflorescence terminal, crowded; fruits red, hairy; sandy woods and openings, LP (to 2 m high; summer)—**Fragrant Sumac**, *Rhus aromatica*
5b. Terminal leaflet on a definite petiole, rounded or acute at base; inflorescence axillary, loose; fruits whitish, glabrous (spring to early summer)—6

6a. Climbing vine, the hairy stems with aerial roots; SLP—**Poison-ivy**, *Toxicodendron radicans* var. *radicans*
6b. Erect shrub, lacking aerial roots; NM & WM (30–80 cm high)—**Poison-ivy**, *Toxicodendron radicans* var. *rydbergii* [*T. rydbergii*—CQ]

Toxicodendron spp. are infamous for the skin rashes and blisters which their resins can produce. Even wading in swamp water containing these resins or exposure to droplets of sap suspended in air can produce blisters in sensitive individuals. **Toxicodendron radicans**, Poison-ivy, is found in Michigan, while the true Poison-oak, **Toxicodendron pubescens** (southeastern USA) and **Toxicodendron diversilobum** (southeastern USA) do not occur in Michigan.

AQUIFOLIACEAE, The Holly Family

Shrubs with alternate, simple leaves. Flowers regular, mostly unisexual (the plants then dioecious with occasional perfect flowers), small, axillary; sepals 4–8 or 0; petals 4–8, white or greenish, often united below; stamens 4–8, often attached below to the petals; pistil 1, style 1 or 0, ovary superior, 4–8-celled. Fruit a red (rarely yellow) berry-like drupe. Spring.

1a. Leaves entire or nearly so, 2–5 cm long; petals yellowish, linear; petioles purple; fruiting pedicel longer than the fruit; bogs and wet woods (to 3 m high)—**Mountain Holly**, *Nemopanthus mucronatus*
1b. Leaves sharply serrate, 5–8 cm long; petals white, obovate; petioles green; fruit longer than its pedicel; bogs and shores (to 5 m high)—**Michigan Holly** or **Black-alder**, *Ilex verticillata*

> Both of the native Michigan hollies, ***Ilex verticillata*** and ***Nemopanthus mucronatus***, are protected by Public Act 182 of 1962, commonly called the "Christmas-tree law". Under this law, it is illegal to remove or cut these plants from any area without a bill of sale or written permission from the owner. Michigan hollies are deciduous, with bright red berries displayed on bare branches in the winter (on female plants). The familiar evergreen Christmas hollies are other species of ***Ilex*** which are garden introductions.

CELASTRACEAE, The Bittersweet Family

Erect or trailing shrubs, small trees, or woody twining vines with alternate or opposite, simple leaves. Flowers regular, perfect or unisexual (the plants then dioecious sometimes with occasional perfect flowers); sepals 4 or 5; petals 4 or 5; stamens 4 or 5, attached to a nectar-secreting disk which fills much of the center of the flower; pistil 1, style 1, ovary superior, embedded in the disk, 3–5-celled. Fruit a capsule, the seeds each covered by an orange or red aril.

1a. Woody twining vines; leaves alternate; flowers mostly unisexual (***Celastrus*** spp., **Bittersweet**)—2
1b. Trees or shrubs, the stems not twining; leaves opposite; flowers perfect; SLP (***Euonymus*** spp.)—3

2a. Flowers in terminal panicles or racemes; native (late spring)—**American Bittersweet**, *Celastrus scandens*
2b. Flowers in axillary cymes; escape from cultivation, SLP—**Oriental Bittersweet**, *Celastrus orbiculata*

The American Bittersweet, **Celastrus scandens**, is protected by Public Act 182 of 1962, commonly called the "Christmas-tree law". Under this law, it is illegal to remove or cut these plants from any area without a bill of sale or written permission from the owner. This native Michigan plant has ovate to oblong leaves and the bright orange fruit is borne at the ends of branchlets. In contrast, the Oriental Bittersweet, **Celastrus orbiculata**, an invasive garden escape, has suborbicular to obovate leaves and the flowers and fruit are in axillary clusters (cymes).

3a. Twigs with several conspicuous corky wings; flowers four-parted; escape from cultivation, SLP (to 2.5 m high; late spring)—**Burning-bush or Winged Euonymus**, *Euonymus alata*
3b. Twigs lack corky wings; flowers four- or five-parted—4

4a. Prostrate or climbing shrub—5
4b. Erect shrub (to 6 m high; late spring)—6

5a. Prostrate shrub with short erect branches; leaves deciduous; young branches smooth; flowers five-parted, greenish-purple; deciduous woods, SLP (to 30 cm high; spring)—**Running Strawberry-bush**, *Euonymus obovata*
5b. Prostrate or climbing (with roots along the stem) shrub; leaves evergreen; young branches warty; flowers four-parted, greenish; escape from cultivation, LP (summer)—**Wintercreeper**, *Euonymus fortunei*

6a. Leaf blades pubescent beneath; flowers brownish-purple; river banks, SLP—**Wahoo**, *Euonymus atropurpurea*
6b. Leaf blades glabrous beneath; flowers greenish-white; escape from cultivation to shores or margins of woods, LP—**Spindle Tree**, *Euonymus europaea*

STAPHYLEACEAE, The Bladdernut Family

Shrubs with opposite, trifoliolate leaves. Flowers regular, perfect, small, in drooping panicles; sepals 5; petals 5, white; stamens 5; pistil 1, styles 3, apically united, ovary superior, embedded in the nectar-secreting disk, 3-celled. Fruit a large, inflated, papery, 3-celled capsule.

One species; moist woods, river banks, SLP (to 5 m high; spring)—**Bladdernut**, *Staphylea trifolia*

ACERACEAE, The Maple Family

Trees or shrubs with opposite, palmately lobed (rarely entire) or pinnately compound leaves. Flowers regular, perfect or unisexual (the
plants then often dioecious with occasional perfect flowers), small;
sepals usually 5, sometimes united below; petals usually (0)5; stamens
often 8 (sometimes 4 or 5 or 10–12), often with a nectar-secreting
disk; pistil 1, styles (1)2, ovary superior, 2-celled. Fruit a double
samara, sometimes splitting to individual winged mericarps. Spring.

1a. Leaves pinnately compound, leaflets usually 3–5; tree; flowers apetalous
and unisexual; moist banks, disturbed areas (to 20 m high)—**Box-elder**,
Acer negundo
1b. Leaves simple, palmately lobed (rarely entire); trees or shrubs; flowers
with or without petals, often perfect—2

2a. Sinuses (angles between the leaf blade lobes) rounded, basally entire;
flowers greenish-yellow, appearing with the leaves—3
2b. Sinuses acute or basally toothed; flowers purple, red, or yellowish, appearing before, with, or after the leaves—5

3a. Inflorescence erect or ascending; flowers with petals; petioles with milky
sap; escape from cultivation—**Norway Maple**, *Acer platanoides*
3b. Inflorescence drooping; flowers apetalous; petioles lack milky sap; rich
woods—4

4a. Leaf blades glabrous beneath, or minutely pubescent on the veins; petioles glabrous (to 40 m high)—**Sugar Maple**, *Acer saccharum*
4b. Leaves downy beneath; petioles pubescent; SLP (to 40 m high)—**Black
Maple**, *Acer saccharum* subsp. *nigrum* [*A. nigrum*—CQ]

5a. Shrubs or small trees; leaves three- to obscurely five-lobed (rarely entire); the lobes with regularly serrate margins; flowers greenish-yellow,
appearing with or later than the leaves—6
5b. Trees; leaves three- to seven-lobed; margins of the lobes entire or incised, but never regularly serrate; flowers red, appearing before the
leaves—8

6a. Leaves three-lobed, some young leaves entire or nearly so; central, terminal lobe about twice as large as the lower lobes; flowers greenish-
white; bud scales more than two, overlapping; roadsides (to 6 m high)—
Amur Maple, *Acer ginnala*
6b. Leaves three- to obscurely five-lobed; all lobes about equal in size;
flowers yellowish green; bud scales two, not overlapping—7

7a. Leaves finely and sharply doubly-serrate; twigs smooth; bark conspicuously striped with white lines; inflorescence a pendant raceme; rich woods, NM (to 12 m high)—**Striped Maple**, *Acer pensylvanicum*

7b. Leaves coarsely and bluntly serrate; young twigs pubescent; bark not striped; inflorescence an ascending or erect panicle; moist woods (to 10 m high)—**Mountain Maple**, *Acer spicatum*

8a. Middle leaf lobe more than half the length of the leaf, narrowed at its base; broken twigs with a strong odor; samaras more than 3 cm long; moist woods and banks (to 30 m high)—**Silver Maple**, *Acer saccharinum*

8b. Middle leaf lobe usually about half the length of the leaf, its sides parallel or broadened at the base; broken twigs without strong odor; samaras less than 3 cm long (to 35 m high)—**Red Maple**, *Acer rubrum*

HIPPOCASTANACEAE, The Buckeye Family

Trees with opposite, palmately compound leaves. Flowers irregular, perfect or unisexual (the plants then monoecious with occasional perfect flowers), showy, in terminal panicles; sepals 5, united below; petals 4 or 5; stamens 6–8; pistil 1, style 1, ovary superior, often 3-celled. Fruit a prickly capsule enclosing a single large brown seed. Spring.

1a. Leaflets seven; buds viscid; corolla of five white or pink petals marked with red or yellow; escape from cultivation, LP (to 25 m high)—**Horsechestnut**, *Aesculus hippocastanum*

1b. Leaflets five; buds smooth; corolla of four greenish-yellow petals; moist woods and banks, SLP (to 15 m high)—**Ohio Buckeye**, *Aesculus glabra*

BALSAMINACEAE, The Touch-me-not Family

Annual herbs with alternate (rarely opposite or whorled), simple leaves. Flowers bilaterally symmetrical, perfect; sepals 3, the lowermost petaloid and prolonged backwards into a spur; petals 3; stamens 5, the anthers united; pistil 1, style 1, ovary superior, 5-celled. Fruit a capsule that, when ripe, opens explosively when touched. Summer.

1a. Leaves opposite or whorled; leaf blade teeth sharp; flowers red to purple; rarely escapes from cultivation (to 2 m high)—*Impatiens glandulifera*

1b. Leaves alternate; leaf blade teeth rounded; flowers yellow or orange; swamps, ditches, wet woodland openings (0.5–1.5 m high)—2

2a. Flowers pale-yellow, with a few red-brown spots; spur 4–6 mm long; LP—**Pale Touch-me-not**, *Impatiens pallida*

2b. Flowers orange, thickly spotted with red-brown or not; spur 7 mm or longer—**Spotted Touch-me-not**, *Impatiens capensis*

RHAMNACEAE, The Buckthorn Family

Shrubs or trees with alternate or opposite, simple, leaves. Flowers regular, perfect or unisexual (the plants then dioecious, sometimes with occasional perfect flowers); sepals 4 or 5; petals 4, 5, or 0; stamens 4 or 5; pistil 1, style(s) 1 or 2–4, united below, ovary superior (although sometimes embedded in the nectar-secreting disk), 2–4-celled. Fruit a drupe, sometimes 3-lobed.

1a. Leaves alternate or opposite, with a single basal mid-vein; flowers axillary, solitary in small umbels, greenish; fruit a fleshy drupe (***Rhamnus*** spp., **Buckthorn**)—2

1b. Leaves alternate, with three principal veins at the base; flowers in dense terminal or axillary umbellate panicles, white; fruit a dehiscent, three-lobed, capsule-like drupe (***Ceanothus*** spp.)—4

2a. Margin of leaf blades entire, sometimes with marginal glands near the tip; flowers perfect; bogs, fens, low woods (to 7 m high; late spring)— **Glossy Buckthorn**, *Rhamnus frangula*

One example of the effects of dissemination by birds on plant distribution is the invasion of wild communities by **Rhamnus frangula** or Glossy Buckthorn. This vigorous shrub or small tree is still sold in the nursery trade as "tallhedge". It is meanwhile being dug out of nature preserves by volunteers. The small black fruits are eaten by birds. The seeds are then excreted over the birds' entire territory, resulting in thickets which shade native vegetation, especially in wet areas. The Nature Conservancy rates Glossy Buckthorn as one of the five most dangerous plant pests in Michigan.

2b. Margin of leaf blades with minute, often rounded teeth—3

3a. Leaves mostly appearing opposite, three to four pairs of lateral veins per leaf blade; thorns terminating the branches; disturbed areas (to 6 m high)—**Common Buckthorn**, *Rhamnus cathartica*

3b. Leaves alternate, six to nine pairs of lateral veins per leaf blade; thorns absent; fens, swamps, wet meadows (to 1 m high)—**Alder-leaved Buckthorn**, *Rhamnus alnifolia*

4a. Shrub over 1 m tall; inflorescences arise from leafless branches of last year; rocky woods, Keweenaw County (1–3 m high; late spring)—**Wild-lilac**, *Ceanothus sanguineus*

4b. Shrub to 1 m tall; inflorescences arise from leafy current-year branches (sandy woods and fields—5

5a. Inflorescence axillary; leaves ovate, mostly 2.5–5 cm wide (summer)— **New Jersey Tea**, *Ceanothus americanus*
5b. Inflorescence terminal; leaves elliptical-lanceolate, mostly 2 cm wide or less; NM (late spring)—**New Jersey Tea**, *Ceanothus herbaceus*

VITACEAE, The Grape Family

Vines climbing by tendrils opposite the palmately lobed or compound leaves. Flowers regular, perfect or unisexual (the plants then monoecious or dioecious); sepals 5, small or 0; petals 5, greenish, the tips sometimes touching; stamens 5; pistil 1, style 1, ovary superior, 2-celled. Fruit a berry. Late spring.

1a. Leaves compound; petals separate (*Parthenocissus* spp., **Virginia Creeper**)—2
1b. Leaves simple; tips of the petals touching; often in woods or thickets (*Vitis* spp., **Grape**)—3

2a. Branches of the tendrils chiefly ending in adhesive disks; inflorescence with strongly unequal branches; woods, thickets—*Parthenocissus quinquefolia*
2b. Branches of the tendrils twining, but not forming disks; inflorescence with dichotomous branches; rocky areas, woods—*Parthenocissus inserta* [*P. vitacea*—CQ]

3a. Leaf blades conspicuously pubescent beneath; a tendril or flower cluster opposite each leaf at 3 or more adjacent nodes; SLP—**Fox Grape**, *Vitis labrusca*
3b. Leaf blades glabrous beneath when mature, or pubescent on the veins only; a tendril or panicle opposite the leaves at no more than two adjacent nodes—4

4a. Leaf blades glaucous beneath; leaf blades often shallowly lobed with obtuse to rounded sinuses between the lobes; LP—**Summer Grape**, *Vitis aestivalis*
4b. Leaf blades green beneath; leaf blades either shallowly or deeply lobed with acute sinuses between the lobes—5

5a. Leaf blades prominently lobed, the marginal teeth acute and ciliate— **River-bank Grape**, *Vitis riparia*

5b. Leaf blade unlobed or slightly three-lobed, the marginal teeth obtuse and
    glabrous; rare, SLP—**Frost Grape**, *Vitis vulpina*

Although the European species **Vitis vinifera** is the major source of wine and table grapes, many of the grapes cultivated in Michigan are derived at least in part from the North American native **Vitis labrusca**.

TILIACEAE, The Linden Family

Trees with alternate, simple, palmately veined leaves. Flowers regular, usually perfect, in axillary cymes that appear to arise from the middle of a leaf-like bract; sepals 5; petals 5; stamens numerous, but united into 5 bundles and with 1 staminode per bundle; pistil 1, style 1, ovary superior, 5-celled. Fruit an indehiscent nut.

One species in Michigan; leaves with asymmetrical cordate bases; rich woods (to 40 m high; summer)—**Basswood**, *Tilia americana*

MALVACEAE, The Mallow Family

Herbs with alternate, simple, mostly palmately veined leaves. Flowers regular, perfect, sometimes subtended by an epicalyx of sepal-like bracts; sepals 5, often united below; petals 5; stamens many, united by their filaments to form a tube surrounding the styles; pistil 1, styles 1–many, ovary superior, 1- or 5–many-celled. Fruit a capsule or separate mericarps.

1a. Epicalyx of bracts immediately subtending the calyx absent; leaves
    broadly heart-shaped; petals orange or yellow; disturbed ground, esp.
    SLP (1–1.5 m high; summer-fall)—**Velvet-leaf**, *Abutilon theophrasti*
1b. Epicalyx of bracts, often 3 or 6, immediately subtending the calyx—2

2a. Calyx subtended by six to many bracts which are sometimes united at
    base; petals white to purple or yellow with a purple base (summer)—3
2b. Calyx subtended by three bracts; petals pink or white; roadsides, gar-
    dens, weedy areas, etc.—6

3a. Epicalyx bracts mostly six to nine, triangular; styles fifteen or more;
    fruit a ring of separate mericarps—4
3b. Epicalyx bracts mostly twelve, linear; styles five; fruit a capsule;
    SLP—5

4a. Flowers 2–4 cm wide, petals pink; stem hairs dense; damp weedy areas, SLP (0.5–1.2 m high)—**Marsh Mallow**, *Althaea officinalis*
4b. Flowers often 10 cm wide, petals white or pink to purple; stem hairs scattered; escapes from cultivation to roadsides, weedy areas, etc. (1.5–3 m high)—**Hollyhock**, *Alcea rosea* [*Althaea rosea*—CQ]

5a. Petals pink to nearly white, over 4 cm long; leaves shallowly (or not) lobed; wet places; wet areas (1–2 m high)—**Rose Mallow**, *Hibiscus moscheutos*
5b. Petals pale yellow with a purple base, less than 3 cm long; leaves deeply lobed; fields, roadsides, etc. (30–50 cm high)—**Flower-of-an-hour**, *Hibiscus trionum*

6a. Plants prostrate or ascending; leaf blades entire or shallowly lobed, the lobes not extending beyond halfway to the base of the blade (summer and late summer)—7
6b. Plants erect; leaf blades deeply lobed or cleft, the lobes extending more than halfway to the base of the blade (40–100 cm high; summer)—8

7a. Petals about twice as long as the sepals; surface of mature mericarps smooth with short hairs—**Common Mallow** or **Cheeses**, *Malva neglecta*
7b. Petals about the same length as the sepals; surface of the mature mericarps wrinkled and glabrous—*Malva rotundifolia*

8a. Lobes of the leaf blade dentate or incised; epicalyx bracts and sepals either glabrous or with simple hairs—**Vervain Mallow**, *Malva alcea*
8b. Lobes of the leaf blade pinnately cleft into linear or narrowly oblong divisions; epicalyx bracts and sepals covered with stellate hairs—**Musk Mallow**, *Malva moschata*

GUTTIFERAE (CLUSIACEAE), The St. John's-wort Family

Herbs or shrubs with opposite entire leaves dotted with translucent glands. Flowers regular, perfect; sepals mostly 5; petals 5, often yellow; stamens 5–many, often in 3 or 5 bundles, the filaments sometimes united below; pistil 1, styles 3–5(6), ovary superior, 1–5(6)-celled. Fruit a capsule. Summer.

1a. Bushy shrubs; petals yellow—2
1b. Herbs; petals yellow or pink to red—3

2a. Styles five (rarely four or six); all inflorescences terminal; moist calcareous areas, esp. along Great Lakes shores (to 1 m high)—**Kalm's St. John's-wort**, *Hypericum kalmianum*

2b. Styles three; inflorescences both terminal and axillary; swamps and fields, SLP (to 2 m high)—**Shrubby St. John's-wort**, *Hypericum prolificum*

3a. Petals pink to red, the flowers normally closed; leaf blades oblong to elliptic with a subcordate base; bogs, shores, wet meadows (*Triadenum* spp., **Marsh St. John's-wort**)—4
3b. Petals yellow, the flowers open; leaf blades mostly narrower and without a subcordate base—5

4a. Styles persisting on capsules, 2–3 mm long; sepals 5 mm or more long with an acute apex; SLP (30–60 cm high)—*Triadenum virginicum*
4b. Styles up to 1.5 mm long; sepals 5 mm or less long with a blunt or rounded apex (30–60 cm high)—*Triadenum fraseri*

5a. Flowers 4 cm or more wide; styles five; stamens many, more than one hundred; wet areas (70–200 cm high)—**Giant St. John's-wort**, *Hypericum ascyron* [*H. pyramidatum*—CQ]
5b. Flowers less than 3 cm wide; styles three; stamens often many, but fewer than 100—6

6a. Leaf blades minute, awl-shaped, 1–3 mm long; sandy areas, SLP (10–50 cm high)—**Orange-grass**, *Hypericum gentianoides*
6b. Leaf blades linear or lanceolate to ovate, 1 cm or more long—7

7a. Petals dotted with black dots—8
7b. Petals without black dots—9

8a. Black spots on petals occur along the edge and tip; sepals with few black dots; stem angled; roadsides, fields, etc. (40–80 cm high)—**St. John's-wort** or **Klamath-weed**, *Hypericum perforatum*
8b. Black spots on entire surface of the petals; sepals conspicuously black-dotted; stem round in cross section; wet areas, fields, esp. SLP (50–100 cm high)—**Spotted St. John's-wort**, *Hypericum punctatum*

9a. Leaf blades pinnately veined, the lateral veins clearly seen; flowers 8–25 mm wide; stamens twenty or more; wet areas, UP (20–50 cm high)—*Hypericum ellipticum*
9b. Leaf blades with one to several parallel veins, the lateral veins absent or obscure; flowers 1–10 mm wide; stamens fewer than twenty—10

10a. Sepals broadest below the middle, tapering to acute tips; leaf blades lanceolate; damp sandy areas (10–60 cm high)—11
10b. Sepals broadest near the middle, not strongly tapered to the obtuse tips; leaf blades broadly elliptic—12

11a. Leaf blades oblanceolate, tapered to a narrow base, with one to three principal veins—***Hypericum canadense***

11b. Leaf blades lanceolate to elliptic-ovate, the base broadly rounded, with three to seven principal veins—***Hypericum majus***

12a. Uppermost bracts linear; damp sandy areas, LP (10–80 cm high)—***Hypericum mutilum***

12b. Uppermost bracts resembling the leaves in shape, but smaller; damp sandy areas, bogs (10–40 cm high)—***Hypericum boreale***

ELATINACEAE, The Waterwort Family

Small creeping aquatic herbs with opposite, simple leaves lacking translucent dots. Flowers regular, perfect, inconspicuous and axillary; sepals 2; petals 2; stamens 2; pistil 1, styles usually 2, ovary superior, 2-celled. Fruit a capsule. Summer.

One species; sandy lake edges, SW & UP (to 5 cm high)—**Waterwort**, *Elatine minima*

TAMARICACEAE, The Tamarisk Family

Shrubs or small trees with alternate, scale-like leaves. Flowers regular, perfect, in racemes; sepals 4; petals 4, pink; stamens 4, filaments attached to a nectar-secreting disk; pistil 1, styles 3 or 4, ovary superior, 1-celled. Fruit a capsule.

One species; rarely escaping from cultivation, LP (to 15 m high; late spring)—**Tamarisk**, *Tamarix parviflora*

CISTACEAE, The Rock-rose Family

Small herbs or shrubs with alternate (sometimes opposite or whorled), simple, entire leaves. Flowers regular (but not the calyx), perfect; sepals 5, the outer 2 smaller than the inner 3; petals 3 or 5 (or 0 in cleistogamous flowers); stamens 3–many; pistil 1, style 1 or 0 (the 1 or 3 stigma(s) then sessile), ovary superior, 1-celled. Fruit a capsule.

1a. Small, bushy or matted shrub; leaves crowded, closely appressed to the branches; petals yellow; sand dunes and forests, esp. along Great Lakes shores (spring and early summer)—**False Heather**, *Hudsonia tomentosa*

1b. Erect herb; leaves flat, spreading; petals yellow or reddish (absent in the later, cleistogamous flowers)—2

2a. Leaves covered with stellate hairs; style present; petals five, yellow, falling soon after the flower opens; open, sandy areas (early summer) (*Helianthemum* spp., **Rock-rose** or **Frostweed**)—3
2b. Leaves glabrous or hairs simple; style absent; petals three, red, often hidden by the sepals (late summer) (*Lechea* spp., **Pinweed**)—4

3a. Petal-bearing flowers solitary at the stem tip, overtopped later by leafy branches; capsules from cleistogamous flowers (along the branches) with five or more seeds (30–60 cm high)—*Helianthemum canadense*
3b. Petal-bearing flowers few, racemose at the stem tip, not overtopped by leafy branches; capsules from cleistogamous flowers with one to three seeds; mostly SLP (20–50 cm high)—*Helianthemum bicknellii*

4a. Stem and leaf blade pubescence of spreading hairs; sandy woods and shores, LP (20–80 cm high)—*Lechea villosa* [*L. mucronata*—CQ]
4b. Stem and leaf blade pubescence of appressed hairs—5

5a. Outer sepals longer than the inner sepals; sandy shores, SLP (20–50 cm high)—*Lechea minor*
5b. Outer sepals shorter than the inner sepals; open sandy fields and woods—6

6a. Fruiting calyx obovoid; leaves with a conspicuous brown tip; SLP (20–80 cm tall)—*Lechea pulchella*
6b. Fruiting calyx subglobose; leaves lack a brown tip; NM (20–60 cm high)—*Lechea intermedia*

VIOLACEAE, The Violet Family

Herbs with alternate or basal, simple leaves. Flowers irregular, perfect; sepals 5; petals 5, the lowermost either swollen near the base or with a backward-projecting *spur*, lateral and spur petals with a *beard* of hairs near the center of the flower or not, lower petals often with one to few colored stripes (*nectar-guides*); stamens 5, converging near the ovary, filaments rarely united; pistil 1, style 1, ovary superior, 1-celled. Fruit a capsule.

1a. Flowers slightly irregular, greenish-white, axillary; lowermost petal swollen basally; erect plant with leafy stem; rich woods, SLP (to 1 m high; spring)—**Green Violet**, *Hybanthus concolor*

1b. Flowers irregular, blue, yellow, or white, conspicuous; lowermost petal
with a backward-projecting spur; plant with or without a leafy stem
(*Viola* spp., **Violet**)—2

Hybridization is common between some **Viola** species, especially among many of the so-called "stemless blue" and "stemless white" violets. This may make some individuals difficult to identify, since hybrids often exhibit various intermediate combinations of the characters of the parental species. Habitat information is important in identification of violets since morphologically similar species often inhabit quite different habitats. For additional information on violets in Michigan, please consult Voss (1985) or Ballard (1994, 1995).

2a. Plant appears stemless, the flowers all on leafless stalks and the leaves
all basal; petals blue or white (spring or early summer)—3
2b. Stems leafy; petals blue, white, or yellow (to 45 cm high; spring and
summer)—16

3a. Leaf blades narrowly lanceolate, not lobed, tapering to the base; damp
sandy shores and fields—**Lance-leaved Violet**, *Viola lanceolata*
3b. Leaf blades sometimes lobed, often with a cordate base, not tapered to
the petiole—4

4a. Ovary pubescent; style tip hook-shaped; green, leafy stolons evident at
the base of the plant; escape from cultivation, often in lawns—**English**
or **Sweet Violet**, *Viola odorata*
4b. Ovary glabrous; style tip not hooked; stolons absent or white; native
species—5

5a. Principal leaves at time of flowering deeply lobed; petals blue (or white
in occasional albino forms)—6
5b. Leaves heart-shaped or kidney-shaped (reniform), not lobed; petals blue
or white—8

6a. Lateral petals glabrous; stamens conspicuously exserted; open sandy
woods and plains, LP—**Birdfoot Violet**, *Viola pedata*

The Bird's-foot Violet, **Viola pedata**, is a dry-land plant characterized by deeply cleft leaves. It is protected by Public Act 182 of 1962, commonly called the "Christmas-tree law". Under this law, it is illegal to remove or cut these plants from any area without a bill of sale or written permission from the owner.

6b. Lateral petals bearded; stamens not exserted—7

7a. Leaves divided to the base into linear segments; prairies, esp. SW—
Prairie Violet, *Viola pedatifida* [*V. palmata* var. *pedatifida*—CQ]

7b. Leaves deeply lobed, the central lobe broader than the lateral lobes; dry woods and prairies, SLP—**Wood Violet**, *Viola palmata*

8a. Flowers violet or blue (rarely white-flowered plants are found with the typical blue-flowered ones)—9
8b. Flowers white, the three lower petals marked with purple stripes—14

9a. Lateral petals glabrous; spur 4 mm or more long; deciduous woods, NM—**Great-spurred Violet**, *Viola selkirkii*
9b. Lateral petals bearded; spur shorter than 4 mm—10

10a. Sepal apex acute or blunt; leaf blades longer than broad with an acute apex—11
10b. Sepal apex rounded; leaf blades about as long as broad with an obtuse apex—13

11a. Leaf blades pubescent, the base truncate sometimes with basal lobes; sandy fields and woods—**Arrow-leaved Violet**, *Viola sagittata*
11b. Leaf blades glabrous or virtually so, the base cordate—12

12a. Beard of the lateral petals with a knob at the tip of each hair; spur petal bearded; moist woods and swamps—**Marsh Violet**, *Viola cucullata*
12b. Beard of the lateral petals not knobbed; spur petal glabrous; moist woods and swamps—**Wood Violet**, *Viola affinis* [incl. in. *V. sororia*—CQ]

13a. Spur petal bearded; sepals glabrous; fens, calcareous shores and swamps—*Viola nephrophylla*
13b. Spur petal glabrous; sepals ciliate; woodlands, meadows—**Common Blue Violet**, *Viola sororia*

14a. Leaf blades ovate, glabrous; wet woods and bogs—**Smooth White Violet**, *Viola macloskeyi*
14b. Leaf blades ovate to reniform, usually pubescent—15

15a. Leaf blade reniform; stolons absent; coniferous swamps, NM—**Kidney-leaved Violet**, *Viola renifolia*
15b. Leaf blades ovate; stolons present; woods and wet areas—**Sweet White Violet**, *Viola blanda*

16a. Stipules large and leaf-like, deeply pinnately divided and nearly or quite as long as the petioles; escapes from cultivation to roadsides, fields, etc. (10–30 cm high)—17
16b. Stipules small, inconspicuous, entire or toothed, and much shorter than the petiole—18

17a. Petals with a dark blue or purple apex and often a yellow base, longer than the sepals—**Johnny-jump-up**, *Viola tricolor*
17b. Petals yellowish-white, sometimes with purple tips, shorter than or equalling the sepals—**Field Pansy**, *Viola arvensis*

18a. Petals yellow; foliage usually villous-pubescent; woods (10–45 cm high)—**Yellow Violet**, *Viola pubescens*
18b. Petals blue or white; foliage pubescent or not—19

19a. Margin of stipules entire; petals white inside, bluish outside; deciduous woods (20–40 cm high)—**Canada Violet**, *Viola canadensis*
19b. Margin of stipules with a fringe of tiny teeth; petals blue or cream—20

20a. Lateral petals glabrous; petals blue, with a darker center (eye); deciduous woods, LP (5–25 cm high)—**Long-spurred Violet**, *Viola rostrata*
20b. Lateral petals pubescent ("bearded"); petals blue or white—21

21a. Petals white or cream-white; deciduous woods, SLP (6–30 cm high)—**Cream Violet**, *Viola striata*
21b. Petals blue—22

22a. Leaves glabrous, yellowish-green; moist areas, often in woods (to 20 cm high)—**Dog Violet**, *Viola conspersa*
22b. Leaves finely pubescent, bluish-green; open, sandy areas, NM (2–15 cm high)—**Sand Violet**, *Viola adunca*

CACTACEAE, The Cactus Family

Fleshy, jointed, leafless herbs armed with numerous spines and/or bristles in groups known as *areoles*. Flowers regular, perfect, large (4–8 cm wide); sepals several to many; petals many, yellow, the petals and sepals both united below, forming a hypanthium; stamens many; pistil 1, style 1, ovary inferior, 1-celled. Fruit a berry. Summer.

1a. Stem joints flattened; areoles with at most one spine and several bristles; sandy areas, WM—**Prickly-pear**, *Opuntia humifusa*
1b. Stem joints not flattened; areoles with three or more spines; rocky areas, NM—**Prickly-pear**, *Opuntia fragilis*

THYMELAEACEAE, The Mezereum Family

Shrubs with alternate, simple, entire leaves. Flowers regular, perfect, in small clusters, opening before the leaves; sepals 4, tiny; petals 0;

hypanthium tubular, yellowish (resembling a calyx); stamens 8, arising from the hypanthium, exserted; pistil 1, style 1, ovary superior, 1-celled. Fruit a drupe.

One species; deciduous woods (1–2 m high; spring)—**Leatherwood**, *Dirca palustris*

ELAEAGNACEAE, The Oleaster Family

Shrubs or small trees with opposite, simple, entire leaves covered with silvery and/or rusty scales. Flowers regular, perfect or unisexual (the plants then monoecious or dioecious), axillary; sepals 4, often yellowish, as lobes on the tubular (or flat) hypanthium; petals 0; stamens 4 or 8, arising from the hypanthium; pistil 1, style 1, ovary superior, 1-celled. Fruit resembling a berry or a drupe.

1a. Leaves opposite; flowers unisexual, stamens eight; shrub (1–3 m high; spring)—**Buffalo-berry** or **Soapberry**, *Shepherdia canadensis*
1b. Leaves alternate; flowers perfect (some may be staminate), stamens four; shrub or tree (*Elaeagnus* spp., **Oleaster**)—2

2a. Small tree; leaf blades covered with silver scales above and beneath; fruit silvery; SLP (to 10 m high; summer)—**Russian-olive**, *Elaeagnus angustifolia*
2b. Shrub or small tree; leaf blades with silver and brown scales beneath, green above; fruit red; LP (to 5 m high; spring)—*Elaeagnus umbellata*

Many plants have been introduced into Michigan through cultivation. Not all cultivated plants are capable of "escaping" to become established in the wild, but many have. Some of these escapes become invasive and become a threat to native populations. ***Elaeagnus umbellata***, an Asian species deliberately planted for "wildlife habitat", is becoming a pest in Michigan, especially in prairies and wetlands. ***Lythrum salicaria*** (Purple Loosestrife) has been cultivated for its attractive flowers, but is now a serious invader of wetlands, overtopping and crowding out native plants such as ***Typha***; it is very difficult to eradicate. It is now illegal in Michigan (P.A. 182) to sell or distribute this species and other nonnative ***Lythrum*** (exception is made for some sterile hybrids).

LYTHRACEAE, The Loosestrife Family

Herbs or shrubs with opposite or seldom alternate, simple, entire leaves. Flowers regular, perfect; sepals 4–7, short, alternating with hypanthium lobes; petals 4–7 or 0, within or at the summit of the cup-shaped, tubular, or globose hypanthium; stamens 4–12, the filaments often of two or three lengths; pistil 1, style 1, ovary superior, 1–5-celled. Fruit a capsule. Summer.

1a. Stem woody below, high-arching; flowers axillary, the petals pink; shores (1–3 m long)—**Whorled** or **Swamp Loosestrife**, *Decodon verticillatus*
1b. Stem not woody below, the plant erect or sprawling; flowers axillary or terminal, petals pink to purple or white—2

2a. Plant erect, usually more than 40 cm high; petals six, red-purple; marshes and shores (*Lythrum* spp., **Loosestrife**)—3
2b. Plant often sprawling, usually less than 40 cm high; petals four (pink or white) or absent—4

3a. Flowers many in terminal clusters resembling spikes; stems tomentose above; leaves opposite or whorled (50 cm–1.5 m high)—**Purple Loosestrife**, *Lythrum salicaria*
3a. Flowers solitary in the axils; stems not tomentose above; leaves mostly alternate (40–80 cm high)—**Loosestrife**, *Lythrum alatum*

4a. Base of the leaf blade tapered; flowers solitary in the axils; triangular appendages alternate with the sepals; shores and marshes, SLP, esp. SW (10–40 cm high)—**Tooth-cup**, *Rotala ramosior*
4b. Base of the leaf blade clasping the stem; flowers few in the axils; awl-shaped appendages alternate with the sepals; mud flats, SE (20–40 cm high)—*Ammannia robusta*

NYSSACEAE, The Tupelo Family

Trees with alternate, simple leaves. Flowers regular, unisexual (the plants then dioecious with or without occasional perfect flowers), axillary; sepals 5; petals 5–10, shorter than the sepals; stamens 8–15 or fewer; pistil 1, style 1, ovary superior, 1-celled. Fruit a drupe.

One species, wet woods and swamp margins, SLP (to 30 m high; spring)—**Sour-gum** or **Black-gum**, *Nyssa sylvatica*

MELASTOMATACEAE, The Melastome Family

Perennial herbs with opposite, simple leaves, with 3–5 principal veins. Flowers mostly regular, perfect, in terminal cymes; sepals 4; petals 4, pink-purple; stamens 8, the anthers conspicuous, curved, yellow; pistil 1, style 1, ovary superior, 4-celled. Fruit a capsule.

One species, open moist areas, SW (20–100 cm high, late summer)—
Meadow-beauty, *Rhexia virginica*

ONAGRACEAE, The Evening-primrose Family

Herbs with opposite or alternate, simple leaves. Flowers regular (rarely irregular), perfect; sepals (2 in *Circaea*) 4; petals (2)4, sometimes 0; stamens (2)4 or 8, attached to the summit or inside of a tubular hypanthium; pistil 1, style 1 (sometimes 4-lobed), ovary inferior, 1-,2-, or 4-celled. Fruit a capsule or an indehiscent nut.

1a. Sepals two; petals two, white, each two-lobed; stamens two; fruit pubescent, the hairs hooked (summer) (*Circaea* spp., **Enchanter's-nightshade**)—2
1b. Sepals four; petals four or sometimes absent; stamens four or eight; hairs on fruit (if present) not hooked—3

2a. Flowers widely spaced along a pubescent raceme axis; fruit ridged; dry or (mostly) moist deciduous woods (30–70 (100) cm high)—*Circaea lutetiana*
2b. Flowers crowded at the end of a glabrous axis; fruit not ridged; swamps and damp woods (10–30 cm high)—*Circaea alpina*

3a. Sepals at the summit of the slender tubular receptacle, which is prolonged beyond the ovary; leaves alternate (spring and summer)—4
3b. Sepals at or slightly above the summit of the ovary; leaves opposite or alternate (summer)—10

4a. Petals white, turning pink when old; fruit an indehiscent nut; fields and disturbed areas, SLP (to 2 m high; summer)—*Gaura biennis*
4b. Petals yellow; fruit a capsule (*Oenothera* spp.)—5

5a. Stamens all equal in length; ovary and fruit not strongly four-angled (*Oenothera* sects. *Oenothera* and *Raimannia*, **Evening-primrose**)—6
5b. Stamens of two lengths; ovary and fruit strongly four-angled (*Oenothera* sect. *Kneiffia*, **Sundrops**)—8

6a. Cauline leaves deeply dentate or pinnately divided; sandy disturbed areas, LP (10–40, rarely to 80 cm high; spring and summer)—*Oenothera laciniata*
6b. Cauline leaves entire, undulate, or finely toothed—7

7a. Stem and foliage densely but closely appressed-pubescent; capsules linear, not thicker near the base; leaves to 1 cm wide; sandy disturbed areas and woods, LP (0.4–1 m high; summer)—*Oenothera clelandii*

7b. Stem and foliage glabrous, or with sparse spreading hairs; capsule broadest near the base; leaves 1 cm or wider, sandy shores, fields, and disturbed areas (0.5–2 m high; summer)—*Oenothera* sect. *Oenothera*, *Oenothera biennis* complex

The **Oenothera biennis** complex as defined here includes four species recognized by Voss (1985). They are **O. biennis**, **O. oakesiana**, **O. parviflora**, and **O. villosa**. While separate taxa can be recognized, the most consistent differences include characters that cannot be used in the field, such as genetic analysis. Other authorities may recognize different taxa within this complex. See Dietrich et al. (1997) for a detailed presentation on this complex.

8a. Petals less than 10 mm long; anthers less than 3 mm long; moist, sandy shores and fields (20–60 cm high; summer)—*Oenothera perennis*

8b. Petals 11–13 mm or longer; anthers 4 mm or longer—9

9a. Stems glabrous or with short glandular hairs; dry fields, SLP & western UP (to 1 m high; summer)—*Oenothera fruticosa*

9b. Stems covered with long, spreading hairs; fields and wooded clearings, LP & Straits (to 80 cm high; spring and early summer)—*Oenothera pilosella*

10a. Stamens 4; petals yellow or greenish, shorter than the sepals; flowers terminal or axillary (longer in *L. alternifolia*) or absent; flowers axillary (*Ludwigia* spp., **False Loosestrife**)—11

10b. Stamens 8; petals white, pink, purple, or red, longer than the sepals (*Epilobium* spp., **Willow-herb**)—13

11a. Aquatic plant of shallow water or muddy ground with prostrate stem and opposite, ovate leaves; flowers minute, apetalous—*Ludwigia palustris*

11b. Plants with erect or ascending stems and alternate, lanceolate leaves; petals present; swamps and marshes—12

12a. Petals minute, greenish; mostly SLP (20–100 cm high)—*Ludwigia polycarpa*

12b. Petals yellow, SLP (40–120 cm high)—*Ludwigia alternifolia*

13a. Flowers in terminal racemes; all leaves alternate; dry woods and disturbed areas, esp. burned areas (1–3 m high)—**Fireweed**, *Epilobium angustifolium*

13b. Flowers axillary; most leaves opposite—14

14a. Leaf blades entire, the margins usually somewhat revolute—15
14b. Leaf blades toothed, the margins flat; wet areas—17

15a. Stems densely pubescent with spreading hairs; fens and swamps (30–60 cm high)—*Epilobium strictum*
15b. Stems pubescent with appressed or incurved hairs—16

16a. Upper surface of leaf blades finely hairy; wet areas (20–100 cm high)—*Epilobium leptophyllum*
16b. Upper surface of leaf blades glabrous except near the midrib; bogs, NM (10–50 cm high)—*Epilobium palustre*

17a. Stems densely pubescent with spreading hairs; stigma four-lobed—18
17b. Stems pubescent with incurved, sometimes glandular hairs; stigma entire—19

18a. Base of leaf blade clasping the stem; petals longer than 10 mm; LP (50–200 cm high)—**Great Hairy Willow-herb**, *Epilobium hirsutum*
18b. Base of leaf blade not clasping the stem; petals up to 10 mm long; northern LP (50–80 cm high)—*Epilobium parviflorum*

19a. Seeds tipped with a tuft of reddish-brown hairs; sepal tips project from buds (to 1 m high)—*Epilobium coloratum*
19b. Seeds tipped with a tuft of white hairs; sepal tips not projecting from buds (to 1.5 m high)—*Epilobium ciliatum*

HALORAGACEAE, The Water-milfoil Family

Aquatic or marsh herbs with alternate, opposite, or whorled, simple, mostly submerged leaves. Flowers regular, perfect or unisexual (the plants then monoecious), inconspicuous, in emergent terminal spikes or axillary; sepals 3 or 4, sometimes 0; petals 3 or 4, sometimes 0; stamens 3, 4, or 8; pistil 1, styles 3 or 4, ovary inferior, 3- or 4-celled. Fruit a small nut, sometimes splitting into 4 mericarps. Summer.

1a. Leaves all alternate, the blades toothed, dissected, or sometimes extremely reduced and scale-like—2
1b. Leaves all or mostly opposite or whorled, all blades finely pinnately divided (most *Myriophyllum* spp., **Water-milfoil**)—3

2a. Leaves entire, very small and scale-like; WM, UP—*Myriophyllum tenellum*

Species of the genus **Myriophyllum** are very difficult to distinguish without flowers; unfortunately, flowers are seldom seen. The key provided here is based on vegetative features. Please refer to Voss (1985) for a key including floral features and additional tips on dealing with sterile plants. The only introduced water-milfoil in Michigan, **Myriophyllum spicatum**, has extremely finely divided leaves. **M. spicatum** is a major weedy pest in recreational lakes. It spreads easily by fragmentation and produces lush growth near the water's surface, crowding out native species.

2b. Leaves toothed (emergent) or pinnately divided into linear segments (submerged), 2 cm or longer (flowering stems 10–40 cm high)—**Mermaid-weed**, *Proserpinaca palustris*

3a. Leaves irregularly whorled, some being alternate; UP—*Myriophyllum farwellii*
3b. Leaves regularly whorled—4

4a. Leaves closely spaced along the stem, internodes less than 10 mm long—5
4b. Leaves less crowded, internodes 10 mm or longer—6

5a. Leaves 12 mm or shorter; UP—*Myriophyllum alterniflorum*
5b. Leaves 15 mm or longer—*Myriophyllum heterophyllum*

6a. Each side of the leaf blade divided into 13 or more segments; LP—**Eurasian Water-milfoil**, *Myriophyllum spicatum*
6b. Each side of the leaf blade divided into 12 or fewer segments—7

7a. Leaves sessile, each side with 9–12 segments—*Myriophyllum verticillatum*
7b. Leaves with distinct petioles, each side with 5–8 segments—*Myriophyllum sibiricum* [CQ, Crow & Hellquist, 2000]

HIPPURIDACEAE , The Mare's-tail Family

Aquatic herbs with whorled, simple, entire, sessile leaves on erect stems. Flowers perfect or unisexual (the plants then monoecious), inconspicuous, sessile and solitary in leaf axils; sepals reduced to a small rim of tissue surrounding the ovary; petals 0; stamen 1; pistil 1, style 1, ovary inferior, 1-celled. Fruit an achene or small drupe.

One species in NM (20–60 cm high; summer)—**Mare's-tail**, *Hippuris vulgaris*

ARALIACEAE, The Ginseng Family

Perennial herbs, sometimes woody below, with alternate, basal, or whorled compound leaves or prickly shrubs with alternate, simple, lobed leaves. Flowers mostly regular, perfect or seldom unisexual (the plants then dioecious), usually in umbels; sepals 0; petals 5; stamens 5; pistil 1, styles usually (1)2, 3 or 5, ovary inferior, usually 2-, 3-, or 5-celled. Fruit a berry-like drupe.

1a. Large, prickly shrub with simple, palmately lobed leaves; ravines, Isle Royale (1–3 m high; late spring)—**Devil's-club**, *Oplopanax horridus*

Oplopanax horridus (Devil's-club) is an example of a group of plants known as "western disjuncts". Marquis and Voss (1982) listed 28 plant species which are found in the vicinity of the Rocky Mountains (or west of them) and in the Great Lakes region, but rarely between these regions. *Oplopanax horridus* is common in moist woods from Alaska south to Oregon and Montana; in the Great Lakes region, it is known only from the Isle Royale archipelago and from nearby Ontario islands. While it is probable that such disjunctions arose following the retreat of the last (Wisconsin) glaciation, there are competing theories as to just how they occurred.

1b. Perennial herbs with or without a woody base; leaves compound—2

2a. Leaves palmately compound, in a single whorl; umbel one, terminal; rich woods (*Panax* spp., **Ginseng**)—3
2b. Leaves often twice or thrice pinnately compound, either alternate or basal; umbels several (*Aralia* spp.)—4

3a. Leaflets sessile; petals white, styles three; fruit yellow (10–20 cm high; spring)—**Dwarf Ginseng**, *Panax trifolius*
3b. Leaflets stalked; petals greenish-white, styles two; fruit red (20–60 cm high; summer)—**Ginseng**, *Panax quinquefolius*

4a. Stem underground, the leaf petioles and flower peduncles arising from the ground; rich woods (to 50 cm high; spring)—**Wild Sarsaparilla**, *Aralia nudicaulis*
4b. Stem erect, with cauline leaves (summer)—5

5a. Stem and petioles spiny or bristly at the base; umbels in terminal cluster; fruit black; sandy woods, disturbed areas (to 1.5 m high)—**Bristly Sarsaparilla**, *Aralia hispida*
5b. Stem and petioles smooth, not spiny; umbels many in a large terminal panicle; fruit purple; rich woods (to 2 m high)—**Spikenard**, *Aralia racemosa*

UMBELLIFERAE (APIACEAE), The Parsley Family

Herbs with alternate, usually compound or dissected leaves, the petioles expanded into a sheathing base. Flowers regular or seldom irregular, perfect or sometimes unisexual (the plants then monoecious), small, in simple or compound umbels (flowers in small *umbellets* terminating the primary inflorescence branches, or *rays*) or rarely in heads; sepals 5, minute or 0; petals 5, often white or yellow; stamens 5; pistil 1, styles 2, ovary inferior, 2-celled. Fruit a schizocarp, splitting into 2 mericarps.

1a. Leaf blades linear, sword-shaped with parallel veins; flowers (petals greenish-white) in a dense head; prairies, marsh edges, SW (to 1 m high; summer)—**Rattlesnake-master**, ***Eryngium yuccifolium***
1b. Leaf blades elliptic or broader with net veins;flowers in umbels—2

Figure 28: Umbelliferous flowers and fruit: A, flower, ovary dissected; B, fruit (schizocarp); C, inflorescence (compound umbel); 1, ray

2a. All leaves simple, the blades kidney-shaped or almost circular; stems creeping or floating; flowers white or more or less greenish (summer) (*Hydrocotyle* spp., **Water-pennywort**)—3

2b. Most leaves compound or at least deeply cleft, the blades variously shaped; stems mostly erect—4

3a. Leaves peltate, the petiole attached to the center of the blade; lakeshores, bogs, SLP—*Hydrocotyle umbellata*

3b. Leaves not peltate, the petiole attached to the margin of the blade; damp woods, stream banks—*Hydrocotyle americana*

4a. Ovary and fruit hairy, spiny, or warty and/or upper stem clearly pubescent; flowers white or greenish, rarely purplish—5

4b. Ovary and fruit glabrous; upper stem glabrous, rarely with tiny hairs; flowers white or yellow, rarely greenish-white—16

5a. Hairs on ovary and fruit distinctly hooked at the tip—6

5b. Hairs on ovary and fruit with straight or curved, but not hooked tips—11

6a. Leaves deeply palmately divided into three to five (rarely seven) segments; greenish flowers in head-like umbels; woods (summer) (*Sanicula spp.*, **Black Snakeroot**)—7

6b. Leaves pinnately divided or dissected; white (rarely purplish) flowers often in spreading umbels—10

7a. Styles short, not projecting beyond the bristles of the mature fruit; deciduous woods—8

7b. Styles long, projecting beyond the bristles of the fruit, and recurved—9

8a. Staminate flowers on pedicels 3–4(or up to 8) mm long, overtopping the fertile flowers and equalling or barely exceeding the fruit; mostly LP (30–80 cm high)—*Sanicula trifoliata*

8b. Staminate flowers short-pedicelled, shorter than the fertile flowers and concealed among the fruits; SLP (30–75 cm high)—*Sanicula canadensis*

9a. Staminate flowers shorter than the fruits; fruit short-stalked, 4 mm long or less; deciduous woods (30–80 cm high)—*Sanicula gregaria*

9b. Staminate flowers longer than the fruits; fruit sessile, longer than 4 mm; deciduous or coniferous woods, swamp borders (30–120 cm high)—*Sanicula marilandica*

10a. Stem hairs retrorse (downward-pointing); bracts subtending an umbel simple, not divided; petals white; wooded disturbed areas, LP (to 1 m high; summer)—**Hedge-parsley**, *Torilis japonica*

10b. Stem hairs spreading, not retrorse; bracts subtending an umbel pinnately divided into linear segments; petals of central flower of an
umbel often purple; roadsides, fields, etc. (40 cm–1 m high; summer &
fall)—**Wild Carrot** or **Queen-Anne's-lace**, *Daucus carota*

Some species of the family Umbelliferae are important food plants (including the cultivated carrot, a variety of **Daucus carota**) and others provide some of the basic seasonings in the
cuisines of the world (such as Caraway, **Carum carvi**, and Dill, **Anethum graveolens**). However, other members of this family are *extremely poisonous*. Any culinary sampling of umbelliferous plants collected in the wild should be done only after consulting references on edible
plants and careful unambiguous identification of the plant.

11a. Leaves pinnately divided or dissected—12
11b. Leaves ternately lobed, divided, or compound; petals all white or
greenish-white—13

12a. Leaflets or segments of the blade linear, less than 1 cm broad; petals of
central flower of an umbel often purple; roadsides, fields, etc. (0.4–1 m
high; summer & fall)—**Wild Carrot** or **Queen-Anne's-lace**, *Daucus
carota*
12b. Leaflets or segments of the blade oblong to lanceolate, 2 cm or more
broad; petals white to greenish white; dry woods and prairies, SLP (to
2 m high; summer)—*Angelica venenosa*

13a. Rays of the umbel more than 15; open woods, stream banks (1–2 m
high; summer)—**Cow-parsnip**, *Heracleum maximum* [*H. lanatum*—
CQ]
13b. Rays (branches) of the umbel fewer than 10; woodlands (40–80 cm
high; spring) (*Osmorhiza* spp., **Sweet-cicely**)—14

14a. Umbellet of flowers not subtended by bracts; mature styles curve outward; NM—*Osmorhiza chilensis*
14b. Umbellet of flowers subtended by bracts; mature styles straight—15

15a. Umbellet usually with eight or fewer flowers; styles less than 2 mm
long; dry woods—*Osmorhiza claytonii*
15b. Umbellet usually with nine or more flowers; styles 2 mm or longer;
moist woods—*Osmorhiza longistylis*

16a. Cauline leaves finely dissected into linear to oblong segments less than
1 cm broad—17
16b. Cauline leaves composed of distinct leaflets, usually variously compound, mostly over 2 cm broad—25

Figure 29: **Heracleum maximum**: *note ternately compound leaves*

17a. Axils of upper bracts bearing small bulblets; umbels few, sometimes none; wet areas (30–100 cm high; summer)—**Water-hemlock**, *Cicuta bulbifera*

17b. Axils of bracts lack bulblets, instead subtending normal umbels—18

18a. Petals yellow; umbellets lack subtending bracts; escaped to roadsides, railroads, fields, etc. (summer)—19

18b. Petals white; umbellets rarely lack subtending bracts—20

19a. Sheathing base of the petiole 3 mm or longer; primary umbel rays fewer than 30; SLP (1–2 m high)—**Fennel**, *Foeniculum vulgare*

19b. Sheathing base of the petiole less than 3 mm long; primary umbel rays 30 or more (to 1.5 m high)—**Dill**, *Anethum graveolens*

20a. Principal branches of the umbel one to four; leaves ternately compound; short plants of rich woods, SLP (early spring or spring)—21

20b. Principal branches of the umbel seven or more; leaves pinnately compound or dissected; tall plants of disturbed areas or swamps (summer)—22

21a. Anthers maroon; fruit elliptic (5–20 cm high; early spring)—**Harbinger-of-spring**, *Erigenia bulbosa*

21b. Anthers yellow; fruit linear-oblong (20–60 cm high; spring)—**Wildchervil**, *Chaerophyllum procumbens*

22a. Plants growing in swamps; SLP (0.4–1.5 m high)—**Hemlock-parsley**, *Conioselinum chinense*

22b. Plants of roadsides, fields, etc.—23

23a. Stems conspicuously spotted with purple (to 3 m high)—**Poison-hemlock**, *Conium maculatum*

23b. Stems not spotted with purple (to 1 m high)—24

24a. Rays of the umbel of unequal lengths; fruit prominently ribbed—**Caraway**, *Carum carvi*

24b. Rays of the umbel of similar lengths; fruit not ribbed—**Chervil**, *Anthriscus sylvestris*

25a. Leaves once-pinnate (or the submerged leaves multiply compound, if present), leaflets five or more (summer)—26

25b. Leaves ternately, palmately, or two to three times pinnately compound—30

26a. Plants of roadsides, fields, sometimes in woods—27

26b. Plants of wet areas—28

27a. Petals yellow; stem pubescent (to 1.5 m high)—**Wild Parsnip**, *Pastinaca sativa*

27b. Petals white; stem glabrous; UP (30-60 cm high)—**Burnet-saxifrage**, *Pimpinella saxifraga*

28a. Margin of leaflets entire, or with a few low teeth near the apex; fruit surrounded by a wing-like margin; SLP (to 1.5 m high)—**Cowbane**, *Oxypolis rigidior*

28b. Margin of leaflets finely toothed to irregularly incised; fruit lacks an expanded margin—29

29a. Leaflets mostly ovate or ovate-lanceolate, the margin irregularly incised; fruit ribs obscure; WM (to 80 cm high)—**Water-parsnip**, *Berula erecta*

29b. Leaflets linear to oblong, finely but sharply serrate; fruit ribs prominent (to 2 m high)—**Water-parsnip**, *Sium suave*

30a. Axils of upper bracts bearing small bulblets; umbels few, sometimes none; leaflets linear; wet areas (30–100 cm high; summer)—**Water-hemlock**, *Cicuta bulbifera*

30b. Axils of bracts lack bulblets, instead subtending normal umbels—31

31a. Most cauline leaves with three leaflets—32

31b. Most cauline leaves with five or more leaflets, often multiply compound—35

32a. Petals white—33

32b. Petals yellow (30–80 cm high; spring)—34

33a. Umbel symmetrical, the rays of one length, more than fifteen; leaflets pubescent; open woods, streambanks (1–2 m high; summer)—**Cow-parsnip**, *Heracleum maximum* [*H. lanatum*—CQ]

33b. Umbel unsymmetrical, the rays of different lengths, fewer than ten; leaflets glabrous; rich woods (30–100 cm tall; early summer)—**Honewort**, *Cryptotaenia canadensis*

34a. Central flower of each umbellet sessile; fruits ribbed, not winged; often in bogs and fens—**Golden-alexanders**, *Zizia aurea*

34b. Central flower of each umbellet pedicellate; fruit with evident lateral wings; oak woods and prairies, SLP—***Thaspium trifoliatum***

35a. Margin of leaflets entire; petals yellow; dry, sandy woods (40–80 cm high; spring)—**Yellow-pimpernel**, *Taenidia integerrima*

35b. Margin of leaflets finely or coarsely toothed; petals white, greenish-white, or yellow—36

36a. Margin of leaflets with few coarse teeth or lobes; woods and wetland margins, SLP (to 1 m high; spring)—***Thaspium barbinode***

36b. Margins of leaflets with many fine teeth—37

37a. Plant robust; stems purple; sheathing petiole bases conspicuous, often over 8 cm long; umbel spherical; petals white or greenish-white; wet areas (to 3 m high; summer)—***Angelica atropurpurea***

37b. Plant smaller; stems mostly green; sheathing petiole bases smaller; umbel not spherical, most rays ascending; petals white or yellow—38

38a. Lateral veins of the leaflets end between the marginal teeth; petals white; wet areas (to 2 m high; summer)—**Water-hemlock**, *Cicuta maculata*

38b. Lateral veins of the leaflets end in the marginal teeth; petals white or yellow—39

39a. Petals white; escapes from cultivation to roadsides, fields, etc. (40–100 cm high; early summer)—**Goutweed** or **Bishop's-weed**, *Aegopodium podagraria*

39b. Petals yellow; woods or wet areas (30–80 cm high; spring)—40

40a. Central flower of each umbellet sessile; fruits ribbed, not winged; often in bogs and fens—**Golden-alexanders**, *Zizia aurea*

40b. Central flower of each umbellet pedicellate; fruit with evident lateral wings; oak woods and prairies, SLP—***Thaspium trifoliatum***

CORNACEAE, The Dogwood Family

Trees, shrubs, or rarely herbs with opposite or seldom alternate or whorled, simple leaves. Flowers regular, perfect, small, in cymes or sometimes heads which are subtended by conspicuous petal-like bracts; sepals 4, minute; petals 4; stamens 4; pistil 1, style 1, ovary inferior, (1)2-celled. Fruit a drupe.

1a. Plant a small herb with one apparent whorl of leaves; flowers in a small head subtended by four white, petal-like bracts; fruits red; mixed or coniferous woods (10–20 cm high; early summer)—**Bunchberry**, *Cornus canadensis*

1b. Plant a shrub or small tree with opposite or alternate leaves; flowers mostly in cymes, seldom in heads subtended by petal-like bracts; fruit red, blue, or white—2

2a. Leaves alternate; woods and thickets (to 6 m high; spring & early summer)—**Alternate-leaved** or **Pagoda Dogwood**, *Cornus alternifolia*

2b. Leaves opposite—3

3a. Flowers in a small head subtended by four petal-like bracts; fruits red; deciduous woods, SLP (to 10 m; spring)—**Flowering Dogwood, *Cornus florida***

3b. Flowers in open flattened cymes not subtended by petaloid bracts; fruits white or blue—4

Flowering Dogwood, **Cornus florida**, is protected by Public Act 182 of 1962, commonly called the "Christmas-tree law". Under this law, it is illegal to remove or cut these plants from any area without a bill of sale or written permission from the owner. While all dogwoods produce flowers, these small trees are notable for the four large white (or in some cultivars, pink) bracts surrounding the head of small inconspicuous flowers.

4a. Lateral veins along each side of the leaf six to eight; young branches greenish with purple spots; woodland borders and thickets (1–4 m high; spring & early summer)—**Round-leaved Dogwood, *Cornus rugosa***

4b. Lateral veins along each side of the leaf three to five (rarely six); young branches grayish or red, not spotted—5

5a. Leaves rough above; river banks, SE (to 6 m high; spring)—**Dogwood, *Cornus drummondii***

5b. Leaves smooth or finely soft-hairy above (spring & early summer)—6

6a. Leaves distinctly pubescent beneath with woolly or spreading hairs; damp areas, esp. shores and dunes (1–3 m high)—**Red-osier, *Cornus stolonifera* [*C. sericea*—CQ]**

6b. Leaves smooth beneath, or pubescent with short appressed hairs—7

7a. Older branches grayish; dry woods and shore thickets (1–5 m high)—**Gray Dogwood, *Cornus foemina* [*C. racemosa*—CQ]**

7b. Older branches reddish—8

8a. Young branches covered with reddish hairs; fruits blue; wet areas, often in thickets (1–5 m high)—**Pale Dogwood, *Cornus amomum***

8b. Young branches glabrous; fruits white; damp areas, esp. shores and dunes (1–3 m high)—**Red-osier, *Cornus stolonifera* [*C. sericea*—CQ]**

Many species of dogwood display brightly colored young branches. **Cornus stolonifera** (= **C. sericea**) includes varieties in which young shoots are either red or yellow.

PYROLACEAE, The Wintergreen Family

Perennial herbs or subshrubs, often with basal, opposite, or whorled, simple, evergreen leaves. Flowers regular, perfect; sepals 5, sometimes united below; petals 5; stamens 10; pistil 1, style 1, ovary superior, 4 or 5-celled. Fruit a capsule. Summer.

1a. Low, trailing evergreen plant with opposite or whorled cauline leaves; flowers white to greenish, in umbels or corymbs; often in dry, sandy woods (***Chimaphila*** spp., **Wintergreen**)—2
1b. Herbs with basal leaves; flowers white to greenish, in racemes or solitary; often in damp woods, bogs—3

2a. Leaves broadest above the middle, dark green (10–30 cm high)— **Prince's-pine** or **Pipsissewa**, ***Chimaphila umbellata***

Pipsissewa, **Chimaphila umbellata,** is protected by Public Act 182 of 1962, commonly called the "Christmas-tree law". Under this law, it is illegal to remove or cut these plants from any area without a bill of sale or written permission from the owner.

2b. Leaves broadest below the middle, spotted with white; mostly WM (10–20 cm high)—**Spotted Wintergreen**, ***Chimaphila maculata***

3a. Flower white, solitary and terminal; NM (3–10 cm high)—**One-flowered Shinleaf**, ***Moneses uniflora***
3b. Flowers white or pinkish, in terminal racemes (scapes 5–30 cm high)—4

4a. Style straight—5
4b. Style bent near the base (most ***Pyrola*** spp., **Shinleaf**)—6

5a. Raceme regular, the white flowers not all pointing in the same direction; style 1.5 mm or shorter; UP—***Pyrola minor***
5b. Racemes one-sided, the white or greenish-white flowers all turned in one direction; style 2.5 mm or longer—**One-sided Shinleaf**, ***Orthilia secunda*** [***Pyrola secunda***—CQ]

6a. Flowers pink or purple; leaf bases often heart-shaped—***Pyrola asarifolia***
6b. Flowers white or greenish; leaf bases often wedge-shaped or broadly rounded—7

7a. Leaves less than 3 cm long, the blades mostly shorter than their petioles; dry woods—***Pyrola chlorantha***
7b. Leaves longer than 3 cm long, the blades usually equalling or longer than their petioles—8

8a. Sepals about as long as broad; dry woods—***Pyrola elliptica***
8b. Sepals longer than broad; dry to moist woods, bogs—***Pyrola rotundifolia***

MONOTROPACEAE, The Indian-pipe Family

Perennial forest herbs which lack chlorophyll, the plants often white, red, or brown with alternate scale-like leaves. Flowers regular, perfect, solitary or in racemes; sepals 0, 4, or 5; petals often 4 or 5, separate or united (the corolla then urn-shaped); stamens 8 or 10, sometimes united below; pistil 1, style 1, ovary superior, 4 or 5-celled. Fruit a capsule. Summer.

1a. Flower solitary; plant often white; rich woods (10–20 cm high)—**Indian-pipe**, *Monotropa uniflora*
1b. Flowers in racemes; plant usually yellow, brown, or pink to red—2

2a. Stems glandular-pubescent; petals united into a bell-shaped corolla; coniferous woods, NM (30–100 cm high)—**Pine-drops**, *Pterospora andromedea*
2b. Stems pubescent (but not glandular) or not; petals all separate; woods (10–30 cm high)—**False Beech-drops** or **Pinesap**, *Monotropa hypopitys*

The Monotropaceae are notable for their lack of the photosynthetic pigment chlorophyll; the plants are either a ghostly white or various colors of red and brown. Unlike the vast majority of vascular plants, which are self-supporting via photosynthesis, members of this family are dependent on the photosynthetic output of other plants. They are *mycotrophic*, associated with a mycorrhizal fungus which in turn is associated with the roots of a tree. The fungus transfers nutrients from the tree to the parasitic plant. Other families of parasitic plants, such as Viscaceae, Cuscutaceae and Orobanchaceae, form *haustoria* (structures which drill directly into the phloem of host plants).

ERICACEAE, The Heath Family

Shrubs, frequently with alternate, opposite, or whorled, simple, evergreen leaves. Flowers regular, perfect (rarely unisexual, the plants then dioecious), often in corymbs or racemes; sepals 4 or 5; petals 4 or 5, united, the corolla often urn- or bell-shaped (or rarely separate); stamens mostly 4, 5, 8, or 10, sometimes attached to the corolla tube; pistil 1, style 1, ovary superior or less often inferior, 4, 5, 8, or 10-celled. Fruit a capsule, berry, or drupe.

1a. Petals separate or the corolla deeply four-lobed, the lobes longer than the tube and often reflexed; bogs—2
1b. Petals united into a tubular, bell-shaped, or urn-shaped corolla, the tube of which is as long as or longer than the lobes—4

2a. Erect shrub; leaf blades 2–5 cm long, densely woolly beneath; petals white; mostly NM (to 100 cm high; early summer)—**Labrador-tea**, *Ledum groenlandicum*

2b. Creeping shrub; leaf blades less than 2 cm long, whitened but not woolly beneath; petals white to pink, often reflexed (*Vaccinium* sect. *Oxycoccus*, **Cranberry**)—3

3a. Leaf blades acute; pedicel with two small red bracts (spring & early summer)—**Small Cranberry**, *Vaccinium oxycoccos*

3b. Leaf blades obtuse; pedicel with two small green bracts (summer)—**Large Cranberry**, *Vaccinium macrocarpon*

4a. Leaves opposite or whorled; corolla purple, saucer-shaped; bogs (to 100 cm high)—5

4b. Leaves alternate; corolla mostly white or pink, bell-shaped or salverform and expanded lobes—6

5a. Branches and twigs cylindrical, not angled; northern LP (early summer)—**Sheep-laurel**, *Kalmia angustifolia*

5b. Branches and twigs with two sharp angles; NM (spring)—**Bog-laurel**, *Kalmia polifolia*

6a. Plants prostrate, with at most a few ascending branches; flowers white or pink—7

6b. Plants erect or ascending—9

7a. Leaf blades 3 cm or more long; flowers over 8 mm long, the corolla not apically constricted; fruit a capsule (early spring)—**Trailing-arbutus**, *Epigaea repens*

Trailing arbutus, ***Epigaea repens***, is protected by Public Act 182 of 1962, commonly called the "Christmas-tree law". Under this law, it is illegal to remove or cut these plants from any area without a bill of sale or written permission from the owner.

7b. Leaf blades 3 cm or less long; flowers 2–5 mm long, the corolla apically constricted; fruit a berry or drupe (late spring)—8

8a. Leaf blades spathulate, broadest beyond the middle; flowers five-parted; fruit red; sandy or rocky areas, NM and WM—**Bearberry**, *Arctostaphylos uva-ursi*

8b. Leaf blades oval, broadest at the middle; flowers four-parted; fruit white; wet woods—**Creeping-snowberry**, *Gaultheria hispidula*

9a. Leaf blades covered with rusty scales or white hairs beneath; bog shrubs—10

9b. Leaf blades smooth, pubescent, or resinous beneath; but neither covered with rusty scales nor whitened—11

10a. Leaf blades linear, white beneath, their margins strongly revolute (to 50 cm high; late spring)—**Bog-rosemary**, *Andromeda glaucophylla*
10b. Leaf blades mostly oblong, covered with rusty scales beneath (to 1.5 m high; spring)—**Leatherleaf**, *Chamaedaphne calyculata*

11a. Leaves evergreen; stem erect from a creeping root-stock; ovary superior, fruit red; woods (10–20 cm high; summer)—**Wintergreen** or **Teaberry**, *Gaultheria procumbens*
11b. Leaves deciduous; plant bushy; ovary inferior, fruit blue or black—12

12a. Leaf blades dotted beneath with yellowish resinous dots; fruits black; dry woods (to 1 m high; spring)—**Huckleberry**, *Gaylussacia baccata*
12b. Leaf blades not resinous-dotted beneath; fruits blue or black—13

13a. Calyx deciduous, the fruit lacks a crown of teeth; UP (*Vaccinium* sect. *Vaccinium*, **Bilberry**)—14
13b. Calyx persistent as a crown of small teeth on the fruit (*Vaccinium* sect. *Cyanococcus*, **Blueberry**)—16

Vaccinium macrocarpon, a native Michigan plant found in bogs, is the cranberry of commerce. Many cultivated strains of blueberries have been bred in part from *Vaccinium corymbosum* (Highbush Blueberry), also a native Michigan plant.

14a. Most full-grown leaf blades less than 2.5 cm long; flowers one to four; low, much-branched shrubs mostly less than 30 cm high; dry areas (early summer)—**Dwarf Bilberry**, *Vaccinium caespitosum*
14b. Most full-grown leaf blades more than 2.5 cm long; flowers solitary; tall, bushy shrubs 0.5–1.5 m high; moist woods and borders (spring)—15

15a. Leaf blades serrulate or entire below the middle, green beneath, acute—**Tall Bilberry**, *Vaccinium membranaceum*
15b. Leaf blades entire or serrulate only below the middle, pale beneath, obtuse—**Oval-leaved Bilberry**, *Vaccinium ovalifolium*

16a. Tall erect shrubs, often 1–4 (rarely 5) m high; bogs and swamps, SLP (spring & early summer)—**Highbush Blueberry**, *Vaccinium corymbosum*
16b. Lower bushy shrubs, 10–80 cm tall—17

17a. Leaf blades pubescent beneath; sand dunes, woods, bogs (20–50 cm high; spring & early summer)—**Canada Blueberry**, *Vaccinium myrtilloides*

17b. Leaf blades glabrous and/or glaucous, at most with a few hairs along the midvein—18

18a. Leaf blades pale green and glaucous, entire or nearly so; fields and dry woods, SLP (20–80 cm high; spring)—**Hillside Blueberry**, *Vaccinium pallidum*

18b. Leaf blades bright green, glabrous beneath or with hairs along the midvein, distinctly serrulate; fruits blue or black; sand dunes, woods, bogs (10–60 cm high)—**Low Sweet Blueberry**, *Vaccinium angustifolium*

PRIMULACEAE, The Primrose Family

Herbs with alternate, opposite, or basal, simple leaves. Flowers regular, perfect; sepals 5 (– 7), united or not; petals 5(–7), united; stamens 5(–7), attached one in front of each petal, sometimes with alternating staminodes; pistil 1, style 1, ovary superior, 1-celled. Fruit a capsule.

1a. Leaves in a basal rosette; inflorescence an umbel subtended by leaf-like bracts on one or more separate peduncles (spring) (*Primula* spp., **Primrose**)—2

1b. Cauline leaves present; inflorescence a panicle, raceme, or flowers solitary—3

2a. Petals entirely yellow; calyx pale, inflated; escape from cultivation, Straits (scapes 10–40 cm high)—**Cowslip**, *Primula veris*

2b. Petals pink, pale purple, or rarely white, the central tube yellow; calyx not inflated; calcareous areas near Great Lakes shorelines, rare in inland fens and moist shores and ledges (scapes to 35 cm high)—**Bird's-eye Primrose**, *Primula mistassinica*

3a. All cauline leaves in one whorl just below the flower cluster; petals and sepals often seven; rich woods, bogs (10–20 cm high; spring)—**Starflower**, *Trientalis borealis*

3b. Cauline leaves several or many, scattered over the stem; petals and sepals often five or six—4

4a. All or the uppermost leaves alternate; petals white—5

4b. All leaves opposite or whorled, rarely a few alternate; petals yellowish, yellow, or red—6

5a. Calyx lobes triangular, shorter than the tube; shores and stream banks, SLP (10–30 cm high; spring and summer)—**Water Pimpernel**, *Samolus parviflorus* [*S. floribundus*—CQ]

5b. Calyx deeply divided, the lobes lanceolate; escape from cultivation (to 1 m high; summer)—**Gooseneck Loosestrife**, *Lysimachia clethroides*

6a. Petals red; fruit a capsule, the upper half opening like a lid; stems prostrate or ascending; lawns and roadsides, LP (10–30 cm high; summer)—**Scarlet Pimpernel**, *Anagallis arvensis*

6b. Petals yellow; fruit a capsule which opens along vertical lines; stems erect, rarely prostrate (summer) (most *Lysimachia* spp., **Loosestrife**)—7

7a. Stem creeping; moist, disturbed areas—**Moneywort**, *Lysimachia nummularia*

7b. Stem erect—8

8a. Leaves punctate with dark dots; staminodes absent (*Lysimachia* sect. *Lysimachia* and sect. *Naumbergia*)—9

8b. Leaves not dark-dotted; small staminodes alternating with the stamens; woods and prairies (*Lysimachia* sect. *Steironema*)—12

9a. Corolla yellow, not streaked or dotted with brown or black; escape from cultivation (to 1 m high)—**Garden Loosestrife**, *Lysimachia punctata*

9b. Corolla streaked or dotted with black or brown—10

10a. Flowers all axillary; leaves whorled; woods, mostly LP (30–90 cm high)—**Whorled Loosestrife**, *Lysimachia quadrifolia*

10b. Flowers in one or several many-flowered racemes; swamps—11

11a. Flowers mostly six-parted, in several axillary racemes; all leaves opposite (30–70 cm high)—**Tufted Loosestrife**, *Lysimachia thyrsiflora*

11b. Flowers mostly five-parted, in a terminal raceme; leaves mostly opposite, rarely a few alternate (40–80 cm high)—**Swamp-candles**, *Lysimachia terrestris*

Plants with mostly whorled leaves and flowers in a terminal raceme are probably **Lysimachia** x***producta**, a fertile hybrid between **L. quadrifolia** and **L. terrestris** which has been collected in the southern Lower Peninsula.

12a. Leaf blades lanceolate to ovate, with conspicuously ciliate petioles; rich woods, wooded shores (40–130 cm high)—**Fringed Loosestrife**, *Lysimachia ciliata*

12b. Leaf blades linear to oblong, the petiole (if present) glabrous or barely ciliate—13

13a. Leaf blades lanceolate, over 7 mm wide, with pinnate venation and flat margins; sandy woods, mostly SLP (20–90 cm high)—***Lysimachia lanceolata***

13b. Leaves linear, 7 mm wide or narrower, with one midvein and revolute margins; fens, marshes, wet prairies, LP (30–100 cm high)—***Lysimachia quadriflora***

OLEACEAE, The Olive Family

Trees or shrubs with opposite, often pinnately compound leaves. Flowers regular, perfect or unisexual (the plants then monoecious); sepals 4, united, or 0; petals 4, united, or 0; stamens usually 2, attached to the corolla tube (when present); pistil 1, style 1, ovary superior, 2-celled. Fruit a capsule, drupe, or a single-seeded samara.

1a. Shrubs; leaves simple, entire; showy racemes of flowers with blue, pinkish, or white petals; persisting from cultivation—2

1b. Trees; leaves compound; flowers inconspicuous, greenish, the petals absent (spring) (***Fraxinus*** spp., **Ash**)—4

2a. Leaf petioles 1 cm or longer; leaf blades broadly ovate with a truncate or cordate base; flowers often blue or pinkish, sometimes white; fruit a capsule (to 6 m high; spring)—**Lilac, *Syringa vulgaris***

2b. Leaf petioles absent or up to 1 cm long; leaf blades ovate to elliptical, the base tapered; flowers white; fruit a black drupe (late spring) (***Ligustrum*** spp., **Privet**)—3

3a. Leaf blades glabrous beneath; corolla tube about 3 mm long, about as long as the lobes; LP (to 5 m high)—***Ligustrum vulgare***

3b. Mid-vein of leaf blade pubescent beneath; corolla tube 5 mm or longer, about twice as long as the lobes; SLP—***Ligustrum obtusifolium***

4a. Twigs sharply four-angled; calcareous woods and river bottoms, SLP (to 30 m high)—**Blue Ash, *Fraxinus quadrangulata***

4b. Twigs not distinctly angled—5

5a. Lateral leaflets sessile; moist woods (to 25 m high)—**Black Ash, *Fraxinus nigra***

5b. Lateral leaflets stalked—6

6a. Leaflets pale green beneath, entire or nearly so; rich woods (to 40 m high)—**White Ash, *Fraxinus americana***

6b. Leaflets bright green beneath, the margin often serrulate; moist or wet woods—7

7a. Wing of samara 7 mm or wider; wet woods, river bottoms, SLP (to 40 m high)—**Pumpkin Ash**, *Fraxinus profunda*

7b. Wing of samara less than 7 mm wide; moist woods (to 25 m high)—8

8a. Leaflets pubescent beneath; young twigs pubescent—**Red Ash**, *Fraxinus pennsylvanica* [*F. pennsylvanica* var. *pennsylvanica*—CQ]

8b. Leaflets glabrous beneath; young twigs glabrous—**Green Ash**, *Fraxinus pennsylvanica* [*F. pennsylvanica* var. *subintegerrima*—CQ]

MENYANTHACEAE, The Buckbean Family

Aquatic herbs with basal trifoliolate leaves. Flowers regular, perfect, in racemes at the summit of a naked pedicel; sepals 5, united or not; petals 5, united; stamens 5, attached to the corolla tube, alternate with the corolla lobes; pistil 1, style 1, ovary partly inferior, 1-celled. Fruit a capsule.

One species in Michigan, flowers white or bluish; cold swamps and bogs (5–30 cm high; late spring)—**Buckbean**, *Menyanthes trifoliata*

GENTIANACEAE, The Gentian Family

Herbs with opposite or whorled, simple, entire leaves (reduced, scale-like, and sometimes alternate in *Bartonia*). Flowers mostly regular, perfect; sepals 4 or 5 (–12), united, the calyx tube often deeply divided; petals 4 or 5 (–12), united; stamens 4 or 5 (–12), attached to the corolla tube, alternate with the corolla lobes; pistil 1, style 1, ovary superior, 1-celled. Fruit a capsule.

> Michigan gentians are protected by Public Act 182 of 1962, commonly called the "Christmas-tree law". Under this law, it is illegal to remove or cut these plants from any area without a bill of sale or written permission from the owner. Gentians include many plants once considered part of the large genus **Gentiana**, but now placed in the genera **Gentianopsis** or **Gentianella**.

1a. Leaves reduced to small scales less than 5 mm long; stem filiform, green or brownish; flowers small, often greenish-yellow; bogs (5–40 cm high; late summer)—*Bartonia virginica*

1b. Leaves not reduced, the blades 1 cm or longer—2

2a. Leaves whorled, often four per whorl; flowers yellowish-green; plant over 1 m tall; rich woods, mostly SW (1–2 m high; summer)—**American Columbo**, *Frasera caroliniensis*

2b. Leaves opposite; plants up to, but often less than, 1 m tall—3

3a. Corolla saucer-shaped, with a short tube and four to twelve spreading lobes up to 2 cm long, pink or rarely white; moist sands, shores, SLP (30–80 cm high; summer)—**Rose-pink** or **Rose Gentian**, *Sabatia angularis*

3b. Corolla bell-shaped, tubular, funnelform, or salverform; tube longer than or equalling the four or five lobes—4

4a. Corolla bell-shaped, each petal with a spur at the base, purplish or whitish (to greenish or yellowish), and about 1 cm long; moist woods and bogs, NM (20–90 cm high; summer)—**Spurred Gentian**, *Halenia deflexa*

4b. Petals lack a basal spur; corolla often longer—5

5a. Corolla pink to reddish-purple, salverform (with spreading lobes), the tube narrow, less than 2.5 cm broad; fields, roadsides, etc. (summer) (*Centaurium* spp., **Centaury**)—6

5b. Corolla mostly blue or white, lobes erect, incurved, or sometimes spreading, the tube wider, more than 2.5 cm broad—7

6a. Basal rosette of leaves present; flowers nearly sessile, the subtending bracts and calyx barely separated, the corolla red-purple; WM (20–50 cm high)—*Centaurium erythraea*

6b. Basal rosette of leaves absent; flowers on short pedicels, the corolla pink; moist saline soil along highways, SE (to 20 cm high)—*Centaurium pulchellum*

7a. Corolla lobes usually four, fringed, bright blue (late summer, early autumn) (*Gentianopsis* spp., **Fringed Gentian**)—8

7b. Corolla lobes usually five, entire, mostly blue or white—9

8a. Leaf blades lanceolate to ovate, over 1 cm wide; wet ditches and moist sandy meadows, LP (30–80 cm high)—*Gentianopsis crinita*

8b. Leaf blades linear or nearly so, less than 1 cm wide; fens, wet calcareous shores—*Gentianopsis procera*

9a. Corolla 2 cm long or a little less, the lobes short and erect, additional tissue between the lobes absent; calcareous meadows and open woods, SLP (20-80 cm high; late summer)—**Stiff Gentian**, *Gentianella quinquefolia*

9b. Corolla 2.5–5 cm long, additional tissue, often as a fold or pleat, present between adjacent lobes (*Gentiana* spp., **Gentian**)—10

10a. Calyx lobes and/or leaf blades rough or ciliate at the margin—11
10b. Calyx lobes and leaf blades smooth—12

11a. Corolla lobes blue, spreading; leaves narrowly lanceolate, indistinctly veined; prairies, dry woods, SLP (20–60 cm high; late summer and early autumn)—**Prairie Gentian**, *Gentiana puberulenta*
11b. Corolla lobes blue above, white below, erect with incurved apices, the mature flower "closed"; leaves ovate to ovate-lanceolate, with three to seven principal veins; prairies, wet meadows (30–100 cm high; summer and early autumn)—**Bottle Gentian**, *Gentiana andrewsii*

12a. Leaf blades ovate or ovate-lanceolate, somewhat cordate at base; flowers greenish-white or yellowish-white; prairies and dry open woods, SLP (30–100 cm high; late summer and early autumn)—**White Gentian**, *Gentiana alba* [*G. flavida*—CQ]
12b. Leaf blades lanceolate or nearly linear, not cordate; flowers blue or white with a bluish apex; NM (late summer)—13

13a. Leaf blades dark green; involucral leaves spreading, not enclosing the calyces of adjacent flowers; acidic shores and meadows, UP (20–80 cm high)—**Bog Gentian**, *Gentiana linearis*
13b. Leaf blades pale green; involucral leaves erect, enclosing the calyces of adjacent flowers; wet, calcareous meadows, NM (30–70 cm high)—**Red-stemmed Gentian**, *Gentiana rubricaulis*

APOCYNACEAE, The Dogbane Family

Perennial herbs with opposite, simple, entire leaves and milky juice. Flowers regular, perfect; sepals 5, united, the calyx tube deeply divided; petals 5, united; stamens 5, attached to the corolla tube, alternate with the corolla lobes; pistil 1, style 1, ovaries 2, each superior, 1-celled. Fruit is 2 follicles.

1a. Stems creeping or trailing; flowers solitary and axillary; petals blue, their lobes 2-3 cm broad; escapes to woods and roadsides (spring)—**Periwinkle**, *Vinca minor*
1b. Stems erect or essentially so; flowers in terminal cymes; petals white or pink, their lobes 1 cm broad or less (spring and summer) (*Apocynum* spp., **Dogbane**)—2

2a. Corolla pinkish, the tube 6–10 mm long; inflorescence overtops sterile branches; petioles present; dry woods (20–80 cm high)—**Spreading Dogbane**, *Apocynum androsaemifolium*

2b. Corolla white or greenish, the tube 3–6 mm long; sterile branches often overtop the inflorescence; petioles present or leaves sessile; fields (50–150 cm high)—**Indian-hemp**, *Apocynum cannabinum*

Since the two species of **Apocynum** (**A. androsaemifolium** and **A. cannabinum**) are known to hybridize in southern Michigan, plants with intermediate characters should be expected. These are sometimes given the name **Apocynum ×floribundum**.

ASCLEPIADACEAE, The Milkweed Family

Perennial herbs or sometimes twining vines with opposite (seldom alternate or whorled), simple, entire leaves and often with milky juice. Flowers regular, perfect, often in cymes or cyme-like umbels; sepals 5; petals 5, united, and often spreading or reflexed, hiding the calyx; a *corona* arises from the base of each stamen forming a colored *hood* and often including a slender *horn*, which together is frequently the most conspicuous part of the flower and may be mistaken for the corolla; stamens 5, attached to the corolla tube, united with each other and with the stigmas to form the *gynostegium* in the center of the flower; pistil 1, styles 2, ovaries 2, each superior, 1-celled. Fruit a follicle. Summer.

1a. Corolla lobes spreading to erect; stems often twining; disturbed woods and thickets, SLP (*Vincetoxicum* spp.)—2

1b. Corolla lobes reflexed; stems erect, not twining; mostly in prairies and open fields (*Asclepias* spp., **Milkweed**)—3

2a. Corolla purplish-black, the upper surface pubescent—**Black Swallowwort**, *Vincetoxicum nigrum*

2b. Corolla pale purple to maroon or pinkish, the upper surface glabrous; SE—**Dog-strangling Vine**, *Vincetoxicum rossicum* [*V. hirundinaria*—CQ]

3a. Leaves in whorls of four to seven, linear; flowers greenish-white (20–50 cm high)—**Whorled Milkweed**, *Asclepias verticillata*

3b. Leaves opposite or alternate; flowers of various colors—4

4a. Most or all leaves alternate, the blades linear to oblanceolate; sandy areas—5

4b. Most or all leaves opposite, the blades lanceolate to ovate—6

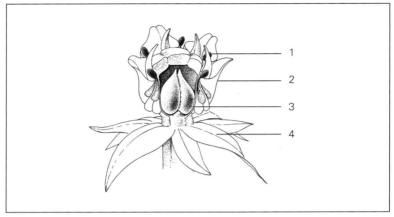

Figure 30: **Asclepias syriaca**, *dissected flower: 1, horn or corona;
2, hood of corona; 3,ovary, 4, petal*

5a. Flowers brilliant orange; LP (30–70 cm high)—**Butterfly-weed**, *Asclepias tuberosa*

5b. Flowers greenish or purplish; SLP (40–100 cm high)—**Prairie Milkweed**, *Asclepias hirtella*

6a. Umbels lateral, mostly sessile; flowers greenish; horns absent; LP (30–80 cm high)—**Green Milkweed**, *Asclepias viridiflora*

6b. Umbel(s) terminal or lateral, peduncled; flowers variously colored; horns present—7

7a. Leaf blades pubescent beneath—8
7b. Leaf blades glabrous beneath or nearly so—9

8a. Corolla lobes green with a purple tinge; fields and roadsides (1–2 m high)—**Common Milkweed**, *Asclepias syriaca*

8b. Corolla lobes bright red or purple; margins of dry woods, SLP (to 1 m high)—**Purple Milkweed**, *Asclepias purpurascens*

9a. Leaves broadly rounded and almost sessile at base—10
9b. Leaves narrowed at the base, distinctly petioled—11

10a. Leaf margins wavy; umbel solitary, terminal and erect on a long peduncle; pedicels pubescent; corolla lobes green with a purple tinge; LP (30–80 cm high)—*Asclepias amplexicaulis*

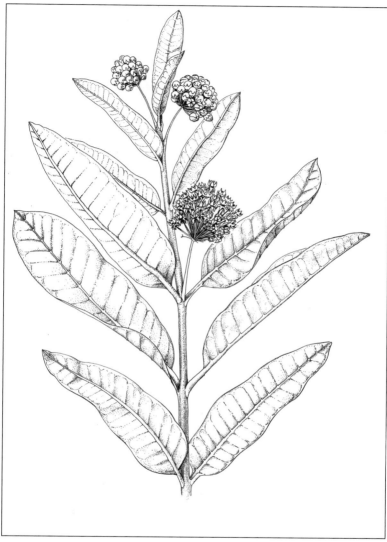

Figure 31:Asclepias syriaca

10b. Leaf margins not wavy; umbels one or several, terminal and/or lateral, on shorter peduncle(s); pedicels glabrous; corolla lobes rose or purple; moist prairies, SE (to 1 m high)—**Smooth Milkweed**, *Asclepias sullivantii*

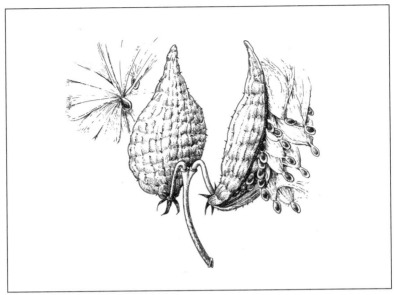

Figure 32: ***Asclepias syriaca****, fruit (follicles)*

11a. Corolla lobes red; wet areas (to 1.5 m high)—**Swamp Milkweed**, *Asclepias incarnata*
11b. Corolla lobes white to greenish; moist woods (80–150 cm high)—**Poke Milkweed**, *Asclepias exaltata*

CONVOLVULACEAE, The Morning Glory Family

Herbs, mostly twining or trailing, with alternate, simple leaves and often with milky juice. Flowers regular, perfect, axillary; sepals 5, united below or not; petals 5, united, the corolla often funnelform; stamens 5, attached to the corolla tube; pistil 1, style 1, ovary superior, 2–4-celled. Fruit a capsule.

The Convolvulaceae and the closely related family Cuscutaceae are distinctive in their twining habits. The herbaceous stems wind tightly around other plants, which is often injurious (the bindweeds are considered major weed pests). One exception is the Low Bindweed, ***Calystegia spithamaea***, which neither binds nor trails.

1a. Style divided at the top into two linear or oblong stigmas; corolla white or pink—2
1b. Style not divided at the top, the stigmas sessile, capitate and entire or at most 3-lobed; corolla white (with or without a red throat), blue, or a variegated blue and white; SLP (summer) (*Ipomoea* spp., **Morning Glory**)—4

2a. Bracts at base of calyx small or absent, the five sepals not concealed: fields, roadsides (spring and summer)—**Field Bindweed**, *Convolvulus arvensis*
2b. Calyx almost concealed by two large heart-shaped bracts (*Calystegia* spp., **Bindweed**)—3

3a. Stem entirely twining or trailing; leaves hastate or sagittate, with basal lobes; fields (summer)—**Hedge Bindweed**, *Calystegia sepium*
3b. Stem erect (portion above the flower may recline); leaf blades rounded or somewhat cordate at base, lacking narrow lobes; sandy woods, rock ledges (late spring and early summer)—**Low Bindweed**, *Calystegia spithamaea*

4a. Stems smooth or nearly so; sepals glabrous; corolla white with a purple-brown throat; stigma entire or two-lobed; woods—**Wild Potato-vine**, *Ipomoea pandurata*
4b. Stems covered with reflexed hairs; lower half of the sepals pubescent; corolla blue, white, pink, or blue and white striped; stigma 3-lobed; roadsides, fields, etc.—**Morning Glory**, *Ipomoea purpurea*

CUSCUTACEAE, The Dodder Family

Parasitic herbs which lack chlorophyll, the stems resembling yellowish or white threads with alternate, scale-like leaves. Flowers regular, perfect, tiny; sepals 4 or 5, united or not; petals 4 or 5, united; stamens 4 or 5, attached to the corolla tube; pistil 1, styles 2, ovary superior, 2-celled. Fruit a utricle or capsule. Late summer.

Dodder (**Cuscuta** spp.) can be quite conspicuous as large mats of bright yellow or orange twining stems attached to host plants. These parasitic plants use *haustoria*, specialized structures produced from short branches, which penetrate the phloem of their host. Flowers are necessary for identification. They display some host specificity, which is also useful in identification of species. There are about ten species in Michigan; the six most common are listed here. **Cuscuta gronovii** is by far the most common.

1a. Flowers sessile, in rope-like clusters; sepals entirely separate; flowers subtended by several bracts; mostly on Compositae; SW—*Cuscuta glomerata*

1b. Flowers in panicles or compact heads; sepals united above the base; bracts immediately beneath the flowers absent—2

2a. Flowers five-parted—3

2b. Flowers four-parted; SLP—5

3a. Calyx less than one-half as long as the corolla tube; widespread, often in low wet areas—*Cuscuta gronovii*

3b. Calyx about one-half as long as the corolla tube or longer; SLP—4

4a. Stigma slender, not capitate; on legumes—**Clover Dodder**, *Cuscuta epithymum*

4b. Stigma capitate; often found on *Polygonum* spp.—*Cuscuta polygonorum*

5a. Flowers pedicellate; tips of the petals acute, incurved; woods—*Cuscuta coryli*

5b. Flowers almost sessile, in small heads; tips of the petals rounded, erect—*Cuscuta cephalanthi*

POLEMONIACEAE, The Polemonium Family

Perennial herbs with alternate or opposite, simple or pinnately compound leaves. Flowers regular, perfect, conspicuous; sepals 5, united; petals 5, united; stamens 5, attached to the corolla tube, alternate with the petals; pistil 1, style 1, ovary superior, 3-celled. Fruit a capsule.

1a. Leaves pinnately divided or compound, all or mostly alternate; petals blue or scarlet and yellow; SLP—2

1b. Leaves simple, opposite or appearing whorled; petals pink-purple or blue, sometimes white (*Phlox* spp., **Phlox**)—3

2a. Leaves pinnately compound; petals blue; rich woods (20–50 cm high; spring)—**Greek Valerian** or **Jacob's Ladder**, *Polemonium reptans*

2b. Leaves pinnately divided into narrow segments; outside of corolla scarlet, the inside yellow; escaped to sandy roadsides, etc. (to 1 m high; summer)—**Standing-cypress**, *Ipomopsis rubra* [*Gilia rubra*—CQ]

3a. Leaves narrowly linear and pointed, less than 1 cm long, often in axillary fascicles; often an escape from cultivation (about 10 cm high; spring)—**Moss-Pink**, *Phlox subulata*

3b. Leaves linear or usually broader, over 1 cm long, rarely in axillary fascicles—4

4a. Corolla lobes deeply two-cleft to the middle; sandy areas, SLP (10–30 cm high; spring)—**Sand Phlox** or **Cleft Phlox**, *Phlox bifida*

4b. Corolla lobes entire and rounded, or somewhat notched at the apex—5

5a. Flowers in summer (30–200 cm high)—6

5b. Flowers in spring (30–60 cm high)—7

6a. Inflorescence broad, more or less flattened; stem not red-spotted; often escapes to fields, roadsides, etc. (to 2 m high)—**Garden Phlox**, *Phlox paniculata*

6b. Inflorescence narrow and cylindric; stem red-spotted; calcareous areas, SW (30–80 cm high)—**Wild Sweet-william**, *Phlox maculata*

7a. Apex of leaf blade ending in a sharp point; corolla pink or red-purple; stems erect; prairies, SLP—**Prairie Phlox**, *Phlox pilosa*

7b. Apex of leaf blade acute or obtuse, not ending in a sharp point; corolla often blue-purple (rarely white); stems ascending; rich woods, LP— **Wild Blue Phlox**, *Phlox divaricata*

HYDROPHYLLACEAE, The Water-leaf Family

Woodland herbs with alternate, simple, lobed or divided leaves. Flowers regular, perfect, in terminal cymes; sepals 5, united but often appearing to be separate; petals 5, united, blue to purple varying to white; stamens 5, attached to the corolla tube and projecting beyond it, alternate with the petals; pistil 1, style 1, apically split, ovary superior, 1-celled. Fruit a 2-valved capsule. Spring.

1a. Leaves pinnately veined and lobed; SLP (30–80 cm high)—**Virginia Water-leaf**, *Hydrophyllum virginianum*

1b. Leaves palmately veined and lobed (30–60 cm high)—2

2a. Upper portion of the stem and inflorescence glabrous or with sparse long hairs; SLP—**Canada Water-leaf**, *Hydrophyllum canadense*

2b. Upper portion of the stem and inflorescence densely covered with both short and long hairs—**Great Water-leaf**, *Hydrophyllum appendiculatum*

BORAGINACEAE, The Borage Family

Herbs with mostly alternate, simple, mainly entire leaves. Flowers mostly regular, perfect, often in helicoid cymes; sepals 5 united or not; petals 5, united; stamens 5, attached to the corolla tube, alternate with the petals; pistil 1, style 1, arising amidst the 4 ovary lobes, ovary superior, 2-celled. Fruit is 4 (sometimes fewer) nutlets.

1a. Corolla saucer-shaped, with a very short tube, bright blue, about 20 mm broad; escape from cultivation (20–60 cm high; summer)—**Borage**, *Borago officinalis*

1b. Corolla tubular, funnel-shaped, or salverform (flattened), the tube always distinct—2

2a. Corolla irregular, with lobes of unequal length; stamens protrude from corolla; sandy fields, railroad rights-of-way, etc. (30–80 cm long; summer and early autumn)—**Viper's Bugloss** or **Blueweed**, *Echium vulgare*

2b. Corolla regular, the lobes equal; stamens shorter than or equalling the corolla—3

3a. Corolla tubular, 10 mm long or more, the tube distinctly longer than the calyx—4

3b. Corolla funnelform or salverform, mostly less than 10 mm long, the tube equalling or shorter than the calyx—7

4a. Stems densely covered with spreading hairs; flowers yellowish-white to pink, seldom blue; escapes to fields, railroad rights-of-way, etc. (30–120 cm high; summer) (*Symphytum* spp., **Comfrey**)—5

4b. Stems glabrous or with appressed hairs; flowers blue or purple, rarely pink or white; moist woods (*Mertensia* spp., **Lungwort**)—6

5a. Leaves decurrent, a wing of tissue extending down the stem; stems hairy, but not bristly—**Comfrey**, *Symphytum officinale*

5b. Leaves not or only very briefly decurrent; stems with bristly hairs—**Prickly Comfrey**, *Symphytum asperum*

6a. Stem and leaves glabrous (30–70 cm high; spring)—**Virginia Bluebell**, *Mertensia virginica*

6b. Stem and leaves pubescent; UP (30–100 cm high; summer)—**Tall Lungwort**, *Mertensia paniculata*

7a. Ovary and fruit covered with hooked prickles—8

7b. Ovary and fruit not prickly—12

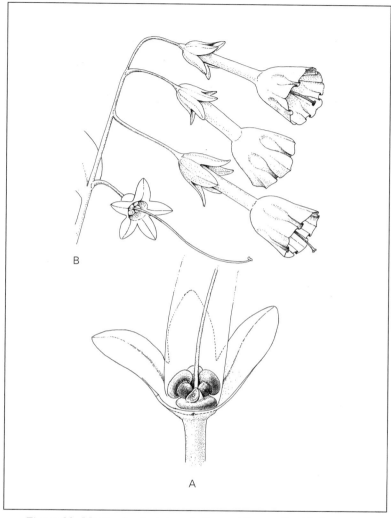

Figure 33: **Mertensia virginica:** *A, flower (dissected) with 4 ovaries (one removed); B, inflorescence*

8a. Principal leaves 2.5 cm wide or more—9
8b. Principal leaves 2 cm wide or less (spring and summer)—11

9a. Corolla white or light blue, to 3 mm wide; woods (to 100 cm or more
    high; summer)—**Beggar's-lice**, *Hackelia virginiana*
9b. Corolla blue or reddish-purple, 8 mm or more wide (spring) (*Cynoglossum* spp.)—10

10a. Leaves chiefly basal, the racemes on long leafless peduncles; corolla
     light blue; woods, NM (40–80 cm high)—**Northern Wild Comfrey**,
     Cynoglossum boreale [*C. virginianum* var. *boreale*—CQ]
10b. Stems leafy; corolla reddish purple; fields and open woods (30–120 cm
     high)—**Hound's-tongue**, *Cynoglossum officinale*

11a. Each flower subtended by a bract; fields and railway rights-of-way, etc.
     (20–80 cm high)—**Stickseed**, *Lappula squarrosa*
11b. Racemes without bracts at the base of each flower; rich woods, shaded
     cliffs, NM—**Stickseed**, *Hackelia deflexa*

12a. Racemes bractless, or bracted only at the base (10–40 cm high)
     (*Myosotis* spp., **Forget-me-not**)—13
12b. Raceme with a bract at the base of each flower—18

13a. Calyx covered with short, appressed hairs which are not apically
     hooked; wet areas (spring and summer)—14
13b. Calyx covered with spreading and/or apically hooked hairs—15

14a. Corolla 5 mm or less broad; style shorter than the nutlets (10–40 cm
     high)—*Myosotis laxa*
14b. Corolla 5 mm or more broad; style longer than the nutlets (20–60 cm
     high)—*Myosotis scorpioides*

15a. Calyx lobes distinctly unequal in length; corolla white; sandy woods
     and banks, SLP and the Keweenaw Peninsula (5–40 cm high; spring
     and early summer)—*Myosotis verna*
15b. Calyx lobes equal in length; corolla blue, rarely white—16

16a. Corolla 5 mm or more broad; escaping cultivation to fields and road-
     sides, etc. (to 50 cm high; spring and summer)—**Garden Forget-me-
     not**, *Myosotis sylvatica*
16b. Corolla 4 mm or less broad—17

Figure 34: **Myosotis sylvatica**: A, flower (dissected); 1, corona,
2, stamen attached to corolla; 3, ovary;
B, inflorescence (helicoid cyme)

17a. Mature (fruiting) pedicels as long or longer than the calyx; roadsides, fields (10–40 cm high; summer)—**Scorpion-grass**, *Myosotis arvensis*

17b. Mature pedicels shorter than the calyx; sandy lawns and fields (to 20 cm high; spring and early summer)—*Myosotis stricta* [*M. micrantha*—CQ]

18a. Corolla blue; escapes to sandy fields, roadsides, etc. (30–80 cm high; summer)—**Bugloss** or **Alkanet**, *Anchusa officinalis*

18b. Corolla not blue (*Lithospermum* spp., **Puccoon** or **Gromwell**)—19

19a. Corolla yellow to deep orange—20

19b. Corolla white, yellowish-white, to greenish—21

20a. Stem and leaves softly pubescent; prairies (10–40 cm high; spring)—**Hoary Puccoon**, *Lithospermum canescens*

20b. Stem and leaves hispid or bristly; sand dunes and barrens, open sandy woods (30–60 cm high; spring and early summer)—**Yellow Puccoon**, *Lithospermum caroliniense*

21a. Corolla white; leaves lacking lateral veins; nutlets brown and wrinkled; sandy fields, LP (10–80 cm high; spring and summer)—**Corn Gromwell**, *Lithospermum arvense*

21b. Corolla yellowish-white to greenish; leaves have lateral veins; fruit white and smooth or somewhat pitted—22

22a. Largest leaf blades less than 2 cm wide, the apex acute; sandy fields (to 1 m high; spring and summer)—**Gromwell**, *Lithospermum officinale*

22b. Largest leaf blades over 2 cm wide, the apex acuminate; rich woods, SLP (40–80 cm high; spring)—**American Gromwell**, *Lithospermum latifolium*

VERBENACEAE, The Vervain Family

Herbs with opposite, simple leaves and often with 4-angled stems. Flowers slightly irregular, perfect, in spikes; sepals 4 or 5, united, the calyx sometimes 2-lipped; petals 4 or 5, united, the corolla sometimes 2-lipped; stamens 4, attached to the corolla tube and rarely projecting beyond it; pistil 1, style 1, terminal, ovary superior, 1-, 2-, or 4-celled. Fruit is 2 or 4 nutlets or an achene. Summer.

1a. Plants prostrate or trailing—2

1b. Plants erect—3

2a. Leaves serrate; flowers pale blue to white, in short dense spikes; damp river banks, SLP (to 60 cm high)—**Fog Fruit**, *Phyla lanceolata*

2b. Leaves pinnately divided or three-lobed; flowers light purple, in loose bracted spikes, the bracts exceeding the calyx; roadsides, railroad rights-of-way, etc.—**Creeping Vervain**, *Verbena bracteata*

3a. Spikes dense, continuous, the flowers overlapping; corolla purple or blue—4

3b. Spikes slender, interrupted, the flowers scattered; corolla white or pale purple—6

4a. Leaves linear to narrowly oblanceolate, tapering at the base; open calcareous areas, railroads (10–70 cm high)—*Verbena simplex*

4b. Leaves broader, not tapering at the base—5

5a. Leaves lanceolate, often with two basal lobes, distinctly petioled; wet meadows and shores (40–150 cm high)—**Blue Vervain**, *Verbena hastata*

5b. Leaves oblong to obovate, without basal lobes, sessile; fields, roadsides, etc. (20–120 cm high)—**Hoary Vervain**, *Verbena stricta*

6a. Flowers in opposite pairs along the axis; fruit an achene, enclosed in the calyx which is reflexed downward; moist woods (50–100 cm high)—**Lopseed**, *Phryma leptostachya*

6b. Flowers not in opposite pairs along the axis; fruit is four nutlets, the calyx erect or ascending; rich woods, wet meadows (40–150 cm high)—**White Vervain**, *Verbena urticifolia*

Plants that resemble **Verbena urticifolia** but which have shorter spikes with more closely spaced purplish flowers and sparse fruit may be **Verbena ×engelmannii**. This hybrid between **V. urticifolia** and **V. hastata** has been collected several times in southern Michigan.

LABIATAE (LAMIACEAE), The Mint Family

Herbs with opposite, simple leaves, square stems, and usually an aromatic odor. Flowers mostly irregular, perfect; sepals 5, united, the calyx sometimes 2-lipped; petals 4 or 5, united, the corolla often 2-lipped; stamens 2 or 4, attached to the corolla tube; pistil 1, style 1, arising from amidst or at the base of the 4 ovary lobes, ovary superior, 4-celled. Fruit is 4 (or fewer) nutlets.

1a. Calyx with a distinct protuberance on the back of the upper side; corolla blue or violet, rarely white or pink (*Scutellaria* spp., **Skullcap**)—2

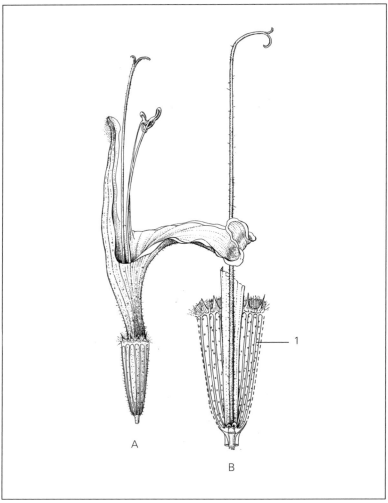

*Figure 35: **Monarda fistulosa**: A, intact flower with 2-lipped corolla; B, flower dissected to show style attachment at base of ovary lobes; 1, calyx*

1b. Calyx without a distinct protuberance—5

2a. Corolla less than 10 mm long—3
2b. Corolla longer than 12 mm—4

3a. Flowers in axillary racemes; marshes (30–70 cm high; summer)—**Mad-dog Skullcap**, *Scutellaria lateriflora*
3b. Flowers axillary, solitary; prairies and limestone plains (10–20 cm high; early summer)—*Scutellaria parvula* [incl. *S. leonardii*—CQ]

4a. Stem leaves sessile or nearly so; flowers axillary; swamps and river-banks (20–80 cm high; summer)—**Marsh Skullcap**, *Scutellaria galericulata*
4b. Stem leaves with petioles 1 cm or more long; flowers in terminal racemes; dry or moist woods, SW (30–60 cm high; late spring and summer)—*Scutellaria elliptica*

5a. Corolla essentially one-lipped, the upper lip barely visible—6
5b. Corolla regular or clearly two-lipped—7

6a. Corolla blue, the lower lip with three to four lobes; creeping plant often in lawns (10–30 cm high; spring)—**Bugle**, *Ajuga reptans*
6b. Corolla pink-purple, the lower lip with five lobes; moist open areas (30–100 cm high; summer)—**Wood-sage**, *Teucrium canadense*

7a. Stamens two—8
7b. Stamens four—22

8a. Corolla regular or nearly so, white; flowers in dense axillary clusters; plants usually of moist open areas (10–100 cm high; summer) (*Lycopus* spp., **Bugleweed** or **Water Horehound**)—9
8b. Corolla distinctly irregular and more or less two-lipped, of various colors—13

9a. Apex of calyx teeth obtuse or acute—10
9b. Apex of calyx teeth acuminate or cuspidate—11

10a. Stamens included in the corolla; center of the summit of the four nutlets depressed—**Bugleweed**, *Lycopus uniflorus*
10b. Stamens exserted beyond the corolla; summit of the four nutlets flat or nearly so, the center area not depressed; SLP—**Bugleweed**, *Lycopus virginicus*

11a. Leaves coarsely incised—**Water Horehound**, *Lycopus americanus*
11b. Leaves evenly serrate—12

12a. Corolla twice as long as the calyx; leaves narrowed at the base; SLP—**Water Horehound**, *Lycopus rubellus*

12b. Corolla barely longer than the calyx; leaves sessile or nearly so; coastal, LP—**Water Horehound**, *Lycopus asper*

13a. Corolla 3–4 mm long, blue; flowers in loose axillary whorled clusters (10–40 cm high; summer)—(*Hedeoma* spp., **Pennyroyal**)—14

13b. Corolla longer than 8 mm, of various colors—15

14a. Leaves lanceolate to ovate, serrate; oak woods, SLP—*Hedeoma pulegioides*

14b. Leaves linear, entire; open sandy areas, prairies, sandy barrens—*Hedeoma hispida*

15a. The lowermost corolla lobe fringed, much longer than the upper; corolla pale-yellow; rich woods, SLP (to 120 cm high; summer)—**Horse-balm**, *Collinsonia canadensis*

15b. The lowermost corolla lobe neither fringed nor much longer than the upper lobe; corolla of various colors, sometimes yellowish—16

16a. Calyx narrowly tubular, the teeth about equal in size; flowers in dense terminal heads (summer) (*Monarda* spp.)—17

16b. Calyx bell-shaped or tubular, two of its teeth different in size from the other three; flowers in loose or dense terminal whorls—20

17a. Corolla scarlet; moist woods (70–150 cm high)—**Oswego-tea** or **Bee-balm**, *Monarda didyma*

17b. Corolla white, lavender, or yellowish—18

18a. Flower-clusters both terminal and axillary; corolla yellowish with purple spots; sandy fields and woods (30–100 cm high)—**Horse Mint**, *Monarda punctata*

18b. Flower-clusters all terminal; corolla lavender or white; fields and dry woods (50–120 cm high)—19

19a. Leaves and stem green, with soft spreading pubescence—**Wild Bergamot**, *Monarda fistulosa* var. *fistulosa*

19b. Leaves and stem grayish, with fine appressed pubescence—**Wild Bergamot**, *Monarda fistulosa* var. *mollis*

20a. Corolla blue or violet; flowers pedicellate, in few-flowered whorls; rarely escapes from cultivation—**Garden Sage**, *Salvia officinalis*

20b. Corolla pink-purple; flowers sessile, in dense terminal whorls; SLP (40–80 cm high; late spring and summer) (*Blephilia* spp.)—21

Figure 36: **Monarda fistulosa,** *inflorescence (head)*

21a. Stems covered with spreading hairs; leaves petiolate; moist woods—
Wood Mint, *Blephilia hirsuta*

21b. Stems covered with recurved hairs; leaves sessile or nearly so; dry
woods, fields—*Blephilia ciliata*

22a. Calyx lobes ten, subulate; woolly plant with whitish flowers in axillary
clusters; barnyards and fields, LP (30–100 cm high; summer)—**Horehound,** *Marrubium vulgare*

22b. Calyx lobes five, all equal or not—23

23a. Calyx two-lipped, one or two lobes different in size and shape from the
other three—24

23b. Calyx regular, all five lobes equal or nearly so at the time of flowering—30

24a. One calyx lobe larger than the other four; flowers light blue; fields,
railroad rights-of-way, rock outcrops (20–80 cm high; spring and early
summer)—**Dragonhead,** *Dracocephalum parviflorum*

24b. Two calyx lobes different in size and shape from the other three—25

25a. Stamens exserted beyond the corolla tube—26

25b. Stamens included in the corolla tube, arching under the upper lip—27

26a. Inflorescence a loose panicle; corolla blue; dry fields and woods (to 70 cm high; late summer)—**Bastard-pennyroyal**, *Trichostoma dichotomum*

26b. Inflorescence spike-like; corolla purple; escape from cultivation (to 25 cm high; summer)—**Wild Thyme**, *Thymus pulegioides* [*T. serpyllum*—CQ]

27a. Inflorescence a terminal spike or head (summer)—28
27b. Inflorescence is one or more axillary clusters—29

28a. Inflorescence bracts leaf-like, the apex acuminate; woodland margins (10–50 cm high)—**Self-heal** or **Heal-all**, *Prunella vulgaris*
28b. Inflorescence bracts narrow, resembling bristles; moist woods (20–50 cm high)—**Wild-basil** or **Dog-mint**, *Clinopodium vulgare* [*Satureja vulgaris*—CQ]

29a. Stems glabrous (nodes sometimes minutely pubescent); calcareous areas; NM (to 60 cm high; late spring and summer)—**Calamint**, *Calamintha arkansana* [*Satureja glabella* var. *angustifolia*—CQ]
29b. Stems pubescent; railroad rights-of-way, etc. (10–20 cm high; summer)—**Basil-thyme**, *Acinos arvensis* [*Satureja acinos*—CQ]

30a. Corolla two-lipped or nearly regular, the upper lip flattened, not conspicuously arched over the stamens—31
30b. Corolla conspicuously two-lipped, the stamens ascending under the concave upper lip—42

31a. Inflorescence of dense terminal spikes—32
31b. Inflorescence of many-flowered whorls, which are axillary, terminal, or aggregated into terminal heads or racemes (summer)—38

32a. Corolla almost regular, the lobes nearly uniform in size (summer) (*Mentha* spp. in part, **Mint**)—33
32b. Corolla distinctly two-lipped, the lower lip longer than the upper—35

33a. Leaf blades densely pubescent beneath; escapes to roadsides, etc., SLP (40–100 cm high)—**Pineapple Mint**, *Mentha suaveolens*
33b. Leaf blades glabrous beneath or pubescent only along the midrib—34

34a. Leaves sessile or with petioles 3 mm or shorter; escapes to roadsides (30-100 cm high)—**Spearmint**, *Mentha spicata*
34b. Leaves with petioles longer than 4 mm (30–80 cm high); moist areas— **Peppermint**, *Mentha* ×*piperita*

35a. Leaves entire, linear-oblong; roadsides, fields, etc. (30–60 cm high; summer and autumn)—**Hyssop**, *Hyssopus officinalis*

35b. Leaves coarsely toothed, ovate-lanceolate to ovate-deltoid (*Agastache* spp., **Giant-hyssop**)—36

36a. Leaves whitened beneath; calyx pubescent; disturbed areas, UP (to 1 m high; summer)—*Agastache foeniculum*

36b. Leaves green beneath; calyx glabrous; open woods, SLP (1–1.5 m high; late summer)—37

37a. Corolla yellowish—*Agastache nepetoides*

37b. Corolla purplish—*Agastache scrophulariifolia*

38a. Inflorescence of axillary whorls or terminal panicles—39

38b. Inflorescence of numerous small terminal heads (summer) (*Pycnanthemum* spp., **Mountain Mint**)—40

39a. All flowers in axillary whorls; corolla almost regular; moist areas (20–80 cm high)—**Wild Mint**, *Mentha arvensis*

39b. All flowers in terminal panicles; corolla two-lipped; occasional escape from cultivation (40–80 cm high; summer)—**Oregano**, *Origanum vulgare*

40a. Leaves and/or inflorescence bracts pubescent above; woods (to 1.5 m high)—*Pycnanthemum verticillatum*

40b. Leaves and inflorescence bracts glabrous above—41

41a. Stems glabrous; sandy fields, LP (50–80 cm high)—*Pycnanthemum tenuifolium*

41b. Angles of the stem pubescent; fens, moist shores, SLP (to 1 m high)—*Pycnanthemum virginianum*

42a. Stems decumbent to diffuse; leaves cordate to nearly circular—43

42b. Stems erect or ascending (summer)—46

43a. Flowers pedicellate, blue; moist woods, lawns (spring)—**Ground-ivy** or **Creeping Charlie**, *Glechoma hederacea*

43b. Flowers sessile, red-purple, rarely white; roadsides, flower beds, etc. (spring to early autumn) (*Lamium* spp., **Dead-nettle**)—44

44a. Upper leaves sessile; LP—*Lamium amplexicaule*

44b. Leaves all petioled—45

45a. Corolla over 1.5 cm long; midrib of leaf blade pale green or white—*Lamium maculatum*

45b. Corolla less than 1.5 cm long; entire leaf blade dark green—***Lamium purpureum***

46a. Inflorescence a loose terminal spike, each inflorescence bract subtending a single flower; corolla rose-colored (to 150 cm high)—**False Dragonhead** or **Obedient Plant**, *Physostegia virginiana*
46b. Inflorescence of axillary or terminal whorls, each inflorescence bract subtending more than two flowers; corolla purple, pink, yellow, or white—47

47a. Apex of calyx lobes spiny—48
47b. Apex of calyx lobes acute to awl-shaped, but not spiny—49

48a. Leaves palmately cleft; flowers pink; shaded gardens, fields, disturbed woods (40–150 cm tall)—**Motherwort**, *Leonurus cardiaca*
48b. Leaves serrate, not lobed; flowers pink, pale-purple, or white; rich woods, sandy shores (20–70 cm high)—**Hemp-nettle**, *Galeopsis tetrahit*

49a. Leaves linear to oblong, entire or nearly so, sessile (summer)—50
49b. Leaves ovate to deltoid, often serrate and/or with long petioles—51

50a. Stem pubescent throughout; flowers purple; escape from cultivation (10–30 cm high)—**Summer Savory**, *Satureja hortensis*
50b. Stem pubescent at nodes (and rarely along stem angles); flowers blue to yellow or white; moist sandy shores and fields (30–50 cm high)—***Stachys hyssopifolia***

51a. Leaves sessile or nearly so; moist sandy shores and woods (30–100 cm high)—**Woundwort**, *Stachys palustris*
51b. At least some leaves with petioles over 8 mm—52

52a. Leaf blades pubescent; fields, roadsides, railroad rights-of-way, etc. (30–100 cm high; summer and autumn)—**Catnip**, *Nepeta cataria*
52b. Leaf blades glabrous; moist woods (to 100 cm high; summer)—***Stachys tenuifolia***

SOLANACEAE, The Nightshade Family

Herbs or sometimes shrubs with alternate, simple leaves. Flowers regular or slightly irregular, perfect; sepals 5, united; petals 5, united; stamens 5, attached to the corolla tube; pistil 1, style 1, ovary superior, 2-celled. Fruit a berry or capsule. Summer.

1a. Plant a climbing or spreading vine or shrub; stem woody or woody only
   near the base—2
1b. Plant herbaceous, erect but not climbing—3

2a. Climbing vine; stem mostly herbaceous, not thorny; leaves mostly lobed;
   corolla saucer-shaped, blue or violet; moist woods and openings—**Nightshade** or **Bittersweet**, *Solanum dulcamara*
2b. Climbing or spreading shrub; stem woody, frequently thorny; leaves entire; corolla tubular, purplish; escape from cultivation, SLP—**Matri-
   mony Vine**, *Lycium barbarum*

3a. Corolla saucer-shaped—4
3b. Corolla tubular or bell-shaped—8

4a. Anthers on long filaments, not clustered in the center of the flower;
   corolla white with yellow center; sandy areas, NM (to 1 m high)—**White
   Ground-cherry**, *Leucophysalis grandiflora*
4b. Anthers on short filaments, clustered and often touching in the center of
   the flower; corolla white, yellow, or bluish (most *Solanum* spp.)—5

5a. Stem and leaves prickly (30–100 cm high)—6
5b. Stem and leaves not prickly; disturbed areas (15–60 cm high)—7

6a. Leaf blades nearly entire, with several large teeth along each margin;
   corolla white or bluish; sandy fields, weedy areas—**Horse-nettle**,
   Solanum carolinense
6b. Leaf blades deeply pinnately lobed; corolla yellow; introduced weed in
   prairies—**Buffalo-bur**, *Solanum rostratum*

7a. Stems glabrous or with short, incurved hairs; fruit black—**Black Nightshade**, *Solanum ptychanthum* [*S. nigrum*—CQ]
7b. Stems covered with long spreading hairs; fruit greenish or yellowish—
   Hairy or **Argentine Nightshade**, *Solanum physalifolium* [*S. sarra-
   choides*—CQ]

8a. Calyx tubular, 3.5 cm or longer; corolla tube white or purple, 7 cm long
   or more; often in cultivated fields (to 1.5 m high)—**Jimson Weed**,
   Datura stramonium
8b. Calyx tubular, 2.5 cm or shorter; corolla tube yellow, yellowish-white, or
   greenish-yellow, the center sometimes purplish, 4.5 cm or shorter—9

9a. Corolla 25 mm wide or more, somewhat irregular; flowers and upper
   leaves sessile; fruit is a capsule, the calyx not inflated; roadsides and
   weedy areas, SLP & Straits (to 1 m high)—**Henbane**, *Hyoscyamus
   niger*

9b. Corolla 20 mm wide or less, strictly regular; flowers and upper leaves not sessile; fruit is a berry enclosed in an inflated calyx; fields and dry woods (*Physalis* spp., **Ground-cherry**)—10

10a. Stem hairs long and spreading, often glandular (20–90 cm high)—*Physalis heterophylla*
10b. Stem glabrous or with short, curved, non-glandular hairs—11

11a. Stem glabrous or with short, upward-curved hairs (40–80 cm high)—*Physalis longifolia*
11b. Stem hairs short and downward-curved (30–60 cm high)—*Physalis virginiana*

Several members of the Solanaceae are often cultivated and only seldom escape. These include the Tomato (**Lycopersicon esculentum**), Potato (**Solanum tuberosum**), Japanese-lantern-plant (**Physalis alkekengi**), Tomatillo (**Physalis philadelphica**), and Tobacco (**Nicotiana tabacum**). Cultivated ornamental hybrid petunias (**Petunia ×atkinsiana**) may also occasionally reseed.

SCROPHULARIACEAE, The Snapdragon Family

Herbs with opposite or alternate (seldom basal), simple leaves. Flowers mostly irregular, perfect; sepals 4 or 5, united or nearly separate; petals 4 or 5, united, corolla sometimes 2-lipped; stamens 2 or 4 (rarely 5), attached to the corolla tube; pistil 1, style 1, ovary superior, 2-celled. Fruit a capsule.

1a. Foliage leaves (not the bracts subtending the flowers) alternate or basal—2
1b. Foliage leaves opposite—16

2a. Corolla with a spur which protrudes between the lower two lobes of the calyx—3
2b. Corolla lacks a spur—7

3a. Flowers in terminal racemes or spikes (*Linaria* spp., **Toadflax**)—4
3b. Flowers solitary, in leaf axils (summer)—6

4a. Corolla blue; sandy areas (10–50 cm high; spring)—**Blue Toadflax**, *Linaria canadensis*
4b. Corolla yellow, the lower lip orange; roadsides, disturbed areas—5

5a. Leaves linear (30–80 cm high; summer and early autumn)—**Butter-and-eggs**, *Linaria vulgaris*
5b. Leaves ovate, sessile (40–120 cm high; summer)—*Linaria dalmatica*

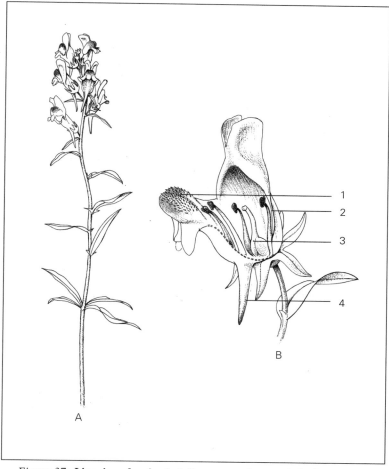

*Figure 37: **Linaria vulgaris***: *A, inflorescence (raceme); B, flower (dissected); 1, lip of corolla (bearded); 2, stamen; 3, pistil; 4, corolla spur*

6a. Stem erect; leaves linear; corolla white or pale blue; railroad ballast, roadsides, etc. (10–50 cm high)—**Dwarf Snapdragon,** *Chaenorrhinum minus*

Chaenorrhinum minus is an introduced plant most often found in cinder railroad ballast. The westward advance of this plant in the United States from the first introductions at coastal ports in the 1800's has been traced to the westward expansion of the railroad system. See Widrlechner (1983) for more details.

6b. Stem trailing; leaves palmately veined and lobed; corolla blue; walls, weedy areas near buildings, etc., LP—**Kenilworth Ivy**, *Cymbalaria muralis*

7a. Anther-bearing stamens five; corolla saucer-shaped; fields, railroads, road-sides, etc. (summer and early autumn) (*Verbascum* spp., **Mullein**)—8
7b. Anther-bearing stamens two or four; corolla tubular or bell-shaped—12

8a. Stem glandular-hairy above; flowers yellow or white, in loose racemes (to 1.5 m high)—**Moth Mullein**, *Verbascum blattaria*
8b. Stem densely white- or gray-woolly throughout; flowers yellow or some-times white, in dense spikes—9

9a. Leaf base appears to extend along stem to next leaf below (1–2 m high)—10
9b. Leaves sessile, clasping, or with only a short extension of the base (to 1.5 m high)—11

10a. Corolla 2.5 cm or less wide; spike dense, usually solitary—**Mullein**, *Verbascum thapsus*
10b. Corolla 2.5 cm or more wide; spike often branched; SW—*Verbascum densiflorum*

11a. Inflorescence much branched, forming a panicle of loose racemes; corolla yellow or white; western UP & SLP—**White Mullein**, *Verbascum lychnitis*
11b. Inflorescence a dense, often simple spike; corolla yellow; SLP—*Verbascum phlomoides*

12a. Anther-bearing stamens two; basal rosette of ovate, palmately veined leaves; sandy areas, SLP (20–40 cm high; late spring)—**Kitten-tail**, *Besseya bullii*
12b. Anther-bearing stamens four, a sterile fifth stamen may or may not be present; basal leaves pinnately lobed (if present)—13

13a. Corolla shorter than twice the length of the calyx (20–60 cm high) (*Castilleja* spp., **Indian Paintbrush**)—14
13b. Corolla much longer than twice the length of the calyx—15

14a. Cauline leaf blades lobed; bracteal leaves subtending the flowers scarlet or sometimes yellow; lakeshores, fens, wet fields, and prairies (late spring and summer)—**Indian Paintbrush**, *Castilleja coccinea*
14b. Cauline leaf blades entire; bracteal leaves subtending the flowers yel-lowish; rocky areas, Lake Superior shoreline (summer)—**Northern Paintbrush**, *Castilleja septentrionalis*

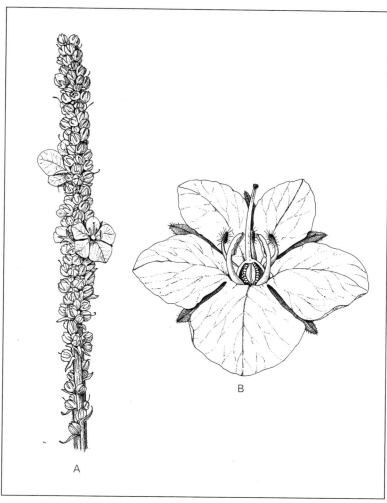

*Figure 38:***Verbascum thapsus***: A, inflorescence (spike);*
B, flower (ovary dissected)

15a. Leaves pinnately lobed or incised; corolla yellow to purple, the upper lip arched, enclosing the anthers; woods and prairies (15–40 cm high; spring)—**Lousewort,** *Pedicularis canadensis*

15b. Leaves entire; corolla white to purple, spotted inside, the upper lip not arched; escape from cultivation, NM (60–180 cm high; summer)—**Common Foxglove,** *Digitalis purpurea*

16a. Anther-bearing stamens four; a sterile fifth stamen may or may not be present—17

16b. Anther-bearing stamens two—37

17a. Upper lip of the corolla arched, enclosing the anthers—18

17b. Upper lip of the corolla not arched, or corolla not two-lipped—20

18a. Leaf blades palmately veined, toothed; corolla purplish; roadsides, weedy areas, UP (10–40 cm high; summer)—**Eyebright,** *Euphrasia stricta* [*E. officinalis*—CQ]

18b. Leaf blades pinnately veined, entire or pinnately lobed—19

19a. Leaf blades pinnately lobed or incised; corolla yellow; swamps (30–80 cm high; late summer)—*Pedicularis lanceolata*

19b. Leaf blades entire, linear to spathulate; corolla white, the lower lip yellow; sandy pine forests, bogs (10–40 cm high; summer)—**Cow-wheat,** *Melampyrum lineare*

20a. Filaments five, four with anthers and one sterile; corolla two-lipped—21

20b. Filaments four, all fertile; corolla two-lipped—26

21a. Corolla reddish-brown or greenish; inflorescence a large terminal panicle; open woods, fencerows (*Scrophularia* spp., **Figwort**)—22

21b. Corolla white to violet; inflorescence a dense spike or terminal panicle—23

22a. Sterile stamen purple; SLP (to 3 m high; summer)—*Scrophularia marilandica*

22b. Sterile stamen yellow (to 2 m high; spring and early summer)—*Scrophularia lanceolata*

23a. Inflorescence a dense terminal or subterminal spike; flowers broad, corolla white, subtended by sepal-like bracts; moist woods (50–80 cm high; summer)—**Turtlehead,** *Chelone glabra*

23b. Inflorescence a loose terminal panicle; flowers narrower, corolla white or pale-violet, not subtended by sepal-like bracts (*Penstemon* spp., **Beard-tongue**)—24

24a. Corolla pale violet, the lower lip of the corolla arched upward, lacking purple stripes; prairies, sandy woods, and roadsides, SLP (40–80 cm high; late spring and early summer)—**Hairy Beard-tongue**, *Penstemon hirsutus*

24b. Corolla white, the lower lip with purple stripes but not arched upward—25

25a. Stems and underside of the leaf blades glabrous; sandy fields and roadsides (to 1.5 m high; late spring and early summer)—*Penstemon digitalis*

25b. Stems and underside of the leaf blades pubescent; sandy fields (30–70 cm high; spring)—*Penstemon pallidus*

26a. Corolla two-lipped, the upper lip very different in size and shape from the lower lip—27

26b. Corolla not two-lipped, all corolla lobes similar in size and shape—32

27a. Middle lobe of lower corolla lip folded downward, enclosing the stamens; corolla blue and white; woods, rocky soil (*Collinsia* spp.)—28

27b. Lower corolla lip spreading or arched upward, not enclosing the stamens; wet areas—29

28a. Corolla 9 mm long or more; rich, moist woods, SLP (20–40 cm high; spring)—**Blue-eyed-Mary**, *Collinsia verna*

28b. Corolla 5–8 mm long; rocky soil, UP (10–40 cm high; spring and early summer)—**Blue-lips**, *Collinsia parviflora*

29a. Leaves pinnately cut into linear segments; lower corolla lip spreading; corolla lavender; shores, SE (10–20 cm high; summer and early autumn)—*Leucospora multifida*

29b. Leaves entire or at most serrate; lower corolla lip arched upward; corolla blue or yellow (summer) (*Mimulus* spp., **Monkey-flower**)—30

30a. Stem erect; corolla blue; leaf blades pinnately veined; wet woods, marshes (20–130 cm high)—*Mimulus ringens*

30b. Stem creeping or spreading, the tips may be ascending; corolla yellow; leaf blades palmately veined—31

31a. Stem and leaves densely covered with long, sticky hairs; wet ditches, NM—**Musk Flower**, *Mimulus moschatus*

31b. Stem and leaves glabrous or with short glandular hairs; wet calcareous soils, cold stream banks—**Yellow Monkey-flower**, *Mimulus glabratus*

Two varieties of Yellow Monkey Flower are found in Michigan. As noted in Voss (1996), the length of the style is the most consistent feature to use in distinguishing plants of **Mimulus glabratus** var. **jamesii** ("styles ca. 3–5 mm long") from **Mimulus glabratus** var. **michiganensis** ("styles ca. (7)8–11 mm long"). The latter variety is the only plant listed by the Federal Government as an endangered species which is endemic to Michigan. It has been the subject of several research studies; see Bliss (1986) for one example.

32a. Corolla yellow; leaves lanceolate to ovate; dry, sandy woods, SLP (*Au-reolaria* spp., **False Foxglove**)—33
32b. Corolla purple, pink, or white; leaves linear (late summer) (*Agalinis* spp., **Gerardia**)—35

33a. Stem glabrous (1–2 m high; summer)—*Aureolaria flava*
33b. Stem pubescent—34

34a. Calyx and pedicels covered with stalked glandular hairs (to 1 m high; late summer)—*Aureolaria pedicularia*
34b. Calyx and pedicels covered with simple, non-glandular hairs (50–150 cm high; summer)—*Aureolaria virginica*

35a. Pedicels equaling or but little longer than the calyx, and conspicuously shorter than the subtending leaf; sandy shores, marshes (30–120 cm high)—*Agalinis purpurea*
35b. Pedicels much longer than the calyx, and generally equaling or exceeding the subtending leaf—36

36a. Corolla pink; sandy areas, SE (20–50 cm high)—*Agalinis gattingeri*
36b. Corolla purple or rarely white; sandy ditches, disturbed areas, SLP (20–60 cm high)—*Agalinis tenuifolia*

37a. Corolla distinctly irregular, two-lipped; sepals five; wet areas—38
37b. Corolla regular or nearly so, four-lobed; sepals four to five—39

38a. Corolla yellowish to white; flowers subtended by a pair of sepal-like bracts; leaves narrowed at the base, with one midvein (10–30 cm high; spring to early summer and autumn)—**Hedge-hyssop**, *Gratiola neglecta*
38b. Corolla pale violet; flowers not subtended by sepal-like bracts; leaves rounded or somewhat clasping at the base, with three to five principal veins (5–30 cm high; summer)—**False Pimpernel**, *Lindernia dubia*

39a. Leaves whorled; flowers white or pale blue, in long spikes; sepals four or five; prairies and open woods (80–200 cm high; summer)—**Culver's-Root**, *Veronicastrum virginicum*

39b. Stem leaves opposite (rarely some in whorls of three), floral bracts may be alternate; flowers white to violet, in terminal or axillary racemes or solitary; sepals four (*Veronica* spp., **Speedwell**)—40

40a. Flowers blue, in racemes, which arise from the axils of the opposite leaves (*Veronica* subg. *Veronica*)—41
40b. Flowers blue or white, solitary in the axils of terminal, often alternate, leaf-like bracts; lawns, gardens, disturbed areas (*Veronica* subg. *Veronicella*)—45

41a. Stem and foliage glabrous or rarely with short glandular hairs; wet areas (*Veronica* sect. *Beccabunga*)(late spring and summer)—42
41b. Stem and foliage covered with spreading hairs; dry areas (*Veronica* sect. *Veronica*) (10–30 cm high)—44

42a. Leaf blades linear or narrowly lanceolate; plant sometimes hairy; swamps (10–40 cm high)—**Marsh Speedwell**, *Veronica scutellata*
42b. Leaf blades elliptic to ovate or ovate-lanceolate; plant glabrous—43

43a. Stem leaves sessile and somewhat clasping; wet ditches and shores (20–100 cm high)—**Water Speedwell**, *Veronica anagallis-aquatica*
43b. Stem leaves on short petioles; swamps, shores (10–100 cm high)— **Brooklime**, *Veronica beccabunga* var. *americana* [*V. americana*—CQ]

44a. Leaf blades narrowed at base into a petiole; subtending bracts longer than flower pedicels; woods and fields (spring and early summer)— **Common Speedwell**, *Veronica officinalis*
44b. Leaf blades sessile, rounded or heart-shaped at base; subtending bracts shorter than flower pedicels; disturbed areas (spring)—**Germander Speedwell**, *Veronica chamaedrys*

45a. All floral bracts much smaller than stem leaves; the inflorescence a spike or raceme; flowers blue—46
45b. Lower floral bracts of similar size to stem leaves, upper bracts gradually smaller; the flowers axillary; flowers blue or white—47

46a. Leaves opposite or in whorls of three, the blades serrate, 4 cm or longer; flowers blue, in a dense terminal spike; escapes from cultivation (to 1 m high; summer)—*Veronica longifolia*
46b. Leaves opposite, the blades entire or weakly toothed, 1.5 cm or shorter; flowers pale blue with darker stripes, in a terminal raceme (10–30 cm high; spring and summer)—**Thyme-leaved Speedwell**, *Veronica serpyllifolia*

47a. Stems erect or ascending; flowers blue or white, about 2 mm wide, nearly sessile, pedicels 2 mm or shorter (5–30 cm high)—48
47b. Stems prostrate or ascending; flowers blue, 4–11 mm wide, on pedicels 5 mm or (often) longer—50

48a. Corolla white (spring and summer)—**Purslane Speedwell**, *Veronica peregrina*
48b. Corolla blue—49

49a. Leaf blades toothed (spring and summer)—**Field Speedwell**, *Veronica arvensis*
49b. Leaf blades toothed, some pinnately cut; NM (spring)—*Veronica verna*

50a. Fruiting pedicels 15 mm or shorter; corolla 8 mm wide or less; SLP (10–30 cm high; spring)—*Veronica polita*
50b. Fruiting pedicels longer than 15 mm; corolla wider than 8 mm—51

51a. Stem erect or ascending, the plant a tufted annual or winter annual; leaf blades 1–2 cm long (10–40 cm high; spring and summer)—**Bird's-eye Speedwell**, *Veronica persica*
51b. Stem trailing, the plant a mat-forming perennial; leaf blades 1 cm long or less—**Creeping Speedwell**, *Veronica filiformis*

> The Scrophulariaceae have provided many garden subjects, some of which escape from time to time. They include Common Foxglove (**Digitalis purpurea**), Yellow Foxglove (**Digitalis grandiflora** = **D. ambigua**), Grecian Foxglove (**Digitalis lanata**), **Digitalis lutea**, and Snapdragon (**Antirrhinum majus**). Many **Veronica** species, including **V. longifolia**, **V. chamaedrys**, and **V. serpyllifolia**, have been introduced from Europe and Asia and are established escapes.

BIGNONIACEAE, The Trumpet-Creeper Family

Trees or woody vines with opposite, simple or compound leaves. Flowers irregular, perfect; sepals 5, united, the calyx 2-lipped or not; petals 5, united, the corolla 2-lipped or not; stamens 2 or 4, attached to the corolla tube; pistil 1, style 1, ovary superior, 2-celled. Fruit an elongate capsule. Both Michigan species are occasional escapes from cultivation.

1a. Tree; leaves simple, the blades broadly ovate; corolla two-lipped, white with brown and yellow spots; SLP (to 30 m high; spring)—**Catalpa**, *Catalpa speciosa*

1b. Woody trailing or climbing vine; leaves compound, the leaflets lanceolate; corolla tubular, reddish-orange; LP (to 10 m high; summer)—
Trumpet-creeper, *Campsis radicans*

OROBANCHACEAE, The Broom-rape Family

Parasitic plants without green color and with alternate scales in place
of leaves. Flowers irregular, mostly perfect; sepals 5, united; petals 4
or 5, united, the corolla 2-lipped; stamens 4, attached to the corolla
tube or 0; pistil 1, style 1, ovary superior, 1-celled. Fruit a capsule.

> These obligately parasitic plants attach to the roots of their hosts by means of *haustoria*. Three
> of the four Michigan species are host-specific. **Epifagus virginiana** (Beech-drops), as the com-
> mon name indicates, parasitizes the roots of American beech, **Fagus grandifolia**. Squaw-root,
> **Conopholis americana**, is found in association with several species of oaks (**Quercus** spp.),
> while **Orobanche fasciculata** is associated with **Artemisia campestris** (Wormwood) on sand
> dunes.

1a. Flowers one to ten, each on a long erect naked pedicel; in a loose
corymb (*Orobanche* spp., **Broom-rape**)—2
1b. Flowers many, sessile, in a branched panicle or a dense spike; rich
woods—3

2a. Stem erect and scaly; flowers four to ten, corollas purple, pedicels
shorter than the stem; sand dunes, northwestern LP (to 30 cm high; late
spring)—*Orobanche fasciculata*
2b. Stem very short, almost below the surface of the ground; flowers one to
three, corollas white or purple, pedicels longer than the stem; sandy
coniferous woods (to 20 cm high; late spring and early summer)—
Orobanche uniflora

3a. Flowers in a branched panicle, the corollas white and purple (10–50 cm
high; late summer)—**Beech-drops**, *Epifagus virginiana*
3b. Flowers in a dense, bracted spike, the corollas pale yellow (5–20 cm
high; late spring and early summer)—**Squaw-root**, *Conopholis americana*

LENTIBULARIACEAE, The Bladderwort Family

Small herbs, growing on rocks, in mud, or (mostly) in water, the
leaves alternate, whorled, or basal, entire or dissected. Flowers irregular, perfect; sepals 5, united, the calyx 2-lipped; petals 5, united, the
corolla 2-lipped with the lower lip extended into a spur; stamens 2, at-

tached to the corolla tube; pistil 1, style 0 or 1, ovary superior, 1-celled. Fruit a capsule. Summer.

> Most members of this family are insectivorous. The leaves of **Pinguicula** act like fly-paper, catching insects which adhere to the leaves and are then digested. **Utricularia** species have bladder-like traps among the leaves which trap small water or soil animals.

1a. Basal rosette of ovate to elliptic, entire leaves present; corolla purple; rocks and calcareous swamps, NM (flowering peduncles 5–15 cm high)—**Butterwort**, ***Pinguicula vulgaris***
1b. Leaves basal or cauline, but not forming a rosette, the blades finely dissected (especially on submerged stems), linear, or absent; corolla purple or yellow; in mud and/or water (flowering peduncles 2–25 cm high) (***Utricularia*** spp., **Bladderwort**)—2

2a. Corolla purple—3
2b. Corolla yellow—4

3a. Peduncle with a single bract near the middle; flower solitary; leaves linear, inconspicuous; sandy or peaty lake bottoms, flowering at the shoreline—***Utricularia resupinata***
3b. Peduncle without bracts, except at the base of the pedicels; flowers 1–4; leaves dissected, whorled; floating in water—**Purple Bladderwort**, ***Utricularia purpurea***

4a. Leaves linear, not evident at the base of the peduncle; bogs, fens, dune wetlands, wet shores—**Horned Bladderwort**, ***Utricularia cornuta***
4b. Dissected leaves present at the base of the peduncle, floating on or submerged in water—5

5a. Stem and numerous dissected leaves floating in water—6
5b. Stem and minute, dissected leaves creeping on the bottom of ponds, while the flowers are on erect stalks, easily detached from the delicate stems—7

6a. One to several scales present along the peduncle below the flowers—**Common Bladderwort**, ***Utricularia vulgaris***
6b. Scales along the peduncle below the flowers absent; bog pools and lakes, NM—***Utricularia geminiscapa***

7a. Upper lip of corolla conspicuous, as long or nearly as long as the lower lip; ultimate leaf segments filiform; often on floating peat mats—***Utricularia gibba***
7b. Upper lip of corolla half as long as the lower lip, or less; ultimate leaf segments flat—8

8a. Spur less than one-half the length of the lower corolla lip; bladder traps and leaves borne on the same branch—***Utricularia minor***

8b. Spur as long as the lower corolla lip; bladder traps and leaves borne on separate branches—**Flat-leaved Bladderwort**, ***Utricularia intermedia***

ACANTHACEAE, The Acanthus Family

Perennial herbs with opposite, simple leaves. Flowers irregular, perfect; sepals 5, united; petals 5, united, the corolla 2-lipped; stamens 2, attached to the corolla tube; pistil 1, style 1, ovary superior, 2-celled. Fruit a capsule.

One species in Michigan. Flowers in dense axillary heads, corolla white with purple lines; shallow waters of lake and river shores, SE (50–100 cm high; summer)—**Water-willow**, ***Justicia americana***

PLANTAGINACEAE, The Plantain Family

Herbs with basal (rarely opposite), simple leaves. Flowers regular, perfect or rarely unisexual (the plants then monoecious), mostly in spikes or small heads; sepals (3)4, united; petals (3)4, united; stamens 4, attached to the corolla tube; pistil 1, style 1, ovary superior, 1- or 2-celled. Fruit a capsule or an achene. Summer.

1a. Flowers unisexual, (3)4-merous, female flowers basal, male flowers solitary, pedicellate; fruit an achene; wet shores and lakes, UP (male flower pedicels to 4 cm high)—***Littorella uniflora***

1b. Flowers perfect, 4-merous, many, in spikes or heads; fruit a capsule; rarely in wet areas (***Plantago*** spp., **Plantain**)—2

2a. Cauline leaves opposite; flowers in axillary heads; railroads, sandy weedy areas (10–60 cm high)—***Plantago arenaria*** [***P. psyllium***—CQ]

2b. All leaves basal; flowers in terminal spikes—3

3a. Leaves linear; plants hairy to densely woolly (flower stalks 30–35 cm high)—4

3b. Leaves broader, lanceolate to broadly ovate or cordate; plants pubescent or not—5

4a. Bracts subtending the flowers several times longer than the flowers; plants hairy; sandy woods and fields, roadsides, SLP—**Bracted Plantain**, ***Plantago aristata***

4b. Bracts about as long as the flowers; plants densely woolly; sandy woods and fields—***Plantago patagonica***

5a. Base of leaf blade cordate or rounded, the blade pinnately veined; streambanks and pond edges in rich deciduous woods; SLP (flower stalks to 60 cm tall)—**Heart-leaved Plantain**, *Plantago cordata*
5b. Base of leaf blade tapered toward the base; the blade with three to many longitudinal veins; dry and/or disturbed areas—6

6a. Sepals and bracts pubescent; leaf blades densely pubescent with grayish hairs; SLP (flower stalks 20–40 cm high)—*Plantago virginica*
6b. Sepals and bracts glabrous or at most with ciliate margins; leaf blades smooth or slightly pubescent; common lawn weeds—7

7a. Leaf blades lanceolate; spikes brownish, not over 10 cm long (flower stalks 25–70 cm high)—**English Plantain**, *Plantago lanceolata*
7b. Leaf blades broadly ovate; spikes greenish, long and slender, usually equaling or longer than the stalk (flower stalks 10–55 cm high)—8

8a. Leaf petioles green; apex of the sepals obtuse—**Common Plantain**, *Plantago major*
8b. Leaf petioles reddish; apex of the sepals acute; LP—*Plantago rugelii*

RUBIACEAE, The Madder Family

Herbs or shrubs, with opposite or whorled, simple leaves. Flowers regular, perfect; sepals 4, united, or minute or 0; petals 3 or 4, united; stamens 4, attached to the corolla tube; pistil 1, styles 1 or 2, ovary inferior, 2- or 4-celled. Fruit a berry, 1 or 2 nutlets, or a capsule.

1a. Shrub; flowers white, in spherical heads; wet areas, LP (1–3 m tall; summer)—**Buttonbush**, *Cephalanthus occidentalis*
1b. Herbaceous plants—2

2a. Leaves opposite—3
2b. Leaves whorled (*Galium* spp., **Bedstraw**)—5

3a. Stems trailing; flowers white, paired; fruit a red berry; rich woods (spring and early summer)—**Partridge-berry**, *Mitchella repens*
3b. Stems erect; flowers white or pale-purple, in small cymes; fruit a capsule (10–30 cm high) (*Houstonia* spp., **Bluet**)—4

4a. Basal leaves present at anthesis, ciliate; sandy or gravelly areas, LP (late spring)—*Houstonia canadensis* [*Hedyotis canadensis*—CQ]
4b. Basal leaves often absent at anthesis, glabrous if present; oak woods (summer)—*Houstonia longifolia* [*Hedyotis longifolia*—CQ]

5a. Flowers yellow; escapes to lawns, roadsides, etc. (40–100 cm high; summer)—**Yellow Bedstraw**, *Galium verum*

> The genus **Galium** (Bedstraw) is easily recognized by the distinctive whorled leaves. Many species of this genus are pubescent, and the lanky stems "scramble" up along other plants, using epidermal hairs for attachment. Identification of species can be difficult, and often requires the presence of mature fruit.

5b. Flowers white, greenish, or purplish—6

6a. Fruit hairy or with hooked bristles—7
6b. Fruit smooth or granular, not hairy nor bristly—12

7a. Leaves in whorls of four; stems erect or ascending (summer)—8
7b. Leaves in whorls of six or eight; stems scrambling over and among adjacent plants or prostrate (spring or summer)—11

8a. Flowers bright white; hairs on fruit straight; roadsides, ditches (20–80 cm high)—**Northern Bedstraw**, *Galium boreale*
8b. Flowers greenish or purplish; hairs on fruit hooked; dry woods—9

9a. Flowers and fruits on pedicels; SLP (20–100 cm high)—*Galium pilosum*
9b. Flowers and fruits all or mostly sessile—10

10a. Leaf blade lanceolate, tapering to an acute or acuminate apex; stem virtually glabrous; corolla purplish, glabrous (30–70 cm high)—*Galium lanceolatum*
10b. Leaf blade ovate to elliptic, the apex obtuse; stem pubescent; corolla greenish, pubescent (20–60 cm high)—*Galium circaezans*

11a. Leaves narrowly oblanceolate to linear, six or (mostly) eight in a whorl; cilia along leaf margin curved toward leaf base; damp woods (spring)—**Goosegrass** or **Cleavers**, *Galium aparine*
11b. Leaves narrowly oval or elliptical, mostly in whorls of six (four on small branches); cilia along leaf margin curved toward leaf apex; woods (summer)—*Galium triflorum*

12a. Apex of leaf blade obtuse or blunt; leaves in whorls of four, seldom to six—13
12b. Apex of leaf blade acute or cuspidate; leaves in whorls of six or eight (sometimes four or five on small branches)—19

13a. Corolla lobes three (rarely four in a few flowers); cilia along leaf margin curved toward leaf base; moist places (summer)—14

13b. Corolla lobes four; cilia along leaf margin straight or curved toward leaf apex—16

14a. Leaves mostly in whorls of four; flowers and fruit on long, filiform, roughened, arching pedicels—***Galium trifidum***
14b. Leaves in whorls of four to six; flowers and fruit on short, glabrous, straight pedicels—15

15a. Plant erect or leaning; pedicels mostly over 4 mm long; flowers about 1.5–2.0 mm in diameter—***Galium tinctorium***
15b. Plant mat-forming; pedicels less than 4 mm long; flowers about 1 mm in diameter or less—***Galium brevipes***

16a. Flowers many, in small many-branched cymes (summer)—17
16b. Flowers in two- to four-flowered cymes—18

17a. Stems mostly erect; woodlands, banks, roadsides, ditches (20–80 cm high)—**Northern Bedstraw**, ***Galium boreale***
17b. Stems scrambling or spreading; wet areas—**Marsh Bedstraw**, ***Galium palustre***

18a. Principal leaves spreading or ascending; swampy woods, SLP (spring and early summer)—***Galium obtusum***
18b. Principal leaves recurved or reflexed; fens and sedge meadows (10–40 cm high; summer)—***Galium labradoricum***

19a. Stems mostly erect, the angles of the stem smooth; leaves in whorls of six or eight; flowers many in terminal panicles; disturbed areas (30–120 cm high; summer)—***Galium mollugo***
19b. Stems scrambling or spreading, the angles of the stem roughened; leaves in whorls of six (four or five on small branches); flowers few, in small loose terminal panicles—20

20a. Stem angles often with sparse bristles; cilia along leaf margin curved toward leaf apex; dry woodlands, SLP (summer)—***Galium concinnum***
20b. Stem angles very bristly; cilia along leaf margin curved toward leaf base; wet woodlands (spring and summer)—**Rough Bedstraw**, ***Galium asprellum***

CAPRIFOLIACEAE, The Honeysuckle Family

Shrubs, woody vines, or herbs, with opposite, simple or pinnately compound leaves. Flowers regular or irregular, perfect; sepals (4)5, united; petals (4)5, united, the corolla sometimes 2-lipped; stamens 4 or 5, attached to the corolla; pistil 1, style (0)1, ovary inferior, 2–5-celled. Fruit a berry, drupe, or capsule.

1a. Leaves pinnately compound; shrubs; inflorescence large terminal cymes (to 3 m high) (***Sambucus*** spp., **Elder**)—2
1b. Leaves simple; shrubs, woody vines, or herbs—3

2a. Inflorescence flattened or convex; fruit usually dark purple; pith of the twigs white; woods and fields (summer)—**Common Elder**, ***Sambucus canadensis***
2b. Inflorescence pyramidal; fruit usually red; pith of the old twigs brown; rich woods (spring)—**Red Elderberry**, ***Sambucus racemosa*** [***S. pubens***—Fassett]

3a. Plant trailing; flowers nodding, in pairs, the petals pink to white; woods and bogs (to 10 cm high; summer)—**Twin-flower**, ***Linnaea borealis*** var. ***longifolia***
3b. Erect herbs, shrubs, small trees, or woody vines—4

4a. Erect herbs; flowers axillary, corolla often dark red; rich woods (60–130 cm high; spring and early summer) (***Triosteum*** **spp.**)—5
4b. Shrubs, small trees, or woody vines—6

5a. Bases of leaves along the middle of the stem somewhat narrowed, the pair connate-perfoliate around the stem; SLP—***Triosteum perfoliatum***
5b. Bases of leaves along the middle of the stem tapered, the pair not perfoliate around the stem—***Triosteum aurantiacum***

6a. Corolla saucer-shaped or shallowly bell-shaped, often white; style very short; inflorescence a cyme, often umbel-like; leaf blades often lobed or margins toothed (***Viburnum*** spp., **Viburnum** or **Arrow-wood**)—7

A number of **Viburnum** species, both native and introduced, are used as landscape shrubs and may be encountered. **Viburnum opulus** var. **americanum** (often called **Viburnum trilobum** in the nursery trade) can be quite tall (up to 5 m), while its European cousin **Viburnum opulus** var. **opulus** is often grown as one of several more compact cultivars. Some **Viburnum** spp., including **V. opulus** and **V. plicatum**, have large showy inflorescences in which sterile flowers at the edge of the cyme have enlarged corollas, usually white. **Viburnum plicatum** cv. **tomentosum**, Double-file Viburnum, is striking for the horizontal arrangement of the cymes.

6b. Corolla tubular at base; style long and slender; inflorescence a cyme, short spike, 6-flowered whorls, or the flowers paired and axillary; leaf margins entire to serrate—14

7a. Leaf blades palmately lobed—8
7b. Leaf blades not lobed—9

8a. Outermost flowers of the inflorescence enlarged (1.5–2.5 cm broad) and imperfect; rich woods (1–5 m high; late spring)—**High-bush Cranberry**, *Viburnum opulus* var. *americanum*
8b. All flowers of the inflorescence alike, 4–5 mm broad; woods (1–2 m high; spring)—**Maple-leaved Viburnum**, *Viburnum acerifolium*

9a. Leaf blades coarsely serrate, all or most of the teeth terminating a prominent lateral vein (spring and early summer)—10
9b. Leaf blades finely serrate or crenate (or rarely entire); the lateral veins obscure or not leading directly to marginal teeth—11

10a. Leaves densely pubescent beneath; dry woods (to 1.5 m high)—**Downy Arrow-wood**, *Viburnum rafinesquianum*
10b. Leaves glabrous beneath, or with tufts of hairs in the forks of the veins; moist woods, SLP (1–5 m high)—**Arrow-wood**, *Viburnum dentatum*

11a. Inflorescence elevated above the bracts on a peduncle 5–30 mm long; shrubs (late spring and early summer)—12
11b. Inflorescence sessile, a peduncle absent; shrubs or small trees; woods and roadsides (spring)—13

12a. Underside of leaf blade glabrous or with tiny reddish-brown scales; fruit bluish-black; woods and swamps (to 4 m high)—**Wild-raisin**, *Viburnum cassinoides* [*V. nudum* var. *cassinoides*—CQ]
12b. Underside of leaf blade covered with star-shaped hairs; fruit red, later darkening; escape from cultivation (to 5 m high)—**Wayfaring Tree** or **Twistwood**, *Viburnum lantana*

13a. Apex of leaf blade distinctly acuminate (to 10 m high)—**Nannyberry**, *Viburnum lentago*
13b. Apex of leaf blade obtuse or barely acute; SLP (to 8 m high)—**Black-haw**, *Viburnum prunifolium*

14a. Leaves serrate; fruit a capsule; woods, roadsides (to 1.2 m high; summer)—**Bush-honeysuckle**, *Diervilla lonicera*
14b. Leaves entire or at most coarsely crenate; fruit a berry or fleshy drupe—15

15a. Corolla tubular, often two-lipped, 1 cm or more long; flowers white to pink, yellow to purplish, in terminal whorls or axillary pairs (*Lonicera* spp., **Honeysuckle**)—16

Two Asian species of Honeysuckle, **Lonicera japonica** and **L. maackii**, are becoming significant threats to native vegetation in the woodlands of southern Michigan. The black berries of **L. japonica** and the red ones of **L. maackii** are eagerly accepted by birds, who disperse these shrubs widely. Wagner (1986), noting that **L. japonica** had become a "serious pest in the southeastern United States", urged that it should not be planted in Michigan. Luken and Thieret (1995) reported that **L. maackii** has become common in the Cincinnati and Chicago areas, which suggests that Michigan woodlands may be similarly threatened.

15b. Corolla bell-shaped, less than 1 cm long; flowers white or pink, in pairs, clusters, or spikes (***Symphoricarpos*** spp., **Snowberry**)—24

16a. Climbing or trailing vines—17
16b. Erect or spreading shrubs or small trees—19

17a. Flowers white aging to yellow, in two-flowered axillary clusters; corolla 3 cm or more long; uppermost leaves petiolate, not connate; aggressive escape in woods, SLP (summer)—**Japanese Honeysuckle**, *Lonicera japonica*
17b. Flowers yellow to purplish, in six-flowered terminal clusters; corolla 2.5 cm or less long; uppermost leaves connate; moist woods—18

18a. Upper surface of leaf blade hairy; corolla yellow or orange; NM (summer)—**Hairy Honeysuckle**, *Lonicera hirsuta*
18b. Upper surface of leaf blade glabrous; corolla yellow to reddish-purple (spring)—**Glaucous Honeysuckle**, *Lonicera dioica*

19a. Ovaries of adjacent flowers united; corolla regular; fruit blue; wet woods, swamps, NM (to 1 m high; spring and early summer)—**Mountain Fly Honeysuckle**, *Lonicera villosa* [*L. caerulea* var. *villosa*—CQ]
19b. Ovaries of adjacent flowers separate; corolla regular or two-lipped; fruit red, rarely yellow—20

20a. Floral peduncles shorter than adjacent leaf petioles, less than 5 mm long; apex of leaf blades long-acuminate; corolla white to yellowish, two-lipped; aggressive invader of disturbed woods, SLP (to 5 m high; spring)—**Amur Honeysuckle**, *Lonicera maackii*
20b. Floral peduncles longer than adjacent leaf petioles, 5–40 mm long; apex of leaf blades rounded or acute, sometimes tapering; corolla white, pink, or yellowish, two-lipped or not (spring)—21

21a. Corolla yellow, two-lipped; swamps (to 2 m high)—**Swamp Fly Honeysuckle**, *Lonicera oblongifolia*
21b. Corolla white, pink, or yellow, regular—22

22a. Branches solid, including a white central pith; bracts subtending the flowers absent or minute; corolla yellow; moist woods, swamps (to 2 m high)—**Fly Honeysuckle**, *Lonicera canadensis*

22b. Branches hollow, central pith absent; bracts subtending the flowers often longer than the ovaries; escapes from cultivation—23

23a. Corolla white, fading to yellow; lower surface of leaf blade pubescent; peduncles 5–15 mm long—***Lonicera morrowii***

23b. Corolla white to pink; lower surface of leaf blade glabrous; peduncles 15–25 mm long (to 3 m high)—**Tartarian Honeysuckle**, *Lonicera tatarica*

Plants with a mixture of the characters in couplet 23 are most likely **Lonicera ×bella,** a vigorous and highly variable hybrid between **L. morrowii** and **L. tatarica**.

24a. Corolla 4 mm or less long; fruit red or reddish; escape from plantings to shores, railways (to 1.5 m high; summer)—**Coralberry**, *Symphoricarpos orbiculatus*

24b. Corolla 5 mm or more long; fruit white or greenish—25

25a. Flowers in short spikes; style exserted beyond the corolla; shores, railways (to 1 m high; summer)—**Wolfberry**, *Symphoricarpos occidentalis*

25b. Flowers usually two in each axil; style included within the corolla; dry woods (to 2 m high; late spring and early summer)—**Snowberry**, *Symphoricarpos albus*

VALERIANACEAE, The Valerian Family

Herbs with opposite, simple leaves. Flowers mostly regular, perfect or unisexual (the plants then monoecious); sepals absent, minute, or expanding into up to 20 plumose segments in fruit; petals 5, united, generally tubular; stamens 3, attached to the corolla tube; pistil 1, style 1, ovary inferior, 1–3-celled. Fruit resembles an achene.

1a. Stem leaves entire or at most dentate at the base; sepals absent or minute; petals white; floodplains, SLP (20–60 cm high; spring)—**Corn-salad**, *Valerianella chenopodiifolia*

1b. Stem leaves pinnately cleft; sepals expanding into up to 20 plumose segments in fruit; petals white to pinkish (*Valeriana* spp., **Valerian**)—2

2a. Leaf blades and/or segments parallel-veined, densely hairy; prairies and fens, SLP (30–120 cm high; spring)—***Valeriana edulis***

2b. Leaf blades and/or segments net-veined, glabrous or sparsely hairy—3

3a. Basal leaf blades entire or with one pair of basal lobes; wet areas (30–100 cm high; spring and early summer)—**Swamp Valerian**, *Valeriana uliginosa*

3b. Basal leaf blades pinnately divided into 10 or more segments; a garden escape to weedy areas (50–150 cm high; spring and summer)—**Common Valerian**, *Valeriana officinalis*

DIPSACACEAE, The Teasel Family

Herbs with opposite, simple leaves. Flowers mostly irregular, perfect, small, aggregated in dense cymose heads subtended by conspicuous bracts; sepals 4, united, or as 8–12 bristle-like teeth; petals 4, united; stamens 4, attached to the corolla; pistil 1, style 1, ovary inferior, 1-celled. Fruit an achene. Summer.

1a. Stem smooth, not prickly; fields, roadsides, UP (30–100 cm high)—**Blue-buttons**, *Knautia arvensis*

1b. Stem prickly; roadsides, fields, etc. (50–200 cm high) (*Dipsacus* spp., **Teasel**)—2

2a. Leaf blades entire, the margin at most toothed—**Wild Teasel**, *Dipsacus fullonum* [*D. sylvestris*—CQ]

2b. Leaf blades pinnately cut into narrow segments; mostly LP—*Dipsacus laciniatus*

CUCURBITACEAE, The Gourd Family

Monoecious herbs, climbing or trailing by tendrils, with alternate, simple, palmately-lobed leaves. Flowers regular, unisexual, the greenish to white staminate flowers in showy racemes; sepals 5 or 6, united; petals 5 or 6, united; stamens (3)5 or 6, the united filaments forming a column; pistil 1, style 1, ovary inferior, 1- or 2-celled. Fruit dry and bristly, inflated or not. Summer.

1a. Sepals and petals 6; fruit inflated, resembling a bladder; disturbed areas, thickets—**Wild Cucumber**, *Echinocystis lobata*

1b. Sepals and petals 5; fruit not inflated, resembles a bur; shores, disturbed areas, mostly SLP—**Bur Cucumber**, *Sicyos angulatus*

A number of commonly cultivated crop plants are members of this family. These plants, which include Watermelon (*Citrullus lanatus*), Pumpkin (*Cucurbita pepo*), Cucumber (*Cucumis sativus*) and Muskmelon (*Cucumis melo*), may escape but do not persist in the wild.

CAMPANULACEAE, The Bellflower Family

Herbs with alternate (rarely basal) simple leaves and milky juice. Flowers regular or irregular, perfect; sepals 5, united below; petals 5, united, tubular or irregularly 1 or 2-lipped; stamens 5, attached at the very base of the corolla or to a disk at the summit of the ovary, the anthers forming a tube surrounding the style; pistil 1, style 1, ovary inferior, 2–5-celled. Fruit a capsule.

1a. Flowers regular, the corolla often bell- or saucer-shaped; anthers touching, but separate (subfamily Campanuloideae)—2
1b. Flowers irregular, one- or two-lipped; anthers attached (subfamily Lobelioideae)—6

2a. Leaf blades circular or nearly so, cordate-clasping at base; flowers axillary and sessile, corolla blue; disturbed areas, LP (10–100 cm high; late spring)—**Venus' Looking Glass**, *Triodanis perfoliata*
2b. Leaf blades linear to ovate, not clasping at base; flowers in a terminal spike or raceme or axillary on long pedicels (summer) (*Campanula* spp., **Bellflower**)—3

3a. Cauline leaf blades linear or nearly so, not over 1 cm wide; corolla blue or white—4
3b. Cauline leaf blades ovate to lanceolate, 2 cm or more wide; corolla blue—5

4a. Stems ascending or erect, round or only obscurely angled; corolla blue; dry areas (10–80 cm high)—**Harebell** or **Bluebell**, *Campanula rotundifolia*
4b. Stems slender, often leaning on other vegetation, three-angled and often rough with reflexed bristles; corolla white or pale blue; wet meadows—**Marsh Bellflower**, *Campanula aparinoides*

5a. Corolla saucer-shaped; flowers in spikes; woods and margins, SLP (50–200 cm high)—**American Bellflower**, *Campanula americana*
5b. Corolla bell-shaped; flowers in one-sided racemes; lawns, roadsides, etc. (40–100 cm high)—**Creeping Bellflower**, *Campanula rapunculoides*

6a. Leaves all basal, tubular; flowers on leafless scapes, corolla blue or white; pond edges, UP (to 100 cm high; summer)—**Water Lobelia**, *Lobelia dortmanna*
6b. Cauline leaves present; flowers on leafy stems; corolla blue, white, or scarlet—7

*Figure 39: **Campanula rotundifolia**: A, habit and inflorescence (raceme), note regular bell-shaped flowers; B, flower (dissected), inferior ovary*

7a. Flowers 2 cm or more long; wet areas (50–150 cm high; summer)—8
7b. Flowers less than 1.5 cm long—9

8a. Flowers scarlet (rarely white), 3 cm or more long—**Cardinal-flower**,
Lobelia cardinalis

Figure 40: **Lobelia siphilitica**: *A, inflorescence (raceme),
note bilaterally symmetrical flowers; B, flower, dissected;
1, stamens (connate); 2, ovary (half-inferior)*

8b. Flowers blue (rarely white), up to 3 cm long—**Great Blue Lobelia**, *Lobelia siphilitica*

9a. Leaf blades linear to narrowly lanceolate; fens, calcareous wet areas (10–40 cm high; summer)—**Brook Lobelia**, *Lobelia kalmii*

9b. Leaf blades lanceolate to ovate; generally in drier areas—10

10a. Flowers in slender, usually unbranched, spike-like racemes; base of calyx tube not inflated in fruit; open areas (30–100 cm high; summer and early autumn)—**Pale Spiked Lobelia**, *Lobelia spicata*

10b. Flowers in branched racemes; base of calyx tube inflated in fruit; woods (to 100 cm high; late spring and summer)—**Indian-tobacco**, *Lobelia inflata*

The Campanulaceae include many popular garden subjects; some specimens may occasionally be collected which are remnants of cultivation. **Campanula persicifolia**, the Peach-leaved Bellflower, is a tall plant (to 1 m) and has bell-shaped flowers in lilac or white. **Campanula portenschlagiana** and **C. posharskyana** are both native to Yugoslavia; the flowers of these creeping plants are lilac-blue and funnelform with pointed corolla lobes (star-shaped). Other garden **Campanula** are **C. medium** (Canterbury bells), **C. carpatica** (Tussock bellflower), and **C. rapunculoides** (Creeping bellflower). **Lobelia cardinalis** (Cardinal-flower) and **L. siphilitica** (Great Blue Lobelia) are present in both native and cultivated populations in Michigan; the latter may have come from other regions in North America. Hybrids between these two **Lobelia** species and with other **Lobelia** species are known. See *Hortus Third* (1976) for more information.

COMPOSITAE (ASTERACEAE), The Composite Family

Annual, biennial, or perennial herbs or rarely small shrubs with alternate, opposite, whorled or basal, simple leaves, sometimes with milky juice. Flowers regular or irregular, perfect or unisexual (the plants then monoecious or dioecious), in one or more heads; sepals as many scales, bristles, or hairs or 0; petals 5, united, tubular below and often modified apically into a single, flattened structure; stamens 5, attached to the base of the corolla tube, the anthers united and forming a tube around the style; pistil 1, style 1, usually divided apically, ovary inferior, 1-celled. Fruit an achene.

The *head* (inflorescence) consists of numerous small flowers attached to a common receptacle; it often resembles a single flower. The head is surrounded by one or more rows of *involucral bracts*, resembling a calyx. Individual flowers may be subtended by a small bract; these bracts collectively are known as the *chaff*.

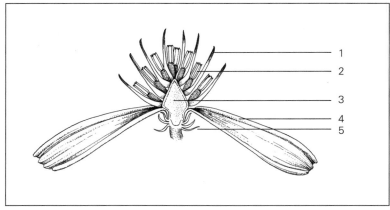

Figure 41: Radiate composite inflorescence: 1, chaff; 2, disk flower (tubular corolla); 3, receptacle; 4, ray flower; 5, involucre (bracts)

The calyx of an individual flower is minute and usually modified into bract-like scales, plumose bristles, or hairs that aid in seed dispersal; these structures collectively are known as the *pappus*, which is best observed on the ripe fruit. Depending on the flower type (see below), the corolla is either entirely tubular, bell-shaped, or tubular near the base and flattened apically and bent to one side of the flower.

The head of a composite may be composed of one or more of three types of flowers. A *discoid* head is composed entirely of *disk* flowers; perfect or sometimes unisexual flowers with a short tubular corolla. A *radiate* head is composed of *disk* flowers in the center of the head surrounded by *ray* flowers; pistillate or sterile flowers with the corolla prolonged into a broad strap-shaped lobe with 1–3 apical teeth. The third type of head is *ligulate*, where each of the flowers is perfect and has the corolla prolonged into a strap-shaped lobe with 5 apical teeth.

When identifying a composite, first determine what flower types are present in the head; disk, disk and ray, or ligulate. The type of pappus present (narrow scales, capillary or feathery bristles, or hairs) should be the next feature examined (a hand lens would be useful), using the oldest heads available.

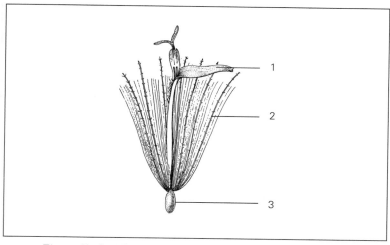

Figure 42: **Sonchus oleraceus**, *flower: 1, corolla; 2, pappus;*
3, ovary (inferior)

The Compositae are the largest family of flowering plants, with about 20,300 species in about 1160 genera worldwide (Zomlefer, 1994). The family can be divided into three subfamilies (Barnesioideae, Asteroideae, and Lactucoideae); the latter two are represented in Michigan. Within each subfamily, the *tribe* is used to group related genera. Nine of the fourteen tribes of Compositae are represented in Michigan. This key is not based on tribal alignment, as the characters defining the tribes are not necessarily easy to use in the field. However, the tribes are indicated wherever appropriate in order to convey some idea of the relationships among genera. See Zomlefer (1994) for a good summary table showing characters of each tribe of the Compositae.

1a. Heads ligulate; flowers perfect; juice milky; leaves alternate or basal (the central flowers must be examined carefully, since they are frequently much smaller than the marginal ones) (Tribe Lactuceae)—3

1b. Heads discoid or radiate; flowers perfect or unisexual; juice watery; leaves alternate, opposite, whorled, or basal—2

2a. Heads discoid, all flowers tubular with regular five-lobed corollas (corollas rarely absent or the corollas of marginal flowers expanded)—48

2b. Heads radiate (in a few species the rays are small and may be overlooked)—123

3a. Pappus absent or consists only of narrow scales—4

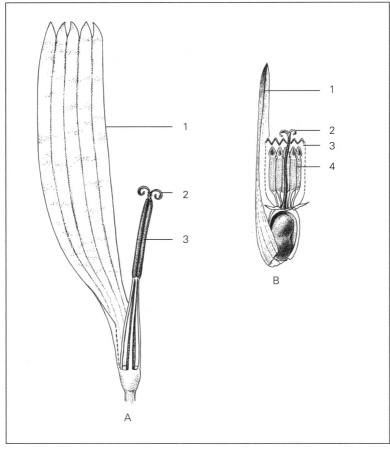

Figure 43: Flower types in the Compositae: A, **Cichorum intybus**
(ligulate flower); 1, corolla; 2, style; 3, stamens; B, **Echinacea purpurea**
(disk flower); 1, chaff; 2, style; 3, corolla; 4, stamens

3b. Pappus consists of capillary (simple) or plumose (feathery) bristles;
    scales sometimes also present—5

4a. Flowers blue; pappus of narrow scales; roadsides, etc. (30–170 cm high;
    summer and autumn)—**Chicory**, *Cichorium intybus*
4b. Flowers yellow; pappus absent; deciduous woods, shaded disturbed
    areas (15–150 cm high; summer)—**Nipplewort**, *Lapsana communis*

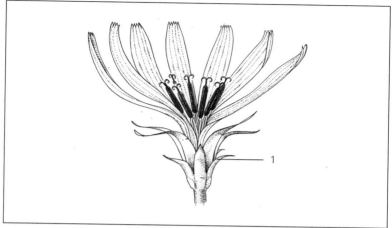

Figure 44: ***Cichorum intybus,*** *inflorescence (ligulate head): 1, involucre*

5a. Pappus bristles plumose (summer)—6
5b. Pappus bristles capillary—12

6a. Leaves basal; flower scape naked or with small scales; flowers yellow; roadsides, etc.—7
6b. Plants with leafy stems; flowers yellow or purple—9

7a. Chaffy bracts subtend the flowers (scape 15–60 cm high)—**Cat's-ear**, ***Hypochaeris radicata*** [***Hypochoeris radicata***—CQ]
7b. Bracts subtending the flowers absent (**Fall-dandelion**, ***Leontodon*** spp.)—8

8a. Heads several; pappus of plumose bristles in all flowers (scape 10–80 cm high)—***Leontodon autumnalis***
8b. Head solitary; pappus of the outer flowers reduced to short scales (scape 10–35 cm high)—***Leontodon taraxacoides***

9a. Leaf blades oblong-lanceolate, the margin sometimes serrate; flowers yellow; roadsides, fields, etc., SLP (20–100 cm high; summer)—**Ox-tongue**, ***Picris hieracioides***
9b. Leaf blades linear to linear-lanceolate, grass-like, the margin entire; flowers yellow or purple; roadsides, open areas (spring and summer) (***Tragopogon*** spp., **Goatsbeard**)—10

10a. Flowers purple (40–100 cm high)—**Salsify**, ***Tragopogon porrifolius***
10b. Flowers yellow—11

11a. Inflorescence bracts shorter than or the same length as the bright yellow corolla (15–80 cm high)—***Tragopogon pratensis***

11b. Inflorescence bracts longer than the pale yellow corolla (30–100 cm high)—***Tragopogon dubius***

12a. Pappus composed of both capillary bristles and short scales; flowers yellow or orange (***Krigia*** spp., **Dwarf-dandelion**)—13

12b. Pappus entirely composed of bristles; flowers yellow to red, blue, or white—14

13a. Plant with leafy stem; heads several; pappus bristles twenty or more per flower; fields, roadsides (20–80 cm high; summer)—***Krigia biflora***

13b. Leaves basal (scape may have a few leaves near the base); head solitary; pappus bristles ten or fewer per flower; sandy areas, LP (scapes 3—20(32) cm high; spring and early summer)—**Dwarf-dandelion**, ***Krigia virginica***

14a. Heads solitary at the summit of leafless scape; flowers yellow, sometimes aging to pink—15

14b. Heads several, on leafy stems, naked or scaly scapes; flowers yellow to red, blue, or white—18

15a. Involucral bracts of two lengths, the outer bracts shorter than the inner; basal leaf blades deeply pinnately-lobed (scape 5–50 cm high; spring to autumn) (***Taraxacum*** spp., **Dandelion**)—16

15b. Involucral bracts all of similar length; basal leaf blades entire or with a few shallow teeth or lobes (summer)—17

16a. Outer involucral bracts reflexed; mature achene tan or brown; lawns, fields, disturbed areas—**Common Dandelion**, ***Taraxacum officinale***

16b. Outer involucral bracts erect or spreading; mature achene red or red-brown; sandy areas—**Red-seeded Dandelion**, ***Taraxacum erythrospermum*** [***Taraxacum laevigatum***—CQ]

17a. Basal leaf blades pubescent; flowers yellow; lawns, NLP [Benzie Co.] (scape 3–25 cm high; spring and summer)—**Mouse-ear Hawkweed**, ***Hieracium pilosella***

17b. Basal leaf blades glabrous; flowers yellow, aging to pink; sandy areas, NLP (scape 10–70 cm high; summer)—**False-dandelion**, ***Agoseris glauca***

18a. Cauline leaves linear; flowers yellow, sometimes aging to blue; fields, roadsides, etc.—19

18b. Cauline leaves absent or lanceolate or broader; flowers yellow to red, blue or white—22

19a. Cauline leaves lack basal auricles; fewer than 15 flowers per head; SLP (30–150 cm high; summer)—**Skeleton-weed**, *Chondrilla juncea*
19b. Cauline leaves with basal auricles—20

20a. Sixteen or fewer flowers per head; flowers yellow, aging to blue; slender beak separates pappus from body of achene; SE (30–100 cm high)—**Willow-leaved Lettuce**, *Lactuca saligna*
20b. Twenty or more flowers per head; flowers yellow; achene beak absent, pappus attached to body of achene (*Crepis* spp., **Hawk's-beard**)—21

21a. Inner side of the inner involucral bracts pubescent; mature achenes dark brown; roadsides (10–100 cm high; summer)—*Crepis tectorum*
21b. Inner side of the inner involucral bracts glabrous; mature achenes light brown; fields (20–90 cm high; summer and autumn)—*Crepis capillaris*

22a. Pappus tawny or brown in color—23
22b. Pappus white—39

23a. Leaf blades mostly pinnately lobed; flowers bluish to white; moist woods (60–200 cm high)—**Tall Blue Lettuce**, *Lactuca biennis*
23b. Leaf blades entire or sometimes toothed; flowers yellow or orange to red (summer or autumn)(*Hieracium* spp., **Hawkweed**)—24

One of the intriguing aspects of plant taxonomy is the dynamic nature of the species composition in any particular local flora. Historical collections and treatises may reveal a very different picture than contemporary plant surveys. Of the fifteen species of **Hieracium** now known in Michigan, eight have been introduced from Europe. Gleason included only one of those eight in the 1918 edition of this book; the three species which are aggressively weedy and common now (**H. aurantiacum**, **H. caespitosum**, and **H. piloselloides**) were barely known in Michigan at that time! These species of **Hieracium** are members of a species complex exhibiting both extensive hybridization and asexual propagation, thus some plants will be found which do not precisely match the characteristics described in the key. See Voss and Böhlke (1978) for a summary of the history of these species in Michigan and a discussion of their taxonomic complexity.

24a. Base of leaf blade coarsely toothed, tapering to a petiole or the base cordate to rounded; disturbed areas—25
24b. Base of leaf blade not coarsely toothed, tapers to a petiole or sessile; disturbed areas or dry woods—27

25a. Base of leaf blade cordate to rounded; disturbed areas, UP (summer)—*Hieracium murorum*
25b. Base of leaf blade tapers to a petiole (late spring and summer)—26

26a. Leaves purple-spotted; mixed woods, Straits—**Spotted Hawkweed**, *Hieracium maculatum*

26b. Leaves not purple-spotted; disturbed areas (15–100 cm tall)—*Hieracium lachenalii*

27a. A rosette of basal leaves conspicuous at flowering time; cauline leaves few (if any) and mostly much reduced; lawns, disturbed areas, less often in prairies, dry woods—28

27b. No rosette of basal leaves at time of flowering; cauline leaves present; sandy woods, fields, beaches—34

28a. Upper leaf blade surface nearly glabrous, the veins red-purple; dry woods, mostly LP (20–80 cm high; spring and early summer)—**Rattlesnake-weed**, *Hieracium venosum*

28b. Upper leaf blade surface pubescent or nearly glabrous, the veins not red-purple—29

29a. Inflorescence a cylindrical panicle; cauline leaves similar to basal leaves; prairies, sandy open woods—30

29b. Inflorescence corymb-like, often crowded; cauline leaves few and much reduced; lawns, disturbed areas—31

30a. Leaves and the lower stem covered with very long hairs (over 1 cm long); forty or more flowers per head; prairies, sandy open woods, SLP (60–200 cm high; summer)—**Prairie Hawkweed**, *Hieracium longipilum*

30b. Leaves and lower stem pubescent, any long hairs shorter than 1 cm long; forty or fewer flowers per head; sandy woods, LP (30–150 cm high; summer and autumn)—*Hieracium gronovii*

31a. Involucral bracts 9–11 mm long; inflorescence of two or three heads; flowers yellow; leafy stolons present; disturbed areas, SLP (15–40 cm tall; late spring)—*Hieracium flagellare*

31b. Involucral bracts 5–8 mm long; inflorescence usually of five or more heads; flowers yellow or orange to red; lawns, disturbed areas (25–90 cm tall; summer)—32

32a. Stem and upper leaf surface glaucous, either glabrous or with only scattered hairs; stolons usually not apparent—**Yellow Hawkweed** or **King Devil**, *Hieracium piloselloides*

32b. Stem and upper leaf surface densely hairy, not glaucous; stolons usually present—33

33a. Flowers yellow—**Yellow Hawkweed** or **King Devil**, *Hieracium caespitosum*

33b. Flowers orange or orange-red—**Orange Hawkweed** or **Devil's Paintbrush**, *Hieracium aurantiacum*

34a. Involucre consists of several rows of similar-sized, overlapping bracts (15–150 cm high; summer)—35
34b. Involucre consists of one row of large bracts subtended by one row of tiny bracts—36

35a. Leaf blades narrowed toward the base; margin of blade has both star-shaped hairs and short, blunt hairs; dunes, sandy areas—**Northern Hawkweed**, *Hieracium umbellatum*
35b. Leaf blades rounded at the sessile base; margin of blade has star-shaped hairs, lacks short, blunt hairs; sandy woods, fields—*Hieracium kalmii*

36a. Inflorescence leafy-bracted; upper stem leafy (summer)—37
36b. Inflorescence bracts much reduced; upper stem naked or with scattered small bracts—38

37a. Inflorescence branches glabrous or with a few scattered glandular hairs; fewer than thirty-five flowers per head; woods, SLP (30–150 cm high)—*Hieracium paniculatum*
37b. Inflorescence branches densely covered with glandular hairs; forty or more flowers per head; sandy woods (20–150 cm high)—*Hieracium scabrum*

38a. Leaves and the lower stem covered with very long hairs (over 1 cm long); forty or more flowers per head; prairies, sandy open woods, SLP (60–200 cm high; summer)—**Prairie Hawkweed**, *Hieracium longipilum*
38b. Leaves and lower stem pubescent, any long hairs shorter than 1 cm long; forty or fewer flowers per head; sandy woods, LP (30–150 cm high; summer and autumn)—*Hieracium gronovii*

39a. Flowers eighty or more per head, yellow; leaves with auriculate, clasping bases and (often) prickly margins; disturbed areas, esp. cultivated ground (10–200 cm high; summer and autumn) (*Sonchus* spp., **Sowthistle**)—40
39b. Flowers five to forty per head, yellow (some aging to blue), or white to purplish; leaves only rarely auriculate-clasping or with prickly margins (summer)—42

40a. Heads 3–5 cm broad; robust, deeply-rooted perennial—**Perennial Sowthistle**, *Sonchus arvensis*
40b. Heads 1.5–2.5 cm broad; annuals—41

41a. Leaf base auricles rounded; leaf margins conspicuously prickly—
Prickly Sow-thistle, *Sonchus asper*
41b. Leaf base auricles acute; leaf margins scarcely prickly; mostly SLP—
Common Sow-thistle, *Sonchus oleraceus*

42a. Heads nodding or in a narrow spike-like cluster; flowers white, cream-
colored, blue, or purplish (40–200 cm high; summer)(***Prenanthes*** spp.,
Rattlesnake-root)—43
42b. Heads in a spreading open panicle; flowers yellow or bluish to white
(most ***Lactuca*** spp., **Wild Lettuce**)—45

43a. Heads pointing in various directions, in a spike-like cluster; involucral
bracts pubescent; lower leaves ovate to spathulate, not lobed; moist
prairies, fens—***Prenanthes racemosa***
43b. Heads nodding, arranged in a panicle; involucral bracts glabrous; lower
leaves usually deltoid, often lobed; woods—44

44a. Five or six flowers per head; SLP—***Prenanthes altissima***
44b. Eight to twelve flowers per head—***Prenanthes alba***

45a. Midrib of underside of leaf blade prickly; fields, roadsides, etc.
(50–150 cm high)—**Prickly Lettuce**, *Lactuca serriola*
45b. Midrib of underside of leaf blade glabrous or pubescent, not prickly—
46

46a. Flowers five per head; sandy shores, Straits (30–100 cm high)—**Wall
Lettuce**, *Lactuca muralis*
46b. Flowers more than twelve per head; woods, fields (30–250 cm high)—
47

47a. Leaves usually glabrous; involucre with fruit 15 mm or less long—
Wild Lettuce, *Lactuca canadensis*
47b. Leaves pubescent; involucre with fruit 15 mm or more long; mostly
LP—***Lactuca hirsuta***

48a. Heads on a scaly peduncle; the whitish flowers opening before the expansion of the basal, broadly triangular to ovate leaves; moist woods,
NM or UP (10–50 cm high, spring) (***Petasites*** spp., **Sweet-coltsfoot**)—
49
48b. Heads on a leafy stem; flowers appear after leaves expand—50

49a. Leaf blades deeply palmately lobed; NM—***Petasites frigidus***
49b. Leaf blades toothed; UP—***Petasites sagittatus***

50a. Leaf blades spiny or prickly and/or the involucral bracts spiny, hooked, fringed, or toothed; leaves alternate or rarely opposite (most are members of the Tribe Cynareae)—51

50b. Leaf blades not spiny or prickly, nor are the involucral bracts spiny, hooked, fringed, or toothed; leaves alternate, opposite, whorled, or basal—73

51a. Leaf blades spiny or prickly—52

51b. Leaf blades lack either spines or prickles—61

52a. Each head one-flowered; heads aggregated in a globular, head-like cluster; flowers blue or white; disturbed areas, sometimes cultivated, LP (1–2.5 m high; summer)—**Globe-thistle**, *Echinops sphaerocephalus*

52b. Each head many-flowered; flowers pink, purple, or rarely white (summer and autumn)—53

53a. Pappus consists of feathery bristles; receptacle covered with bristles (*Cirsium* spp., **Thistle**)—54

53b. Pappus consists of capillary bristles; receptacle bristles present or absent roadsides, fields (*Carduus* spp., **Plumeless Thistle**)—60

54a. Leaf bases decurrent, making the stem appear spiny and winged—55

54b. Leaf bases not decurrent, stem not spiny-winged—56

55a. Involucre 1–2 cm high; involucral bract tips not spiny; NM, wetlands (30–200 cm tall; early summer)—**European Swamp Thistle**, *Cirsium palustre*

55b. Involucre 2–4 cm high; involucral bract tips spiny; roadsides, pastures, etc. (50–150 cm high; summer and autumn)—**Bull Thistle**, *Cirsium vulgare*

56a. Involucre 1–2 cm high; plants forming large colonies; flowers unisexual; fields, ditches, etc. (30–150 cm high; summer)—**Canada Thistle**, *Cirsium arvense*

56b. Involucre 2 cm or longer; plants usually not forming colonies; flowers perfect—57

57a. Leaf blades green on both sides; wet meadows, swamps, fens (50–200 cm high)—**Swamp Thistle**, *Cirsium muticum*

57b. Underside (or both sides) of leaf blades white- or brown-woolly—58

58a. Leaves conspicuously white-woolly on both sides; blades deeply pinnately divided with linear divisions; flowers almost white; Great Lakes sand dunes and beaches (50–100 cm high; late spring and summer)—**Pitcher's** or **Dune Thistle**, *Cirsium pitcheri*

58b. Leaves conspicuously white- or brown-woolly below, not above; flowers purple or pink—59

59a. Stem leaves entire or shallowly lobed; woodland edges, disturbed areas, SLP (1–3 m high; summer and autumn)—**Tall Thistle**, *Cirsium altissimum*

59b. Stem leaves obviously pinnately divided; fields, disturbed areas (1–3 m high; summer and autumn)—**Pasture Thistle**, *Cirsium discolor*

60a. Heads erect, less than 2.5 cm broad; involucral bracts less than 2 mm wide; LP (30–150 cm high)—*Carduus acanthoides*

60b. Heads mostly nodding, more than 2.5 cm broad; involucral bracts 2 mm wide or more; SE (60–200 cm high)—**Nodding** or **Musk Thistle**, *Carduus nutans*

61a. Staminate and pistillate flowers in separate heads, the staminate above in clusters or racemes, the pistillate below and axillary (often inconspicuous); leaves alternate or opposite—62

61b. All heads alike, flowers perfect; leaves alternate—65

62a. Involucral bracts of pistillate flowers thickly covered with sharp hooked spines; leaves alternate; fields, flood plains, disturbed areas (20 cm–200 cm high; summer)—**Cocklebur**, *Xanthium strumarium*

62b. Tips of involucral bracts of pistillate flowers spiny or tuberculate, spines not hooked; leaves alternate or opposite (summer and autumn) (*Ambrosia* spp., **Ragweed**)—63

Ragweed (***Ambrosia*** sp.) and Goldenrod (***Solidago*** sp.) represent two different pollination strategies, with a direct effect on human life. ***Ambrosia*** plants bear small, greenish, unisexual flowers. The male flowers produce large amounts of smooth, light pollen which is carried by the wind to the female flowers; this ragweed pollen is the allergen responsible for most common cases of fall hayfever. ***Solidago*** flowers are bisexual and are pollinated by insects. Their sticky pollen is not airborne and does not contribute to the misery of hay-fever sufferers. Goldenrod has presumably been blamed for hayfever because it blooms at the same time as ragweed.

63a. Leaves all opposite, the blades deeply palmately three- or five-lobed, sometimes entire; flood plains, roadside ditches, etc. (1–5 m high)—**Giant Ragweed**, *Ambrosia trifida*

63b. Leaves alternate above, opposite below, the blades deeply pinnately divided (30–100 cm high)—64

64a. Annual; leaves usually twice-pinnately divided; tips of pistillate involucral bracts spiny; disturbed areas—**Common Ragweed**, *Ambrosia artemisiifolia*

64b. Perennial; leaves usually once-pinnately divided; tips of pistillate involucral bracts tuberculate; old, sandy fields—**Western Ragweed**, *Ambrosia psilostachya*

65a. Tip of involucral bracts hooked; leaf blades ovate, 10 cm or more broad, most with a cordate base; flowers pink or purple (seldom white); marginal corollas not enlarged; roadsides, pastures, etc. (0.5–1.5 m high)—**Common Burdock**, *Arctium minus*

65b. Tip of involucral bracts spiny or deeply fringed, not hooked; leaf blades linear to ovate-lanceolate, sometimes pinnately divided; flowers rose, purple, blue, white, or rarely yellow; marginal corollas sometimes enlarged (most *Centaurea* spp., **Star-thistle** or **Knapweed**)—66

66a. Flowers yellow; cauline leaves ovate-lanceolate, entire; roadsides, UP, sometimes cultivated (to 1 m high; summer)—*Centaurea macrocephala*

66b. Flowers rose, purple, blue, or white; cauline leaves entire, toothed, or pinnately divided; fields, roadsides, disturbed areas—67

67a. Leaves pinnately divided; flowers rose, purple, or white (10–80 cm high)—68

67b. Leaves entire, toothed, or irregularly lobed; flowers blue or purple—69

68a. Flowers rose or purple; tips of involucral bracts deeply fringed; pappus present (summer and autumn)—**Spotted Knapweed**, *Centaurea maculosa*

68b. Flowers white; tips of involucral bracts spiny; pappus absent; esp. NLP (summer)—**White-flowered** or **Tumble Knapweed**, *Centaurea diffusa*

Two of our species of **Centaurea** (*C. maculosa* and *C. diffusa*) have aggressively colonized many roadsides, fields, and disturbed areas in Michigan. *Centaurea diffusa* is more commonly seen in the northern Lower Peninsula while *C. maculosa* is likely to be found throughout the state. Hybrids between the two species are known in the northern LP, thus plants with a mixture of the characters for each species may be found in that area.

69a. Leaf bases decurrent, forming wings along the stem; leaves mostly entire; UP, sometimes cultivated—**Mountain Bluet**, *Centaurea montana*

69b. Leaf bases not decurrent, stem not winged; leaves entire or often toothed or lobed (20–80 cm high; summer and autumn)—70

70a. Leaves linear, less than 1 cm wide; escape from cultivation (20—120 cm high)—**Bachelor's-button**, *Centaurea cyanus*

70b. Leaves over 1 cm wide—71

71a. Involucral bract tips 3 mm or less long, black, deeply fringed—**Short-fringed Knapweed**, *Centaurea nigrescens* [*Centaurea dubia*—CQ]
71b. Involucral bract tips 3 mm or more long, brown, fringed or toothed— 72

72a. Involucral bract tips irregularly toothed—*Centaurea jacea*
72b. Involucral bracts tips fringed—*Centaurea ×pratensis*

73a. Staminate and pistillate flowers in separate heads, the staminate above in racemes, the pistillate below and axillary (summer and autumn) (*Ambrosia* spp., **Ragweed**)—74
73b. All heads alike, flowers perfect or unisexual—76

74a. Leaves all opposite, the blades deeply palmately three- or five-lobed, sometimes entire; flood plains, roadside ditches, etc. (1–5 m high)— **Giant Ragweed**, *Ambrosia trifida*
74b. Leaves alternate above, opposite below, the blades deeply pinnately divided (30–100 cm high)—75

75a. Annual; leaves usually twice-pinnately divided; tips of pistillate involucral bracts spiny; disturbed areas—**Common Ragweed**, *Ambrosia artemisiifolia*
75b. Perennial; leaves usually once-pinnately divided; tips of pistillate involucral bracts tuberculate; sandy, old fields—**Western Ragweed**, *Ambrosia psilostachya*

76a. Principal bracts of the involucre in a single row, sometimes with a few much smaller basal bracteoles (most are members of the Tribe Senecioneae)—77
76b. Principal bracts of the involucre in multiple, overlapping rows—84

77a. Flowers yellow or orange—78
77b. Flowers white or pinkish (summer)—80

78a. Leaf blades large, broadly ovate, to 30 cm long, mostly opposite; upper stem and leaves somewhat viscid-pubescent; moist woods, SLP (60-200 cm high; summer and autumn)—**Leaf-cup**, *Polymnia canadensis*
78b. Leaf blades smaller, ovate or elliptic to 10 cm long, alternate and basal; upper stem and leaves glabrous (*Senecio* spp., in part, **Ragwort**)—79

79a. Leaf blades deeply lobed; tips of involucral bracts often black; gardens, disturbed areas (10–40 cm high; spring to autumn)—**Common Groundsel**, *Senecio vulgaris*

79b. Leaf blades serrate, not deeply lobed; tips of involucral bracts reddish or purple; moist woods, wet meadows, shores, UP (30–80 cm high; summer)—**Squaw-weed**, *Senecio indecorus*

80a. Heads five-flowered; flowers white or pinkish; SLP (*Cacalia* spp., **Indian-plantain**)—81

80b. Heads with more than five flowers; flowers white—82

81a. Leaf blades elliptic, with entire margins and several parallel main veins from base to apex; wet prairies (0.6–1.8 m high)—*Cacalia plantaginea*

81b. Leaf blades broadly triangular or ovate, with toothed or lobed margins and several palmate-diverging main veins; woods, open areas (1–3 m high)—*Cacalia atriplicifolia*

82a. Leaves opposite; woods (30-150 cm high; summer and autumn)— **White Snakeroot**, *Eupatorium rugosum*

82b. Leaves alternate—83

83a. Leaves mostly on the lower half of the stem; leaf blades white-woolly beneath; moist woods, UP (up to 1 m high; summer)—**Trail Plant**, *Adenocaulon bicolor*

83b. Leaves found all along the stem; leaf blades not white-woolly beneath; wet meadows, marshes, shores (10 cm–2.5 m high; late summer)— **Fireweed**, *Erechtites hieraciifolia*

84a. Leaves compound or dissected; flowers yellow or orange (summer and autumn)—85

84b. Leaves entire or merely lobed, never truly compound or dissected—96

85a. Leaves opposite, often trifoliolate; some of the involucral bracts leaf-like, longer than the heads (*Bidens* spp. in part, **Beggar-ticks**)—86

85b. Leaves alternate, often finely dissected; involucral bracts short and not leaf-like (Tribe Anthemideae)—88

86a. Outer involucral bracts three to five, the margins not ciliate; swamp forests, wet thickets, SLP (30–80 cm high; late summer)—*Bidens discoideus* [*Bidens discoidea*—CQ]

86b. Outer involucral bracts five or more; the margins conspicuously ciliate (20–120 cm high; summer and autumn)—87

87a. Outer involucral bracts five to ten; achenes blackish; shores, damp woods, ditches—*Bidens frondosus* [*Bidens frondosa*—CQ]

87b. Outer involucral bracts ten to sixteen; achenes yellowish-brown; shores, damp fields—*Bidens vulgatus* [*Bidens vulgata*—CQ]

88a. Heads up to 5 mm wide, arranged in spikes, racemes, or panicles
(Artemisia spp. in part, Wormwood)—89
88b. Heads 5–20 mm wide, arranged in corymb-like cymes—94

89a. Leaves finely gray-pubescent on both sides; long hairs present between
flowers on the receptacle; roadsides, fields, etc. (40–100 cm high; summer)—**Absinth**, *Artemisia absinthium*
89b. Leaves glabrous or pubescent above, sometimes densely white-woolly
beneath; long hairs between the flowers absent—90

90a. Leaves densely white-woolly beneath—91
90b. Leaves green beneath—92

91a. Leaves 5–10 cm long; clump-forming perennials; fields, etc., SE & UP
(50–200 cm high; summer and autumn)—**Mugwort**, *Artemisia vulgaris*
91b. Leaves 1–3 cm long; perennials forming colonies from rhizomes; sandy
areas, LP (40 cm or more high; late summer)—*Artemisia pontica*

92a. Leaf lobes narrowly linear, strictly entire; open, sandy areas (10–100
cm high; summer)—**Wild Wormwood**, *Artemisia campestris*
92b. Leaf lobes finely toothed (30–300 cm high; late summer and autumn)—93

93a. Heads often nodding, on peduncles in a loose, spreading panicle; disturbed areas, SE—**Sweet Wormwood**, *Artemisia annua*
93b. Heads erect, nearly sessile, crowded in a leafy spike-like panicle; sandy
roadsides, disturbed areas—**Biennial Wormwood**, *Artemisia biennis*

94a. Receptacle cone-shaped, pointed; lawns, roadsides, etc. (5–40 cm high;
late spring and summer)—**Pineapple-weed**, *Matricaria discoidea* [*Matricaria matricarioides*—CQ]
94b. Receptacle flat or at most convex (*Tanacetum* spp., *Tansy*)—95

95a. Heads 5–10 mm wide, more than twenty per plant; roadsides, fields,
etc. (40–150 cm high; late summer and autumn)—**Common Tansy**,
Tanacetum vulgare
95b. Heads 10–20 mm wide, up to fifteen per plant; Great Lakes shores, NM
(10–80 cm high; summer)—**Lake Huron Tansy**, *Tanacetum huronense*

96a. All or most leaves opposite or whorled—97
96b. All leaves alternate—106

97a. Receptacle bracts present subtending the flowers; pappus of three or
four rigid awns or absent—98

97b. Receptacle bracts absent; pappus of capillary bristles (*Eupatorium* spp.)—101

98a. Leaf blades ovate, often broadly so; flowers greenish; small heads sub-sessile, in panicles; pappus absent; barnyards, moist disturbed areas (50–200 cm high; late summer and autumn)—**Marsh-elder**, *Iva xan-thifolia*

98b. Leaf blades lanceolate to lance-ovate; flowers yellowish to orange-yel-low; heads on long peduncles; pappus of two to four stiff awns (late summer and autumn)(*Bidens* spp. in part, **Beggar-ticks**)—99

99a. Heads nodding after flowering; leaves sessile, the bases often connate; wet shores (10 cm–1 m or more high)—**Nodding Beggar-ticks**, *Bidens cernuus* [*Bidens cernua*—CQ]

99b. Heads erect, not nodding after flowering; leaf petioled or sessile, the bases not connate; moist disturbed areas (10–200 cm high)—100

100a. Disk corollas light yellow; achenes flattened or 3-angled; SLP (10–200 cm high)—*Bidens comosus* [*Bidens comosa*—CQ]

100b. Disk corollas yellow-orange; achenes 4-angled (10–200 cm high)—*Bidens connatus* [*Bidens connata*—CQ]

101a. Leaves whorled; flowers pink or purple (60–200 cm high; summer)—102

101b. Leaves opposite; flowers white (summer and autumn)—103

102a. Inflorescence ovoid or pyramidal; four to seven flowers per head; woods, SLP—**Green-stemmed Joe-Pye-Weed**, *Eupatorium pur-pureum*

102b. Inflorescence depressed or flattened; eight or more flowers per head; marshes, fens—**Joe-Pye-Weed**, *Eupatorium maculatum*

103a. Flowers five per head; fields etc., SLP (80–200 cm high)—**Tall Bone-set**, *Eupatorium altissimum*

103b. Flowers nine or more per head—104

104a. Leaves sessile, united at the base; marshes, fens, bogs (40–150 cm high)—**Boneset**, *Eupatorium perfoliatum*

104b. Leaves petioled—105

105a. Involucral bracts overlapping, of two or three lengths; disturbed areas, SW (40–200 cm high)—**Late Boneset**, *Eupatorium serotinum*

105b. Involucral bracts not overlapping, all of similar length; woods (30-150 cm high)—**White Snakeroot**, *Eupatorium rugosum*

106a. Lower surface of the leaf blades (and often other plant parts) white-woolly—107
106b. Lower surface of the leaf blades green, glabrous or pubescent—115

107a. Pappus none; heads in panicled spikes; sandy roadsides, fields, etc. (30–100 cm high; summer and autumn)—**Western Mugwort,** *Artemisia ludoviciana*
107b. Pappus of capillary bristles; heads in flat- or round-topped clusters or slender spikes (Tribe Inuleae)—108

108a. Receptacle bracts present, subtending the flowers; sandy roadsides, NLP & Straits (3–50 cm high; summer)—*Filago arvensis*
108b. Receptacle bracts absent—109

109a. Principal leaves basal; heads sessile or subsessile in small flat-topped clusters; flowers white or purplish; (10–40 cm high; spring and early summer)(*Antennaria* spp., **Everlasting** or **Pussy-toes**)—110

The genus **Antennaria** has always been a taxonomically difficult group. *Apomixis*, in which seed is produced without sexual recombination of male and female gametes, is well known in this genus. Although most **Antennaria** plants are dioecious, only female plants are found in apomictic species. In Michigan, **Antennaria neglecta** and **A. parlinii** are sexual species, while **A. howellii** is apomictic. All apomictic clones arose at one time from sexual plants. With the switch to asexual propagation, one small colony with minor variations from the parent population can be amplified into great numbers of plants. Many of these local variants have been named as species and other taxa, resulting in over 350 names described from North American plants. In a recent examination of the genus, Bayer and Stebbins (1993) accept only 46 taxa. This key has followed their treatment.

109b. Principal leaves cauline; heads in small or large flat-topped clusters or panicles; flowers white; (summer or autumn)—112

110a. Basal leaves over 2 cm wide, with three or five primary veins; open woods, fields—*Antennaria parlinii* [*Antennaria plantaginifolia* var. *parlinii*—CQ]
110b. Basal leaves less than 2 cm wide, with one primary vein—111

111a. Middle and upper cauline leaves with flat, scarious tips; fields, SLP—*Antennaria neglecta*
111b. Middle and upper cauline leaves with awl-shaped tips; open woods, fields—*Antennaria howellii* [incl. in *Antennaria neglecta*—CQ]

112a. Involucral bracts pearly white; perennials forming colonies from rhizomes; open woods, forest edges, fields (30–90 cm high; summer)— **Pearly Everlasting**, *Anaphalis margaritacea*

112b. Involucral bracts yellow-white, greenish, or pale brown; tufted annuals or biennials (*Gnaphalium* spp., **Cudweed**)—113

113a. Leaf bases decurrent along the stem; open sandy fields (30–100 cm high; summer) **Clammy Cudweed**, *Gnaphalium macounii*

113b. Leaf bases not decurrent along the stem—114

114a. Inflorescence of several axillary and terminal clusters of heads subtended by leaves which are conspicuously longer than the heads; wet fields, pastures, shores (5–25 cm high; summer and autumn)—**Low Cudweed**, *Gnaphalium uliginosum*

114b. Inflorescence a panicle of heads, the heads not surpassed by subtending bracts; fields (30–100 cm high; summer)—**Fragrant Cudweed**, *Gnaphalium obtusifolium*

115a. Flowers yellow; leaf blades oblanceolate to elliptic, with rounded teeth, sometimes with basal lobes; roadsides, fields, etc., mostly SLP (50–150 cm high; summer and autumn)—**Costmary**, *Chrysanthemum balsamita*

115b. Flowers white, red, or purple; leaf blades linear to elliptic, lacking basal lobes—116

116a. All or most involucral bracts with a petaloid or scarious margin; flowers pink-purple; heads showy, in a long spike or spike-like raceme (late summer)(*Liatris* spp., **Blazing Star**)—117

116b. Involucral bracts green, entirely herbaceous; flowers white or red to purple; heads in panicles, spike-like racemes, or corymb-like clusters—120

Many Compositae are popular garden plants. In addition to those included in this key, other commonly grown species which occasionally escape from cultivation include **Calendula officinalis** (Pot-marigold), **Chrysanthemum ×superbum** (Shasta Daisy), **Eupatorium coelestinum** (Hardy Ageratum), and **Aster tataricus**.

117a. Pappus plumose, the lateral hairs over 0.5 mm long; prairies, sandy areas, LP (20–60 cm high)—*Liatris cylindracea*

117b. Pappus barbed, the lateral hairs minute, less than 0.3 mm long—118

118a. Heads sessile, five- to fourteen-flowered; moist prairies, SLP (60–200 cm high)—*Liatris spicata*

118b. Heads pedicellate (rarely subsessile), with fifteen or more flowers per head—119

119a. Middle involucral bracts with a scarious, sometimes irregularly cut or wavy, margin; prairies, dry woods, SLP & UP (40–150 cm high)—*Liatris aspera*

119b. Middle involucral bracts lack a scarious margin; open woods, prairies, LP (30–80 cm high)—*Liatris scariosa*

120a. Leaf blades linear; flowers violet; heads in a panicle or spike-like raceme; tap-rooted annual; saline roadsides, etc. (10–70 cm high)—**Rayless Aster**, *Aster brachyactis*

120b. Leaf blades lanceolate to ovate; flowers white, red or purple; heads in corymb-like clusters; clump-forming perennials—121

121a. Flowers white; pappus of plumose bristles; sandy fields, prairies, WM (30–130 cm high; late summer and autumn)—**False Boneset**, *Kuhnia eupatoriodes*

121b. Flowers bright red or purple; pappus of short and long capillary bristles (late summer and autumn)(*Vernonia* spp., **Ironweed**) (Tribe Vernonieae)—122

122a. Heads include fifteen to thirty flowers; hairs along midvein of lower leaf surface short and straight; moist ditches, floodplain forests, SE (1–3 m high)—*Vernonia gigantea*

122b. Heads include thirty to forty-five flowers; hairs along midvein of lower leaf surface long and crooked; fields, prairies (1–2 m high)—*Vernonia missurica*

123a. Heads on a scaly peduncle; the flowers opening before the expansion of the basal, broadly triangular to ovate leaves (spring)—124

123b. Heads on a leafy stem (rarely with only bract-like scales); flowers appear after leaves expand—126

124a. Head solitary; flowers yellow; roadsides (10–50 cm high; spring)—**Coltsfoot**, *Tussilago farfara*

124b. Heads several; flowers whitish; moist woods, NM or UP (10–50 cm high) (*Petasites* spp., **Sweet-coltsfoot**)—125

125a. Leaf blades deeply palmately lobed; NM—*Petasites frigidus*

125b. Leaf blades toothed; UP—*Petasites sagittatus*

126a. Rays yellow or brown—127

126b. Rays white to blue or red, never yellow or brown—190

127a. Leaves basal, the stem merely with bract-like scales; prairies, roadsides, SLP (1–3 m high; summer)—**Prairie Dock**, *Silphium terebinthinaceum*

127b. Leaves cauline—128

128a. Cauline leaves all or mostly opposite or whorled (Tribe Heliantheae)—129
128b. Cauline leaves all or mostly alternate—145

129a. Ray flowers fertile, pistillate (the two-lobed style protrudes from their base)—130
129b. Ray flowers sterile, with neither stamens nor pistil—133

130a. Leaf blades lobed, broadly ovate, to 30 cm long, mostly opposite; upper stem and leaves somewhat viscid-pubescent; fields, woods, SW (1–3 m high; summer)—**Leaf-cup**, *Polymnia uvedalia*
130b. Leaf blades toothed or entire, not lobed—131

131a. Disk flowers perfect; style is divided into two branches; fields, fens, roadsides (50–150 cm high; summer and autumn)—**False Sunflower**, *Heliopsis helianthoides*
131b. Disk flowers staminate; style is undivided (summer) (*Silphium* spp. in part, **Rosin-weed**)—132

132a. Leaves (or petiole bases) united at base into a cup surrounding the square stem; woods, SLP (1–2.5 m high)—**Cup Plant**, *Silphium perfoliatum*
132b. Leaves sessile, often clasping (but not united around) the round stem; roadsides, prairies, SW (0.5–1.5 m high)—**Rosin-weed**, *Silphium integrifolium*

133a. Plant a submerged aquatic; leaf blades dissected into filiform segments; ponds—**Water Marigold**, *Megalodonta beckii* [*Bidens beckii*—CQ]
133b. Plants terrestrial (emergent if growing in water); leaf blades lobed, pinnately divided, or more often entire—134

134a. Cauline leaves lobed or pinnately divided—135
134b. Cauline leaves entire or with the margins at most serrate—141

135a. Pappus consists of three or four barbed scales or two bristly awns; leaf margins serrate (summer and autumn) (*Bidens* spp. in part; **Beggarticks**)—136
135b. Pappus consists of two smooth, short teeth; leaf margins not serrate (*Coreopsis* spp. in part, **Tickseed**)—137

136a. Leaf blades entire or three-lobed; moist disturbed areas (10–200 cm high)—***Bidens connatus*** [***Bidens connata***—CQ]

136b. Leaf blades pinnately divided into three to seven nearly linear segments; marshes, SLP (30–150 cm high)—**Tickseed-sunflower**, ***Bidens coronatus*** [***Bidens coronata***—CQ]

137a. Rays yellow, with a reddish basal spot; disk corollas mostly 4-lobed; escapes to roadsides, SE (40–120 cm high; summer)—**Plains Coreopsis**, *Coreopsis tinctoria*

137b. Rays yellow, without a basal spot; disk corollas mostly 5-lobed—138

138a. Leaves compound or palmately three-lobed; sandy woods, prairies (summer)—139

138b. Leaves entire or pinnately divided into linear segments (spring)—140

139a. Leaf lobes linear-oblong, all about equal; SW (40–90 cm high)—**Finger** or **Prairie Coreopsis**, *Coreopsis palmata*

139b. Leaves mostly divided into three lanceolate leaflets; SLP (1–3 m high)—**Tall Tickseed**, *Coreopsis tripteris*

140a. Leaves mostly entire, crowded near the base of the stem; dry soils (20–40 cm high)—***Coreopsis lanceolata***

140b. Leaves entire or pinnately divided, all along the stem; fields, roadsides (20-100 cm high)—***Coreopsis grandiflora***

141a. Bracts of the involucre all essentially alike in form and texture (flowers in summer and autumn) (***Helianthus*** spp., **Sunflower**)—181

141b. Bracts of the involucre in two distinct sets, differing in form or consistency or both—142

142a. Leaf blades entire; dry soils (20–60 cm; spring)—***Coreopsis lanceolata***

142b. Leaf blades serrate; moist areas (summer and autumn)(***Bidens*** spp. in part, **Bur-marigold**)—143

143a. Heads nodding after flowering; leaves sessile, the bases often connate; wet shores (10 cm–1 m or more)—**Nodding Beggar-ticks**, ***Bidens cernuus*** [***Bidens cernua***—CQ]

143b. Heads erect, not nodding after flowering; leaves petioled or sessile, the bases not connate; moist disturbed areas, shores (10–200 cm high)—144

144a. Disk corollas light yellow, mostly 4-lobed; achenes flattened; SLP (10–200 cm high)—***Bidens comosus*** [***Bidens comosa***—CQ]

144b. Disk corollas yellow-orange, mostly 5-lobed; achenes 4-angled
(10–200 cm high)—***Bidens connatus*** [***Bidens connata***—CQ]

145a. Heads small, seldom more than 1 cm wide, including the rays, bloom-
ing in late summer and autumn; flowers numerous, crowded in spikes,
racemes, corymbs, or panicles (summer and autumn) (***Solidago*** spp.
and ***Euthamia*** spp., **Goldenrod**) (Tribe Astereae)—146

It is easy to recognize a "goldenrod", but much more difficult to determine the species. This
key includes 15 of the 22 species of **Solidago** known in Michigan (Voss, 1996). Recent treatments of this group have segregated eight of the more than 100 species traditionally placed
in the genus **Solidago** into a separate genus, **Euthamia**, based on inflorescence characters
and the presence of glandular-dotted leaves. Two members of this genus are known in Michigan and are included here.

145b. Heads medium size or large, more than 1 cm, and usually exceeding 2
cm in width, including the rays—164

146a. Inflorescence a corymb, the heads crowded at or near the ends of the
branches at about the same distance from the base of the panicle,
forming a rounded or flat-topped infloresence—147

146b. Heads more or less uniformly distributed along the length of the
branches, forming a cylindrical or pyramidal inflorescence (never flat-
topped), or the heads in clusters or short racemes in the upper leaf
axils—151

147a. Leaf blades covered with small resinous dots (***Euthamia*** spp.)—148
147b. Leaf blades glabrous or pubescent, but not covered with resinous
dots—149

148a. Leaf blades 3–12 mm wide, with three to five primary veins; moist
open areas (30–150 cm high)—**Flat-topped Goldenrod**, *Euthamia
graminifolia*
148b. Leaf blades 1–3 mm wide, usually with one primary vein; moist sandy
shores—***Euthamia remota***

149a. Stem rough-hairy; prairies, sandy fields, SLP & UP (25–150 cm
high)—**Stiff Goldenrod**, *Solidago rigida*
149b. Stem glabrous; swamps, wet areas (40–100 cm high)—150

150a. Lower leaf blades flat, with one primary vein—**Ohio Goldenrod**, *Solidago ohioensis*
150b. Lower leaf blades folded, with three or five primary veins; SLP—*Solidago riddellii*

151a. Heads chiefly in clusters or short racemes in the upper leaf axils, or occasionally the upper compacted into a leafy cluster terminating the stem; woods—152

151b. Heads more or less uniformly distributed along the length of the branches, forming a cylindrical or pyramidal inflorescence—156

152a. Stem below the inflorescence pubescent, the hairs spreading or appressed (10–100 cm high)—153

152b. Stem below the inflorescence glabrous; woods (30–120 cm high)—155

153a. Rays whitish; woods, mostly SE—**White Goldenrod** or **Silverrod**, *Solidago bicolor*

153b. Rays yellow—154

154a. Achenes pubescent; sand dunes, rocky areas, WM & NM—**Gillman's Goldenrod**, *Solidago simplex*

154b. Achenes glabrous; woods—**Hairy Goldenrod**, *Solidago hispida*

155a. Basal leaf blades abruptly narrowed to winged petioles; stem angled, often taking a zig-zag appearance—**Zigzag Goldenrod**, *Solidago flexicaulis*

155b. Basal leaf blades tapering to a nearly sessile base; stem not angled; LP—**Bluestem Goldenrod**, *Solidago caesia*

156a. Basal leaves much larger than the greatly reduced or bract-like upper ones—157

156b. Leaves essentially uniform in size from base to summit of stem—161

157a. Racemes or branches of the panicle either short and arranged along a more or less elongated central axis, or elongated and ascending, scarcely recurved, forming a narrow, more or less elongated panicle—158

157b. Racemes or branches of the panicle usually elongated, spreading outwards, usually recurved, forming a broader panicle—160

158a. Petioles of the lowermost leaves have a sheathing base; bogs, fens, sedge meadows (60-150 cm high)—**Bog Goldenrod**, *Solidago uliginosa*

158b. Petioles of the lowermost leaves do not have a sheathing base—159

159a. Achenes pubescent; sand dunes, rocky areas, WM & NM (10–90 cm high, or prostrate)—**Gillman's Goldenrod**, *Solidago simplex*

159b. Achenes glabrous; prairies (30–150 cm high)—**Showy Goldenrod**, *Solidago speciosa*

160a. Stem pubescent; leaf blades pubescent or rough; sandy fields, woods (10–100 cm high)—**Gray Goldenrod**, *Solidago nemoralis*

160b. Stem below the infloresence glabrous; leaf blades glabrous or sometimes with short hairs; sandy fields (30–120 cm high)—**Early Goldenrod**, *Solidago juncea*

161a. Leaf blades with one primary vein; thickets, moist woods (30–150 cm high)—**Rough-leaved Goldenrod**, *Solidago rugosa*

161b. Leaf blades with three or five primary veins (25–200 cm high)—162

162a. Stem glabrous below the inflorescence—**Late Goldenrod**, *Solidago gigantea*

162b. Stem pubescent, sometimes not below the middle—163

163a. Stem glabrous below the middle; involucre 2–3 mm long—**Canadian Goldenrod**, *Solidago canadensis*

163b. Stem pubescent below the middle; involucre 3–5 mm long—**Tall Goldenrod**, *Solidago altissima* [*Solidago canadensis* var. *scabra*—CQ]

164a. Ray flowers fertile, pistillate (the two-lobed style protrudes from their base)—165

164b. Ray flowers sterile, with neither stamens nor pistil (Tribe Heliantheae)—173

165a. Lower leaves more than 20 cm long (summer)—166

165b. Lower leaves less than 15 cm long—167

166a. Leaf blades deeply pinnately lobed; prairies, roadsides, SLP (1–3 m high)—**Compass Plant**, *Silphium laciniatum*

166b. Leaf blades toothed or serrate; moist disturbed areas, esp. SLP (1–2 m high)—**Elecampane**, *Inula helenium*

167a. Principal bracts of the involucre in a single row, sometimes with a few much smaller basal bracteoles; upper cauline leaves pinnately divided (most *Senecio* spp., **Ragwort**)—168

167b. Principal bracts of the involucre in multiple, overlapping rows; upper cauline leaves pinnately divided or not; flowers in late summer and autumn—171

168a. Basal leaf blades cordate at base; swamps, wet woods (30–80 cm high; spring and summer)—**Golden Ragwort**, *Senecio aureus*

168b. Basal leaf blades truncate at base or tapering into the petiole—169

169a. Basal leaves obovate to orbicular; woods, swamps, SE (20–70 cm high; spring)—**Round-leaved Ragwort,** *Senecio obovatus*

169b. Basal leaves oblong to elliptic (spring and early summer)—170

170a. Base of basal leaf blades truncate, the leaf clearly petiolate; moist woods, wet meadows, shore, UP (30–80 cm high; summer)—**Squawroot,** *Senecio indecorus*

170b. Base of basal leaf blades rounded or tapering to the petiole; prairies, meadows (10–70 cm high)—**Northern Ragwort,** *Senecio pauperculus*

171a. Leaf blades pinnately divided; escape to roadsides, disturbed areas, mostly NM (30–70 cm high; summer)—**Yellow Chamomile,** *Anthemis tinctoria*

171b. Leaf blades entire to spinulose-serrate—172

172a. Leaf bases decurrent along the stem, forming wings; leaf blades serrate; involucre pubescent, but not viscid; wet meadows, fens (50–150 cm high; late summer and autumn)—**Sneezeweed,** *Helenium autumnale*

172b. Leaf bases not decurrent, stem wings absent; leaf blades sharply spinulose-serrate; involucre viscid; roadsides, fields, disturbed areas (10–100 cm high; summer)—**Gumweed,** *Grindelia squarrosa*

173a. Receptacle conic or oblong-cylindrical, often much elongated in the mature flower (summer)—174

173b. Receptacle flat, convex, to somewhat conic, not elongated (summer and autumn)—178

174a. Chaffy bracts subtend both the disk and ray flowers; rays spreading or often drooping; leaf blades pinnately divided; prairies, fields , SLP & UP (40–120 cm high; summer)—**Gray-headed Coneflower,** *Ratibida pinnata*

174b. Chaffy bracts subtend only the disk flowers; rays spreading; leaf blades entire, lobed, or pinnately divided (*Rudbeckia* **spp., Coneflower**)—175

175a. Disk (central portion of the head) yellow or greenish; leaf blades pinnately divided; stream banks, flood plains (0.5–3 m high; summer)—**Tall** or **Cutleaf Coneflower,** *Rudbeckia laciniata*

175b. Disk brown or purple; leaf blades entire to serrate or pinnately divided (summer and early autumn)—176

176a. The largest cauline leaves deeply three-lobed or sometimes pinnately divided; moist woods, disturbed areas, LP (50–150 cm high)—***Rudbeckia triloba***
176b. All leaf blades entire or toothed (30–100 cm high)—177

177a. Leaf blades entire or sparingly serrate; disturbed areas—**Black-eyed Susan**, ***Rudbeckia hirta***
177b. Leaf blades sharply serrate; fens, swamps, SLP—***Rudbeckia fulgida***

178a. Receptacle bracts absent; disk corollas brownish-purple; moist sandy areas (20–100 cm high; summer and early autumn)—***Helenium flexuosum***
178b. Receptacle bracts present, either chaffy or bristly; disk corollas brownish-purple or yellow—179

179a. Receptacle bristly; rays yellow with purple bases; disk corollas brownish-purple; escapes to roadsides, fields (20–70 cm high; late spring and summer)—**Blanketflower**, ***Gaillardia aristata***
179b. Receptacle covered with chaffy bracts—180

180a. Leaf bases decurrent along the stem, forming wings; all leaves alternate; disk corollas yellow, the flowers loosely arranged and spreading; thickets, floodplain forests, mostly SE (1–3 m high; late summer and autumn)—**Wing-stem** or **Yellow Ironweed**, ***Verbesina alternifolia***
180b. Leaf bases not decurrent along the stem, wings absent; lowermost leaves often opposite; disk corollas (often) yellow or red-purple, the flowers not spreading (***Helianthus*** spp., **Sunflower**)—181

Sunflowers (***Helianthus*** sp.) combine the well-known inflorescence (broad yellow sterile ray flowers surrounding a flat or convex, chaffy, receptacle, itself surrounded by subequal inflorescence bracts) with simple entire or toothed leaves. Ten of the 15 species known in Michigan (Voss, 1996) are included here; be warned that hybridization is extremely common, blurring the distinguishing characters of each species.

181a. Disk flowers brown or purple—182
181b. Disk flowers yellow—184

182a. Leaves opposite; petioles short; receptacle cone-shaped; roadsides, prairies (1–2 m high; late summer)—**Prairie Sunflower**, ***Helianthus pauciflorus***
182b. All but the lowermost leaves alternate; petioles prominent; receptacle flat—183

183a. Bracts at the center of the receptacle apically white-hairy; roadsides (30 cm–1 m high)—**Plains Sunflower**, *Helianthus petiolaris*

183b. Bracts at the center of the receptacle glabrous or only slightly hairy; shores, disturbed areas (1–3 m high)—***Helianthus annuus***

184a. Basal leaves present and much larger than the smaller cauline leaves; eight or fewer pairs of cauline leaves; prairies, LP (0.5–1.5 m high; late summer and early autumn)—**Western Sunflower**, *Helianthus occidentalis*

184b. Leaves chiefly cauline, basal leaves absent or not much larger than the cauline leaves; cauline leaves mostly more than eight pairs—185

185a. Stems hairy below the inflorescence; upper leaves mostly or all alternate (1–3 m high; late summer and early autumn)—186

185b. Stems glabrous below the inflorescence; upper leaves alternate or opposite—188

186a. Leaf blades broad, exceeding 4 mm wide; moist fields, disturbed areas—**Jerusalem Artichoke**, *Helianthus tuberosus*

186b. Leaf blades narrower, 3 mm wide or less—187

187a. Leaf blades flat, with one primary vein; stem covered with spreading hairs; swamps, fens, wet prairies—**Tall Sunflower**, *Helianthus giganteus*

187b. Leaf blades often folded, with three primary veins (at least at base); stem densely covered with white, often appressed hairs; prairies, railway rights-of-way—**Maximilian Sunflower**, *Helianthus maximilianii*

188a. Leaves sessile or virtually so; leaves all opposite; dry woods (0.5–1.5 m high; summer)—**Woodland Sunflower**, *Helianthus divaricatus*

188b. Leaves petioled, the petiole exceeding 5 mm; uppermost leaves opposite or alternate (50 cm–3 m high)—189

189a. Involucral bracts long, exceeding the height of the disk; leaf blades thin, serrate; dry woods, SLP (0.5–1.5 m high; late summer and autumn)—***Helianthus decapetalus***

189b. Involucral bracts shorter, less than or equal to the height of the disk; leaf blades thick, entire to sparsely serrate; fields, open woods, mostly LP (1–2 m high; summer)—***Helianthus strumosus***

190a. Principal leaves on the stem opposite or whorled (summer and autumn)—191

190b. Principal leaves on the stem alternate, or with smaller ones clustered in the axils—195

191a. Leaf blades pinnately lobed or dissected; flower heads (excluding rays) over 6 mm wide (60–200 cm high)—192

191b. Leaf blades nearly entire to toothed, not lobed; flower heads smaller, the head (excluding rays) 6 mm or less wide—193

192a. Leaf blades pinnately lobed, broadly ovate, to 30 cm long; rays white; moist woods, SLP—**Leaf-cup**, *Polymnia canadensis*

192b. Leaf blades pinnately dissected, to 11 cm long; rays pink, violet or white; escapes to disturbed areas, fields—*Cosmos bipinnatus*

193a. Rays many; disk corollas white; mud banks, Lake Erie region, SE— **Yerba de Tago**, *Eclipta prostrata*

193b. Rays often five; disk corollas yellow; lawns, gardens, disturbed areas (20–70 cm high) (*Galinsoga* spp., **Peruvian Daisy**)—194

194a. Leaf blades entire or with small, shallow teeth; stems glabrous or with sparse, often appressed hairs; mostly LP—*Galinsoga parviflora*

194b. Leaf blades serrate; stems pubescent, the hairs often spreading— *Galinsoga quadriradiata*

195a. Leaves dissected or deeply lobed or pinnately divided; pappus a short crown or absent, never capillary bristles; rays white to pink; lawns, gardens, roadsides, disturbed areas (Tribe Anthemideae)—196

195b. Leaves entire or at most serrate; pappus a short crown of scales, capillary hairs or bristles, or absent; rays white to pink, blue, or purple— 202

196a. Rays often five, white to pink, 3 mm or less long; disk 4 mm or less wide (20–100 cm high; summer and autumn)—**Yarrow**, *Achillea millefolium*

196b. Rays ten or more, white, 4 mm or more long; disk 5 mm or more wide—197

197a. Center of the receptacle chaffy bracted (10–90 cm high; spring and summer) (*Anthemis* spp.)—198

197b. Receptacle naked, lacking chaffy bracts—199

198a. Ray flowers sterile; foliage strongly scented—**Dog-fennel**, *Anthemis cotula*

198b. Ray flowers fertile, pistillate; foliage not strongly scented; SLP— **Corn-chamomile**, *Anthemis arvensis*

199a. Leaf blades pinnately lobed, the segments ovate to rounded (20–80 cm high) (*Chrysanthemum* spp. in part)—200

199b. Leaf blades finely pinnately divided, the segments linear-filiform (summer) (*Matricaria* spp. in part)—201

200a. Disk 1 cm or more wide; fields (spring to early autumn)—**Ox-eye Daisy**, *Chrysanthemum leucanthemum*
200b. Disk less than 1 cm wide (summer)—**Feverfew**, *Chrysanthemum parthenium*

201a. Foliage strongly scented; receptacle cone-shaped (20–80 cm high)—*Matricaria recutita*
201b. Foliage not strongly scented; receptacle rounded, hemispheric (10–70 cm high)—**Scentless Chamomile**, *Matricaria perforata* [*Matricaria maritima*—CQ]

202a. Heads small, 3–6 mm broad, including the white or pinkish rays; flowers many in a long inflorescence (summer and autumn)—203
202b. Heads 7 mm broad or larger, including the rays; rays white to blue or purple; flowers few to many—204

203a. Leaf blades oblanceolate to elliptic; dry open woods, mostly SE (10–100 cm high)—**White Goldenrod** or **Silverrod**, *Solidago bicolor*
203b. Leaf blades linear or narrowly lanceolate; disturbed areas (10–150 cm high)—**Horse-weed**, *Conyza canadensis*

204a. Pappus a short crown, awns, scales, or pappus absent (summer and autumn)—205
204b. Pappus consists only of many capillary hairs (Tribe Astereae)—211

205a. Disk corollas purple or brown—206
205b. Disk corollas yellow or white (rarely pinkish)—208

206a. Receptacle covered with fine bristles; rays purple, with or without a yellow tip, spreading; escapes to roadsides, fields (10–60 cm high; spring and summer)—**Blanketflower**, *Gaillardia pulchella*

Many of the sunny "wildflower seed mixtures" include composites native to dry, sunny prairie areas, such as species of **Liatris**, **Ratibida**, **Echinacea**, and **Gaillardia**. These mixtures have the potential to affect the native Michigan flora. Mixtures produced in other states could include species not native to Michigan but which could become established and spread in Michigan. The two species of **Gaillardia** listed in this key are examples of this effect; while they are indeed "native wildflowers" on the prairies of the central United States, there is no evidence of native populations of either species in Michigan. Another effect may occur when seeds from a species represented in Michigan, but produced from a distant population of that species, is included in a mixture. The genetic makeup of the local native flora may be subtly altered as genes from the distant population are introduced.

206b. Receptacle covered with stiff scales which are longer than the disk corollas; rays reddish-purple, pink, or white, drooping (60–180 cm high; summer and early autumn) (***Echinacea* spp., Purple Cone-flower**)—207

207a. Leaf blades ovate or ovate-lanceolate, most of them serrate; rays mostly reddish-purple, fading to pink with age; prairies, open roadsides, woods, LP—*Echinacea purpurea*

207b. Leaf blades narrowly lanceolate, gradually narrowed at the base, entire; rays pink or white; prairies, dry fields—*Echinacea pallida*

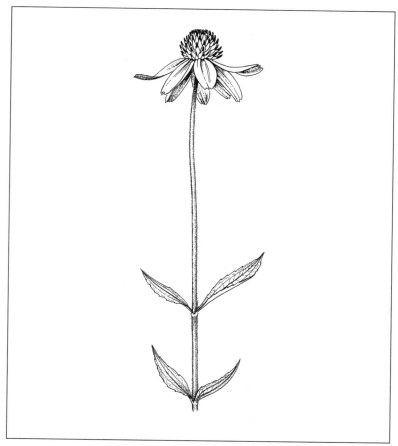

*Figure 45: **Echinacea purpurea**, inflorescence (radiate head)*

208a. Disk corollas white or rarely pinkish; rays white; roadsides, disturbed areas, NM (30–60 cm high; summer)—**Sneezewort**, *Achillea ptarmica*

208b. Disk corollas yellow; rays white or pink—209

209a. Leaves all basal; flower heads solitary on a naked scape; rays pink to white; lawns, disturbed areas (scape 5–15 cm high; spring and summer)—**English Daisy**, *Bellis perennis*

209b. Leaves cauline; flower heads numerous or solitary; rays white or pink—210

210a. Leaf blades serrate; flower heads solitary, terminating branches; rays white; fields (20–80 cm high; spring to early autumn)—**Ox-eye Daisy**, *Chrysanthemum leucanthemum*

210b. Leaf blades entire; flowers in a leafy corymb; rays pink; marsh edges, SE (30–150 cm high; summer and early autumn)—*Boltonia asteroides*

211a. Involucral bracts narrow, all of the same length and arranged in a single row (spring and summer, only a few plants persisting in bloom until autumn)(*Erigeron* spp., **Fleabane**)—212

211b. Involucral bracts of several unequal lengths, the outer successively shorter (or rarely nearly equal), loosely or closely overlapping (late summer and autumn) (*Aster* spp.)—215

Aster is another large genus of composites which is easy to recognize but difficult to identify to species. Most species flower in the late summer and early fall; the inflorescence has many narrow, blue or white rays, a pappus of capillary hairs, and alternate simple leaves. Eighteen of the 28 species known in Michigan (Voss, 1996) are included in this key.

212a. Bases of cauline leaves clasping the stem; rays blue, light purple, or pink—213

212b. Bases of cauline leaves tapering to the stem; rays white, sometimes pale pink or blue; disturbed areas—214

213a. Rays 100 or more per head, light purple or pink; moist meadows (20–70 cm high)—**Common Fleabane**, *Erigeron philadelphicus*

213b. Rays mostly less than 100 per head, blue; roadsides, sandy woods, mostly SLP (15–60 cm high; spring)—**Robin's-plantain**, *Erigeron pulchellus*

214a. Cauline leaves linear, entire (30–70 cm high)—**Daisy Fleabane**, *Erigeron strigosus*

214b. Cauline leaves ovate-lanceolate, most of them toothed (60–150 cm high)—**Daisy Fleabane**, *Erigeron annuus*

215a. Basal leaves petioled, the blades with cordate bases—216
215b. Basal leaves sessile or petioled, the blades with a tapering base—220

216a. Rays white tinged with blue; inflorescence branches glandular; woods
(20–120 cm high)—**Big-leaved Aster**, *Aster macrophyllus*
216b. Rays blue or violet; inflorescence branches not glandular—217

217a. Leaf blades entire; dry woods, sandy areas, esp. SLP (20–150 cm
high)—**Sky-blue Aster**, *Aster oolentangiensis*
217b. Leaf blades serrate (20–120 cm high)—218

218a. Heads few (seldom more than 50), in a loose spreading cluster; inflorescence bracts few; woods, NM—**Lindley's Aster**, *Aster ciliolatus*
218b. Heads many (usually over 100), in an elongate or spreading crowded
panicle; inflorescence bracts many—219

219a. Flowers mostly lilac; green tip of the involucral bracts diamond-
shaped; petioles not or only barely winged; woods, SLP—**Heart-
leaved Aster**, *Aster cordifolius*
219b. Flowers white; green tip of the involucral bracts long and slender;
petioles winged; open woods, prairies, fields, roadsides—**Arrow-
leaved Aster**, *Aster sagittifolius*

220a. Basal leaves not petioled; stem leaves with auriculate-shaped clasping
bases—221
220b. Basal and stem leaves sessile or petioled, but never auriculate-clasping—224

221a. Stem covered with spreading hairs; leaf blades entire or nearly so—
222
221b. Stem glabrous or with lines of short hairs on the upper stem; leaf
blades entire or serrate—223

222a. Rays reddish-purple to deep pink; involucral bracts covered with glandular hairs; moist meadows, fens, mostly SLP (30–200 cm high)—
New England Aster, *Aster novae-angliae*
222b. Rays blue to whitish; involucral bracts not covered with glandular
hairs; swamps, fens (50–250 cm high)—**Swamp** or **Purple-stemmed
Aster**, *Aster puniceus*

223a. Rays blue; leaf blades at least 1 cm wide; dry woods, fields, prairies
(30–100 cm high)—**Smooth Aster**, *Aster laevis*
223b. Rays white to lavender; leaf blades linear, most less than 0.5 cm wide;
fens, sedge meadows (15–100 cm high)—**Rush Aster**, *Aster borealis*

224a. Rays minute or absent; leaf blades linear; flowers violet; heads in a panicle or spike-like raceme; tap-rooted annual; saline roadsides. (10–70 cm high)—**Rayless Aster**, *Aster brachyactis*

224b. Rays conspicuous; leaf blades usually wider; rays blue, violet, or white; rhizomatous or clumping perennials—225

225a. Stems and leaves gray with a silky pubescence; rays violet; prairies, WM (30–70 cm high)—**Silky Aster**, *Aster sericeus*

225b. Stem and leaves green, not silky—226

226a. Involucral bracts narrowed to a pointed, awl-shaped tip; rays white, pink or purple—227

226b. Involucral bracts acute or obtuse at the flattened tip—228

227a. Cauline leaves linear, up to 10 mm long and 1 mm wide; heads many in an open inflorescence; rays white (rarely pinkish); sandy shores, prairies, roadsides (10–150 cm high)—**Frost Aster**, *Aster pilosus*

227b. Cauline leaves linear to oblong, more than 10 mm long and 2 mm wide; heads solitary or few; rays pink or purplish; fens, eastern UP (10–60 cm high)—**Bog Aster**, *Aster nemoralis*

228a. Heads in flat-topped clusters; rays white—229

228b. Heads in panicles, irregular clusters, racemes, or solitary; rays white, pink, or blue—230

229a. Leaf blades rigid, linear to oblanceolate; prairies, shores, dunes (10–70 cm high)—**Sneezewort Goldenrod**, *Solidago ptarmicoides*

This species has long white rays and was long considered to be an aster (***Aster ptarmicoides***). However, recent hybridization studies have shown that it is better considered as a goldenrod (***Solidago ptarmicoides***). It can hybridize with two other species of goldenrods, but not with asters. Some would rather place it outside both genera, using the name ***Unamia ptarmicoides***.

229b. Leaves not rigid, lanceolate to elliptic; bogs, fens, wet meadows (1–2 m high)—**Flat-topped Aster**, *Aster umbellatus*

230a. Underside of leaf blades pubescent, either over the entire surface or along the midvein—231

230b. Underside of leaf blades glabrous—233

231a. Leaf blades linear, most 5 mm or less wide; rays 20 or more, white to lavender; inflorescence a short, broad cluster; fens, sedge meadows (15–100 cm high)—**Rush Aster**, *Aster borealis*

231b. Leaf blades often lanceolate, 5 mm or more wide; rays 15 or fewer, white to purplish; inflorescence a more or less one-sided raceme (30–120 cm high)—232

232a. Surface of the underside of the leaf blade hairy; flood plain forests, SLP—**Lake Ontario Aster**, *Aster ontarionis*
232b. Underside of the leaf blade pubescent only along the midvein; woods, floodplains, fens—**Calico Aster**, *Aster lateriflorus*

233a. Upper portion of the stem pubescence arranged in lines—234
233b. Upper portion of the stem glabrous or if pubescent, the hairs not in lines; esp. SLP (30–100 cm high)—235

234a. Rays white or slightly tinged with blue; inflorescence elongate; leaf blades oblong to narrowly lanceolate, most more than 5 mm wide; fens, sandy shores, ditches, thickets (60–150 cm high)—**Panicled Aster**, *Aster lanceolatus*
234b. Rays white to lavender; Inflorescence short and broad; leaf blades linear, most 5 mm or less wide; fens, sedge meadows (15–100 cm high)—**Rush Aster**, *Aster borealis*

235a. Inflorescence a more or less one-sided raceme; involucral bracts spine-tipped; rays white; prairies, pastures—**Heath Aster** or **White Prairie Aster**, *Aster ericoides*
235b. Inflorescence not one-sided, the heads solitary at the end of minutely leafy peduncles; involucral bracts not spine-tipped; rays lavender to white; moist sandy areas—**Bushy Aster**, *Aster dumosus*

GLOSSARY

Some terms are used in a combined form. In those cases, refer to each of the two individual terms. A combination term may indicate that both characteristics are evident; a leaf which is *cordate-clasping* has heart-shaped leaves (cordate) with leaf bases that nearly encircle the stem (clasping). Some combination terms indicate a range between two characteristics; *ovate-deltoid* leaves may range from oval in outline (ovate) to triangular in outline (deltoid).

Modifiers are used on some terms. In each such case, refer to the root word. For example, *suborbicular* leaves are nearly round in outline (orbicular), but may be slightly flattened or otherwise fall short of a perfect roundness. Also note that some terms may be used as either a noun or an adjective. Thus, an inflorescence which is a *cyme* or a *raceme* may be referred to as a *cymose* or *racemose* inflorescence.

Achene	Dry indehiscent fruit consisting of one seed with a tightly appressed fruit wall
Acorn	Dry indehiscent fruit (a nut) with a hard fruit wall subtended by an involucre or "cup" of scales, as in *Quercus*
Acuminate	Tapers to extended point (*Fig. 7*)
Acute	As in acute angle; apex formed by two straight sides at an angle less than 90°
Adherent	Closely attached
Anther	The portion of the stamen which contains the pollen in flowers (*Fig. 2*)
Anthesis	The period when anthers mature and release pollen; often the stigma is also receptive to pollen
Appressed	Pressed flat against a surface
Aril	A fleshy outgrowth of the seed
Ascending	Growth habit in which stems or branches grow upward and outward
Auricle	Lobe which has rounded ends; "ear-shaped", as in auriculate leaf bases (*Fig. 7*)
Awn	A long bristle, often part of a leaf, bract, or fruit
Axil	The angle immediately above the leaf attachment to the stem (*Fig. 4* shows the axillary bud)

Banner	Large, often showy upper petal in flowers of the Legumi-nosae (*Fig. 23*)
Beak	An elongated, pointed appendage of fruits or other struc-tures
Beard	A stripe or island of hairs on petals
Bell-shaped	Flower with the corolla fused in the shape of a bell (*Fig. 39*)
Berry	Fleshy fruit with few to many seeds; may have several compartments, as in the tomato
Bipinnate	Twice-pinnately compound leaf in which leaflets are them-selves pinnately compound (*Fig. 5*)
Blade	The flat, broad portion of a leaf; modified or absent in some plants (*Fig. 4*)
Bract	Leaflike structures associated with flowers and inflores-cences; they may be reduced, brightly colored, or other-wise modified
Calyx	The outer whorl of leaf-like structures (sepals) in the flower; green or occasionally colored
Capitate	Head-like, often with a rounded end as in a capitate stigma
Capsule	Dry dehiscent fruit, with one to several compartments and often many seeds; splits along several lines at maturity
Carpel	A unit of ovule-bearing tissue within the ovary (pistil). An ovary may consist of one to many carpels, often visible in cross-section.
Catkin	Inflorescence which is a spike or raceme of unisexual re-duced flowers; typical of several families of woody plants (*Fig. 3*)
Cauline	Attached to the stem
Chaff	Small bracts subtending individual flowers in the Com-positae, often dry and hard or membranous (*Fig. 41*)
Ciliate	With fine marginal hairs
Clasping	Leaf base which partially surrounds the stem (*Fig. 7*)
Cleistogamous	Flowers which do not open for pollination and are self-fer-tilized; sometimes near the ground, underground, or pro-duced late in season
Column	A structure formed by the adhesion of the stamens to the style in the Orchidaceae
Cone	Woody or fleshy structure consisting of whorls of scales bearing naked seeds in gymnosperms
Connate	Joined or fused with parts of the same kind (*Fig. 40*, con-nate stamens)
Cordate	Heart-shaped (*Fig. 6*, cordate leaf; *Fig. 7*, cordate leaf base)
Corolla	The inner whorl of leaf-like organs (petals) in the flower; frequently colored, sometimes absent

Corona	A tissue projection between the petals and stamens, often seen as a "crown" near the throat of the corolla tube (*Figs. 18, 34*); or, as in the Asclepiadaceae, a colored structure surrounding each anther (*Fig. 30*)
Corymb	Flat-topped or convex inflorescence in which flower pedicels are inserted at different points and are of different lengths (*Fig. 3*)
Crenate	Leaf margins with rounded medium coarse teeth (*Fig. 8*)
Crenulate	Leaf margins with rounded fine teeth (*Fig. 8*)
Crisped	Crimped or ruffled leaf margins
Cuspidate	Apex which is a sharp, abrupt point (*Fig. 7*)
Cyme	A branched inflorescence in which the central or terminal flower opens first (*Fig. 3*). A helicoid cyme branches only to one side (*Fig. 34*)
Decumbent	Growth habit in which stems trail on the ground, but tips become erect
Decurrent	Extending downward, as when leaf bases extend down along the stem (*Fig. 7*)
Dehiscent	The (dry) fruit wall splits or breaks to release seeds
Deltoid	Triangular in outline (*Fig. 6*)
Dentate	Leaf margins with medium coarse teeth (*Fig. 8*)
Dichotomous	Divided into two equal portions; dichotomous branching often results in Y-shaped branches
Dioecious	Imperfect (staminate and pistillate) flowers on separate male and female plants
Discoid	Inflorescence type in the Compositae consisting only of disk flowers; also, any structure shaped like a disk
Disk	Nectariferous tissue at the base of the ovary or among the stamens in some flowers
Disk flower	Perfect or sometimes unisexual flowers in the Compositae, with a short tubular corolla (*Fig. 43*)
Dissected	Finely divided leaf blade (*Fig. 8*)
Divided	A leaf in which lobes are formed from cuts three-quarters or more of the distance from the margin to the midrib; extremely fine divisions with a feathery appearance are *dissected*
Drupe	Fleshy fruit with a single seed enclosed by a thick fruit wall, as in the plum
Drupelet	A small drupe; sometimes part of an aggregate fruit as in raspberries
Elliptic	Elongate form with rounded edges
Emergent	Foliage of an aquatic plant which is held above the water
Entire	Margin lacks teeth or lobes (*Fig. 8*)

Exserted   Thrust out, not enclosed, as when stamens protrude from the flower

Fascicle   A small cluster or bundle, as of leaves or flowers

Filament   The stalk which bears an anther (*Fig. 2*)

Filiform   Threadlike

Floret   Small individual flower, usually as part of a compact inflorescence

Follicle   Dry fruit opening along single seam (*Fig. 32*)

Funnelform   Flower with fused petals which open out gradually, resembling a funnel

Gamopetalous   Flowers in which the petals are fused to one another

Glabrous   A hairless surface

Glandular   Leaf, stem, or floral surfaces with small glands visible

Glaucous   Waxy, often whitened or bluish

Globose   Globe-shaped

Glume   Bract at the base of the grass spikelet (*Fig. 9*)

Hastate   Leaf base in which the lobes flare outwards (*Fig. 7*)

Head   Inflorescence in which numerous flowers, often small and usually sessile, are clustered tightly on a flat or discoid axis (*Fig. 3*)

Helicoid   Coiled; see *cyme*

Hemiparasitic   Parasitic plants which produce some nutrients via photosynthesis, but obtain some nutrients from other plants

Herbaceous   Non-woody; also refers to green tissue such as leaves

Heterophyllous Plants with more than one size and/or shape of leaf, as in the Polygonaceae

Hirsute   Surface with rough hairs

Hispid   Surface with stiff, sometimes bristly, hairs

Homophyllous   Plants with leaves of one size and shape, in contrast to *heterophyllous,* as in the Polygonaceae

Hood   Part of the corona surrounding the anthers in *Asclepias* (*Fig. 30*)

Horn   A pointed protrusion arising within the hood in *Asclepias* (*Fig. 30*)

Hypanthium   Cup-like structure formed by the fused bases of the sepals, petals, and stamens which often surrounds, but is not attached to the ovary

Imperfect   Flowers which do not have both male and female organs (stamen and pistil). They are either *staminate* (male) or *pistillate* (female).

Indehiscent   The fruit wall does not open at maturity

Inflorescence   The flowering structure of the plant, including a peduncle (stalk), one or more flowers and their pedicels, with any associated bracts (*Fig. 3*)

Internode	The portion of the stem between leaf attachments (nodes)
Inferior ovary	Floral type in which the ovary is fused to the hypanthium and the sepals, petals, and stamens are attached above the ovary (*Fig. 2*)
Involucre	A whorl of bracts, often closely appressed, which subtends an inflorescence, as in the Compositae (*Figs. 41, 44*)
Irregular	Flowers in which one or more parts of a whorl differ in size or shape
Keel	A ridge along an axis, as in leaves or floral parts; also the two lower, united petals of a leguminous flower (*Fig. 23*)
Lanceolate	A long narrow oval with a tapered apex (*Fig. 6*)
Leaflet	One part of the blade of a compound leaf (*Fig. 5*)
Legume	A dry fruit opening by two seams, typical of the Leguminosae
Lemma	The outer bract surrounding a grass floret (*Fig. 9*)
Ligulate	A perfect flower in the Compositae with the corolla prolonged into a strap-shaped lobe with 5 apical teeth (*Fig. 43*); a *ligulate head* consists only of these flowers
Ligule	A distinct ridge at the top of the sheath (petiole) of grasses and sedges (*Fig. 4*)
Lip	One of (often two) segments of an unequally divided corolla, as in the Labiatae (*Fig. 35*). In some flowers, only one highly modified lip is present, as in the Orchidaceae
Lobe	A protuberance, as in a lobed leaf (*Fig. 8*); the free tips of a fused corolla or calyx.
Mericarp	Segment of a schizocarp (Geraniaceae, Umbelliferae) (*Fig. 28*)
Midrib	The major vein through the center of the blade of the leaf (*Fig. 4*)
Monoecious	Imperfect (staminate and pistillate) flowers on the same plant
Mucronate	Apex with a hard, short point (*Fig. 7*)
Nerve	A prominent vein other than the midrib (*Fig. 4, 9*)
Net-veined	Leaf venation where primary veins branch from the midrib or converge near the base of the blade (*Fig. 4*)
Node	The location on the stem where one or more leaves attach
Nut	A dry indehiscent fruit with a hard, even woody, fruit wall surrounding a single seed; smaller fruits are called *nutlets* (*Fig. 33*)
Obcordate	Narrow at the base and cordate at the apex (*Fig. 6*)
Oblanceolate	Tapers from a narrow base to a rounded apex (*Fig. 6*)
Oblong	A long narrow oval with slightly flattened sides
Oblique	Leaf base in which one side of the blade is attached above the other (*Fig. 7*)

Obovate   Narrow at the base but otherwise oval (*Fig. 6*)

Ocrea   Modified, sheath-like stipules at the base of leaves in the Polygonaceae (*Fig. 16*)

Orbicular   Circular in outline (*Fig. 6*)

Ovary   The portion of the pistil which encloses the ovule and later the seed (*Fig. 2*).

Ovate   Shape with a broad base, tapering to an angle at the apex (*Fig. 6*)

Ovoid   Egg-shaped

Palea   The inner bract surrounding the grass floret (*Fig. 9*)

Palmate   Arrangement in which several axes converge to one point, such as palmate venation (*Fig. 4*) or palmately compound leaves (*Fig. 5*)

Panicle   Inflorescence where flowers occur along branches arising from a central unbranched axis; branching is often loose, producing an "open" inflorescence (*Fig. 3*)

Pappus   Plumose (featherlike) bristles, scales, or hairs derived from the sepals of individual flowers in the Compositae, and best seen on the ripe fruit (*Fig. 42*)

Parallel-veined   Leaf venation where primary veins are parallel to the midrib (*Fig. 4*)

Parasite   Plants which derive nutrients from other plants

Pedicel   The stalk which bears the receptacle and thus the flower (*Fig. 2*)

Peduncle   The stalk bearing an entire inflorescence (*Fig. 3*)

Peltate   Leaf with the petiole attached at or near the center of the blade (*Fig. 6*)

Perfect   Flowers which have both male and female organs (stamen and pistil).

Perfoliate   Describes a leaf in which the base surrounds the stem, so that the stem appears to pierce the blade (*Fig. 7*)

Perianth   The leaflike organs in the two outer whorls of the flower (sepals and petals, called tepals when not clearly different) (*Fig. 2*)

Perigynium   A saclike structure (bract) which encloses the pistil of *Carex* (Cyperaceae) (*Fig. 11*)

Petal   A member of the inner whorl of leaflike organs of the flower; frequently colored or sometimes absent (*Fig. 2*)

Petaloid   Structures such as stamens, sepals, or bracts, which are colored and shaped to resemble petals

Petiole   The part of the leaf attached to the stem; may be modified (see *sheath*), or absent (*sessile* leaves) (*Fig. 4*)

Pilose   Covered with fine, soft hairs

Pinnate   Arrangement in which two or more parallel axes arise from

	a central axis, such as pinnate venation (*Fig. 4*) or pinnately compound leaves (*Fig. 5*)
Pistil	The female organ of the flower, composed of ovary, style, and stigma (*Fig. 2*); formed from one or more ovule-bearing carpels
Pistillate	Flowers lacking functional stamens; female flowers
Pome	Fruit in which the receptacle enclosing the ovary swells to form an accessory fruit wall, as in the apple.
Prickle	Sharp epidermal outgrowth (*Fig. 14*)
Prostrate	Growth habit in which stems are flat on the ground
Pubescence	Hairs on a surface
Punctate	Describes a surface with many small impressions resembling shallow punctures
Raceme	Inflorescence in which individual pedicellate flowers are arranged along an unbranched central axis (*Fig. 3*)
Rachilla	The axis in a spikelet of the Gramineae (*Fig. 9*) and the Cyperaceae
Radiate	Inflorescence type in the Compositae consisting of disk flowers surrounded by ray flowers (*Fig. 41*)
Ray flower	Pistillate or sterile flowers in the Compositae with the corolla prolonged into a broad strap-shaped lobe with 1–3 apical teeth; they are typically the outer row of flowers in a radiate head (*Fig. 41*)
Receptacle	The end of a pedicel (stalk) to which the flower is attached (*Fig. 2*); it may be fleshy or enlarged with numerous flowers or floral parts attached to its surface (*Figs. 22, 41*)
Recurved	Curved outward or downward
Reflexed	Bent downward
Regular	Flowers in which all parts of a whorl are the same size and shape
Reniform	Kidney-shaped, as in a reniform leaf (*Fig. 6*)
Resinous	Has a sticky surface
Retrorse	Backward-pointing, as in hairs which point down or toward the axis
Rhizome	Thick fleshy stem at or under the soil surface
Rosette	Plant form in which the internodes are extremely short and leaves form a circle around a short crown close to the ground
Salverform	Flower with a fused corolla which is tubular at the base, then flattens to a plate-like top (*Fig. 34*)
Samara	Dry indehiscent fruit with a wing, as in *Acer* and *Fraxinus*
Saucer-shaped	Flower with a fused corolla which spreads out to form a flat dish
Scale	Small, thin, seldom green, leaflike structures such as those

	covering woody plant buds or reduced leaves in some plants
Scape	A leafless stalk of an inflorescence, often originating in a rosette of leaves (*Fig. 3*)
Schizocarp	Dry dehiscent fruit which splits into segments (mericarps) at maturity (Geraniaceae, Umbelliferae)(*Fig. 28*)
Sepal	A member of the outermost whorl of leaflike organs of the flower; often green, but may be colored (*Fig. 2*)
Serrate	Leaf margins with medium coarse, acutely angled teeth (*Fig. 8*)
Serrulate	Leaf margins with fine, acutely angled teeth (*Fig. 8*)
Sessile	Structure attached at the base without a stalk, as in a sessile leaf (*Fig. 7*) or stigma (*Fig. 13*)
Sheath	A tubular, modified petiole which encircles the stem as in the Gramineae (*Fig. 4*)
Silicle	Dry fruit characteristic of certain Cruciferae; splits at two seams, leaving a membranous center; less than twice as long as wide (*Fig. 19*)
Silique	Dry fruit characteristic of certain Cruciferae; splits at two seams, leaving a membranous center; at least twice as long as wide (*Fig. 20*)
Silky	With long, thin, smooth hairs
Sinus	Area between two lobes of a leaf (*Fig. 8*)
Solitary	Inflorescence with a single flower, sometimes occurring on a scape arising from a rosette of leaves (*Fig. 3*)
Spathe	A large bract which subtends a *spadix*, the typical inflorescence of the Araceae (*Fig. 3*)
Spathulate	Long and narrow, tapering from the apex to the base (*Fig. 6*)
Spike	Inflorescence in which sessile flowers occur along a central axis (*Figs. 3, 38*)
Spikelet	Reduced flowers in small spikes as part of an inflorescence characteristic of the Gramineae (*Figs 9, 10*) and the Cyperaceae (*Fig. 11*)
Spreading	Growth habit in which branches are held off the ground and are more or less horizontal
Spur	A hollow protuberance of one or more sepals or petals of some flowers (*Fig. 37*)
Stamen	The pollen-bearing (male) organ of the flower, composed of filament and anther (*Fig. 2*)
Staminate	Flowers lacking functional pistils; male flowers
Staminode	A stamen without a functional anther; when present, they often differ in appearance from fertile stamens in the same flower

Stellate	Star-shaped
Sterile	Flowers lacking functional sexual organs, or stamens lacking functional anthers
Stigma	The receptive surface (for pollen) of the pistil (*Fig. 2*)
Stipule	Appendages, usually paired, present at the base of the petiole. They can be spines, scales, glands, or resemble leaves (*Figs. 4, 22*)
Stolon	Stems at or under the surface of the soil, usually budding to produce new above-ground stems
Striate	With thin lines over the surface
Style	The portion of the pistil which bears the stigma (*Fig. 2*)
Sub-	Prefix indicating that a characteristic is minimally expressed; *e.g.* "subcordate", only vaguely heart-shaped
Subulate	Long and narrow, tapering from the base to the apex; awl-like (*Fig. 6*)
Superior ovary	Flower type in which the ovary sits on top of the receptacle and sepals, petals, and stamens are attached below the ovary (*Figs. 2, 12*)
Tendril	A portion of a leaf or stem modified for attachment or twining (*Fig. 14*)
Tepal	Perianth segments which cannot be distinguished as petals and sepals
Ternate	Segments in 3's; a ternately compound leaf, often based on palmate venation, has one or more sets of leaflets, each in groups of three (*Fig. 5*)
Thorn	A hard, sharp, shortened branch
Tomentose	Densely hairy
Trifoliolate	Compound leaf with three leaflets, often based on pinnate venation (*Fig. 5*)
Truncate	Abrupt square or broad end, as in truncate leaf base (*Fig. 7*)
Tubercle	A small bump or protrusion
Tuberculate	Describes a surface bearing tubercles
Tubular	Flower with a fused corolla which forms a tube without flaring outward
Twining	Twisting around an axis, as in the stems of some vines twisting around another plant stem
Umbel	Flat-topped or convex inflorescence in which flower pedicels are inserted at the same central point and are of the same length (*Fig. 3*); compound umbels (*Fig. 28*) have secondary umbellets originating at the ends of the first set of peduncles.
Urn-shaped	Flower in which the fused petals (corolla) converge, forming a small narrow opening

Utricle   One-seeded, dry indehiscent fruit with a thin papery wall
Valve   One section of the fruit wall in dehiscent dry fruits; also, persistent tepals enclosing the fruit (achene) in the Polygonaceae (*Fig. 15*).
Venation   The pattern of veins in leaf blades (*Fig. 4*)
Viscid   Sticky to the touch
Whorled   More than two branches or leaves arising from a node (*Fig. 5*)

SUBJECT INDEX

INDEX TO PLANT NAMES

Centimeters

15
14
13
12
11
10
9
8
7
6
5
4
3
2
1
0

QUICK INDEX TO PLANT FAMILIES